Topics in

Functional Analysis and Applications

TOPICS IN FUNCTIONAL ANALYSIS AND APPLICATIONS

S. Kesavan
School of Mathematics
Tata Institute of Fundamental Research
Bangalore, India

JOHN WILEY & SONS
New York Chichester Brisbane Toronto Singapore

Randall Library UNC-W

First published in 1989 by
WILEY EASTERN LIMITED
4835/24 Ansari Road, Daryaganj
New Delhi 110 002, India

Distributors:

Australia and New Zealand:
JACARANDA-WILEY LTD., JACARANDA PRESS
John Wiley & Sons, Inc.
GPO Box 859, Brisbane, Queensland 4001, Australia

Canada:
JOHN WILEY & SONS CANADA LIMITED
22 Worcester Road, Rexdale, Ontario, Canada

Europe and Africa:
JOHN WILEY & SONS LIMITED
Baffins Lane, Chichester, West Sussex, England

South East Asia:
JOHN WILEY & SONS, INC.
05-05 Block B, Union Industrial Building
37 Jalan Pemimpin, Singapore 2057

Africa and South Asia:
WILEY EASTERN LIMITED
4835/24 Ansari Road, Daryaganj
New Delhi 110 002, India

North and South America and rest of the world:
JOHN WILEY & SONS, INC.
605 Third Avenue, New York, N.Y. 10158, USA

Copyright © 1989, WILEY EASTERN LIMITED
New Delhi, India

Library of Congress Cataloging in Publication Data

ISBN 0-470-21050-8 John Wiley & Sons, Inc.
ISBN 81-224-0062-0 Wiley Eastern Limited

Printed in India at Urvashi Press, Meerut, India.

QA
320
.K43
1989

Preface

With the discovery of the theory of distributions, the role of Functional Analysis in the study of partial differential equations has become increasingly important. Not only is a proper functional analytic setting important for the theoretical study of the well-posedness of initial and boundary value problems but also for the construction of good numerical schemes for the computation of approximate solutions. Numerical methods like the Finite Element Method draw heavily upon the results of Functional Analysis both for the construction of the schemes as well as their error analysis.

It is thus clear that applied mathematicians must have a good background in Functional Analysis and its applications to partial differential equations. Such courses are absent from the curricula of most Indian universities. One of the main reasons for this is the lack of suitable textbooks on which such a course could be based and it is this gap that the present book is expected to fill. Indeed the material covered in this book is not original and can be found distributed amongst several treatises, texts or papers. The difficulty faced by teachers is the task of assembling material suitable for an introductory course from an abundant literature. The present book aims to provide such an introductory course on the functional analytic methods used in the study of partial differential equations and is based on lectures given by myself at the Tata Institute of Fundamental Research in the past few years.

The prerequisites for the use of this book are basic courses on Analysis (theory of measure and integration, L^p-spaces, etc.), Topology and Functional Analysis (Banach and Hilbert spaces, strong and weak topologies, compact operators etc.). Apart from these requirements, every effort has been made to keep the treatment as self-contained as possible. The interdependence of the various chapters is as follows:

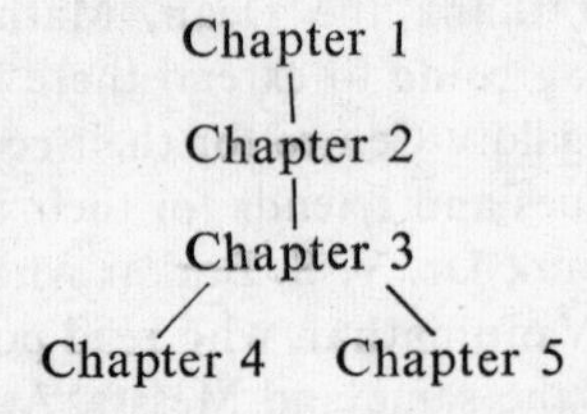

The first chapter covers the main aspects of the theory of distributions and the Fourier transform. The notion of distribution solutions to partial differential equations is introduced. The second chapter studies the impor-

tant properties of Sobolev spaces. These spaces form a natural functional analytic framework for the study of weak solutions of elliptic boundary value problems which is the topic discussed in the third chapter. Here we study the existence, uniqueness and regularity of weak solutions of linear elliptic boundary value problems. The general theory is illustrated by several classical examples from physics and engineering. We also study maximum principles and eigenvalue problems. The fourth chapter is devoted to the study of evolution equations, i.e. initial and initial-boundary value problems, using the theory of semigroups of linear operators on a Banach space. After a brief introduction to the abstract theory, illustrations via some standard partial differential equations of physics are provided. The last chapter provides an introduction to the study of semi-linear elliptic boundary value problems from the point of fixed point theorems, approximation methods and variational principles.

Comments at the end of each chapter provide additional information or results not given in the text and also give important bibliographic references. Each chapter is also provided with a selection of exercises which are designed to fill gaps in the proofs of some theorems or prove additional theorems or to construct examples or counter examples to notions introduced in the text.

As the Tamil poet Avvai put it, one's knowledge and one's ignorance roughly bear the ratio of a fistful of soil to the volume of the Earth. In the same way it must be remembered that this book is just an introduction to the study of the modern theory of partial differential equations and is by no means an exhaustive treatment of the subject. The literature is very vast and references to the important works and the frontiers of current research are indicated wherever possible.

The material covered in the text is more or less a prerequisite for any student aspiring for a good research career in applied mathematics whether it be from the theoretical or computational point of view. This book could thus be used for an M.Phil. or a pre-Ph.D. course in the applied mathematics curriculum in Indian universities. In case the M.Sc. curriculum provides a strong base in Analysis, Functional Analysis and Topology, it could also be used for an elective course at that level.

The preparation of this manuscript was possible due to the excellent facilities available at the Bangalore Centre of the Tata Institute of Fundamental Research and I thank the Dean, Mathematics Faculty of this Institute for generously agreeing to extend these facilities for this purpose and the Staff of the Bangalore Centre for their cooperation. I would also like to thank my colleagues and friends for their help and encouragement. In particular I wish to thank Dr. V. S. Borkar who egged me on to embark on this project, Dr. M. Vanninathan who read portions of the manuscript and helped me improve the same and Messrs. A. Patnaik, B. K. Ravi and A. S. Vasudevamurthy who, in various ways, helped me during the preparation of the manuscript. Finally I thank Ms. N. N. Shanthakumary for her neat and careful typing of the manuscript. I am grateful to the

personnel of Wiley Eastern Ltd. for their cooperation in bringing out this volume. To my family I owe the moral support extended throughout the execution of this project.

Finally, I wish to dedicate this book to the memory of my late father.

Bangalore,
August 1988

S. KESAVAN

Notations

I. Notations in Euclidean Spaces

$\mathbb{R}$ stands for the real line.

$\mathbb{R}^n$ stands for the n-dimensional Euclidean space over $\mathbb{R}$.

$\mathbb{C}$ stands for the set of complex numbers.

e_i stands for the ith standard basis vector in $\mathbb{R}^n$.

x $=(x_1 \ldots x_n)$ is a vector in $\mathbb{R}^n$ with coordinates x_i, $1 \leqslant i \leqslant n$.

$|x|$ is the Euclidean norm of $x \in \mathbb{R}^n$.

dist (x, A) is the Euclidean distance of $x \in \mathbb{R}^n$ from a subset $A \subset \mathbb{R}^n$.

Ω stands for an open set in $\mathbb{R}^n$.

$\bar{\Omega}$ stands for its closure in $\mathbb{R}^n$.

$\Gamma = \partial\Omega$ stands for the boundary of Ω.

$\Omega' \subset \subset \Omega$ means that Ω' is a relatively compact open subset of Ω.

II. Function Spaces

$C(\Omega)$ is the space of continuous functions on Ω.

$C(\bar{\Omega})$ is the space of continuous functions on $\bar{\Omega}$.

$C^k(\Omega)$ is the space of k times continuously differentiable functions on Ω.

$C^k(\bar{\Omega})$ is the space of functions in $C^k(\Omega)$ which together with all derivatives possess continuous extensions to $\bar{\Omega}$.

$C^\infty(\Omega)$ $= \bigcap_{k=0}^{\infty} C^k(\Omega)$.

$C^\infty(\bar{\Omega})$ $= \bigcap_{k=0}^{\infty} C^k(\bar{\Omega})$.

$\mathscr{D}(\Omega)$ is the space of functions in $C^\infty(\Omega)$ with compact support in Ω ($\mathscr{D} = \mathscr{D}(\mathbb{R}^n)$).

$\mathscr{D}'(\Omega)$ is the space of distributions on Ω ($\mathscr{D}' = \mathscr{D}'(\mathbb{R}^n)$).

$\mathscr{E}(\Omega)$ $=C^\infty(\Omega)$ ($\mathscr{E} = \mathscr{E}(\mathbb{R}^n)$).

$\mathscr{E}'(\Omega)$ is the space of distributions with compact support in Ω ($\mathscr{E}' = \mathscr{E}'(\mathbb{R}^n)$).

$\mathscr{S}$ is the Schwartz space of rapidly decreasing functions in $\mathbb{R}^n$.

$\mathscr{S}'$ is the space of tempered distributions on $\mathbb{R}^n$.

$W^{m,p}(\Omega)$ is the Sobolev space of order m for $1 \leqslant p \leqslant \infty$ with norm $\|\cdot\|_{m,p,\Omega}$ and semi-norm $|\cdot|_{m,p,\Omega}$.

$W_0^{m,p}(\Omega)$ is the closure of $\mathscr{D}(\Omega)$ in $W^{m,p}(\Omega)$.

$W^{m,2}(\Omega)$ $= H^m(\Omega)$ with norm $\|\cdot\|_{m,\Omega}$ and semi-norm $|\cdot|_{m,\Omega}$.

$W_0^{m,2}(\Omega)$ $= H_0^m(\Omega)$.

$W^{0,p}(\Omega)$ $= L^p(\Omega)$ with norm $|\cdot|_{0,p,\Omega}$.

$W^{0,2}(\Omega)$ $= L^2(\Omega)$ with norm $|\cdot|_{0,\Omega}$.

III. General Remarks

1. In any normed linear space $B(x; a)$ will stand for the open or closed ball (depending on the context) centered at x and of radius a.
2. In any estimate or inequality the quantity C will denote a generic positive constant and need not necessarily be the same constant as in the preceding calculations.

Contents

ONE

Distributions

1.1 INTRODUCTION

The main aim of this book is to develop the basic tools of functional analysis which will be useful in the study of partial differential equations and to illustrate their use via examples. As a first step towards this, the notion of differentiable functions must be generalized. When we study a partial differential equation, we understand—in the classical sense—that a solution must be differentiable at least as many times as the order of the equation and that it must satisfy the equation everywhere in space (and time). However such a point of view is very restrictive and several interesting equations which model physical phenomena will fail to possess such solutions and thus we will be prevented from studying mathematically such physical situations. Let us consider a few examples.

Example 1.1.1 Consider the following equation:

$$u_t + uu_x = 0,\ x \in \mathbb{R},\ t > 0 \tag{1.1.1}$$

where the subscripts denote differentiation with respect to the corresponding independent variable. This equation is known as *Burger's Equation* and is closely related to a class of partial differential equations known as *hyperbolic conservation laws*. Let $u(x, t)$ be a 'smooth' solution of (1.1.1) satisfying an initial condition of the form

$$u(x, 0) = u_0(x),\ x \in \mathbb{R} \tag{1.1.2}$$

where $u_0(x)$ is a given function of x. Let us now define a curve $x = x(t)$ in the x-t plane by means of the ordinary differential equation

$$\frac{dx}{dt}(t) = u(x(t), t). \tag{1.1.3}$$

Along such a curve we have

$$\frac{du}{dt} = \frac{\partial u}{\partial t} + \frac{\partial u}{\partial x}\frac{dx}{dt} = u_t + uu_x = 0$$

since u satisfies (1.1.1). Hence along each such curve, $u =$ a constant. It then follows from (1.1.3) that such curves are straight lines. These are called 'characteristic curves' and the curve through the point $(x_0, 0)$ on the

real line will have the form

$$x = x_0 + ct,\ c = u_0(x_0) \tag{1.1.4}$$

and all along this curve $u(x, t) = u_0(x_0)$.

Now consider a smooth initial function u_0 as in figure 1.

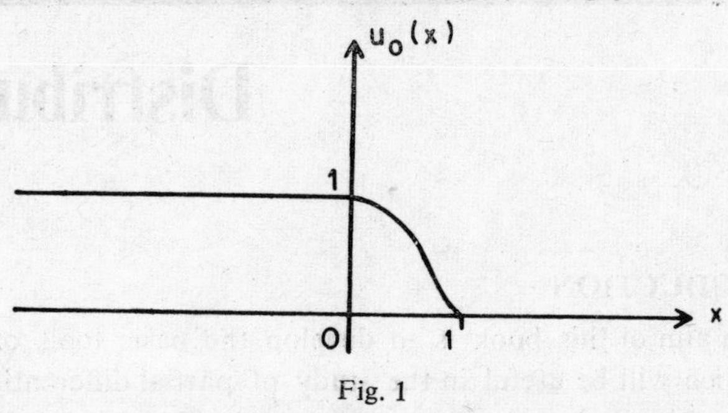

Fig. 1

The corresponding characteristic curves are shown in figure 2.

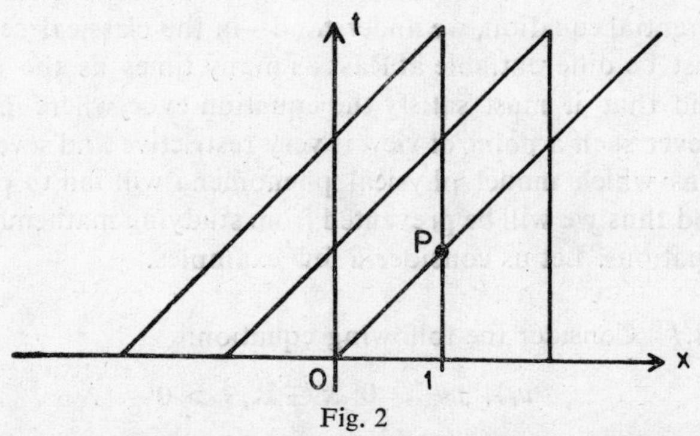

Fig. 2

It is seen from this figure that, for instance, at the point P two characteristic curves meet and hence the value of u at P is not even well defined. Thus except for a short time, we cannot even expect the function to be continuous! If we only wish to study classical solutions, such equations cannot be tackled and we will not be able to study interesting physical phenomena such as *shock waves* (which are very important in aeronautics). We are thus led to the need of generalizing the notion of a solution of partial differential equations which should eventually include discontinuous functions being recognized as solutions (albeit in a weak sense). ∎

Example 1.1.2 Let $\Omega \subset \mathbb{R}^2$ be a bounded open set. If Ω is the region occupied by a thin membrane fixed along the boundary $\partial\Omega$ and acted upon by a vertical force, then the displacement in the vertical direction is given by a function $u(x)$, $x \in \Omega$, which satisfies a partial differential equation of the form

$$\left.\begin{aligned} -\Delta u &= f \text{ in } \Omega \\ u &= 0 \text{ on } \partial\Omega \end{aligned}\right\} \tag{1.1.5}$$

where Δ is the *Laplace operator* defined by

$$\Delta u = u_{xx} + u_{yy}. \tag{1.1.6}$$

This is a partial differential equation involving the second order derivatives of the function u. However, in mechanics, what is more important is that u minimizes the *strain energy* functional

$$J(v) = \tfrac{1}{2}\int_{\Omega}\left[\left(\frac{\partial v}{\partial x}\right)^2 + \left(\frac{\partial v}{\partial y}\right)^2\right] dx\,dy - \int_{\Omega} fv\,dx\,dy \qquad (1.1.7)$$

amongst all 'admissible displacements' v. A fact that strikes us immediately is that no second derivatives are involved in the definition of J! Thus in looking for the equilibrium state of the membrane the space of 'admissible displacements' need not involve functions which are twice differentiable and in fact for several 'reasonable' data f, it is not correct to do so. Nevertheless we would like to know the connection between the problem (1.1.5) and the minimizer u of the functional J. In other words though u may not be twice differentiable in the classical sense we would like to say that it still satisfies (1.1.5) in a weak sense.

Many computational schemes to approximate the solution of (1.1.5) stem from the variational characterization described above. ∎

The above examples are but few instances which motivate the need of generalizing the notion of a solution of a partial differential equation, which in turn motivates the need of generalizing the notion of differentiable functions. In other words, we will study a larger class of objects—called **Distributions**—on which we can define a (generalized) derivative and wherein the usual rules of calculus will hold. Further, for smooth functions, this new notion of a derivative must coincide with the usual one.

A rough idea as to how to set about realizing this class can be obtained from the following discussion.

Let $f \in L^2(\mathbb{R})$, the space of square integrable functions on $\mathbb{R}$. It can be shown that the space $\mathscr{D}$ of infinitely differentiable functions with compact support in $\mathbb{R}$ is dense in $L^2(\mathbb{R})$. (By the support of a function $\phi : \mathbb{R} \to \mathbb{R}$ (or $\mathbb{C}$) we mean the set

$$K = \overline{\{x \in \mathbb{R} \mid \phi(x) \neq 0\}} \qquad (1.1.8)$$

which is always closed by definition.) Thus, since $L^2(\mathbb{R})$ is a Hilbert space, f is completely known once its innerproduct with each element of $\mathscr{D}$ is known, i.e. when all the numbers

$$\int_{\Omega} f\phi,\ \phi \in \mathscr{D}$$

are known. Now assume that f is continuously differentiable, with derivative f'. By integration by parts, we have

$$\int_{\mathbb{R}} f'\phi = -\int_{\mathbb{R}} f\phi'. \qquad (1.1.9)$$

Notice now that the right-hand side of (1.1.9) does not involve the derivative of f ! Also notice that the operations $\phi \to \int_{\mathbb{R}} f\phi$ and $\phi \to \int_{\mathbb{R}} f\phi'$ are linear on $\mathscr{D}$. Hence if we can define a suitable topology on $\mathscr{D}$ which

makes these operations continuous, we can define f as a *continuous linear functional* on $\mathcal{D}$ and define f' via the right-hand-side of (1.1.9) *even when f is not differentiable* as long as the integrals make sense. This is the procedure we will follow in the next few sections.

1.2 TEST FUNCTIONS AND DISTRIBUTIONS

Let ϕ be a real (or complex) valued continuous function defined on an open set in $\mathbb{R}^n$. The **support** of ϕ, written as supp (ϕ), is defined as the closure (in $\mathbb{R}^n$) of the set on which ϕ is non-zero (cf. (1.1.8)). If this closed set is compact as well, then ϕ is said to be of compact support. The set of all infinitely differentiable (i.e. C^∞) functions defined on $\mathbb{R}^n$ with compact support is a vector space which will henceforth be denoted by $\mathcal{D}(\mathbb{R}^n)$ or, simply, $\mathcal{D}$. We will now show that this class of functions is quite rich.

Lemma 1.2.1 Let $f: \mathbb{R} \to \mathbb{R}$ be defined by

$$f(x) = \begin{cases} \exp(-x^{-2}), & x > 0 \\ 0 & , x \leqslant 0. \end{cases} \tag{1.2.1}$$

Then f is a C^∞ function.

Proof: We only need to check the smoothness at $x = 0$. As $x \uparrow 0$ all derivatives are zero. As $x \downarrow 0$, the derivatives are finite linear combinations of terms of the form $x^{-l} \exp(-x^{-2})$, l an integer greater than or equal to zero.

A simple application of l'Hospital's Rule shows that these terms tend to zero as $x \downarrow 0$. ∎

We can use the above lemma to construct examples of elements of $\mathcal{D}$.

Example 1.2.1 Consider the function

$$\phi(x) = \begin{cases} \exp(-a^2/(a^2 - x^2)), & |x| < a \\ 0, & |x| \geqslant a \end{cases} \tag{1.2.2}$$

Then a simple application of the preceding lemma shows that $\phi \in \mathcal{D}(\mathbb{R})$ and supp $(\phi) = [-a, a]$.

More generally, define

$$\phi(x) = \begin{cases} \exp(-a^2/(a^2 - |x|^2)), & |x| < a \\ 0 & , |x| \geqslant a \end{cases} \tag{1.2.3}$$

where $|x|^2 = \sum_{i=1}^{n} |x_i|^2$, $x = (x_1, \ldots, x_n) \in \mathbb{R}^n$. Then $\phi \in \mathcal{D}(\mathbb{R}^n)$ with supp (ϕ) = Ball Centre 0 and radius a (denoted $B(0; a)$). ∎

Example 1.2.2 This is a slight, but very useful, variation of the previous example. Let $\epsilon > 0$ and set

$$\rho_\epsilon(x) = \begin{cases} k\epsilon^{-n} \exp(-\epsilon^2/(\epsilon^2 - |x|^2)), & |x| < \epsilon \\ 0 & , |x| \geqslant \epsilon \end{cases} \tag{1.2.4}$$

where

$$k^{-1} = \int_{|x| \leqslant 1} \exp(-1/(1 - |x|^2))\, dx. \tag{1.2.5}$$

Then by (1.2.3), we know that $\rho_\epsilon \in D(\mathbb{R}^n)$ with supp $(\rho_\epsilon) = B(0; \epsilon)$, the ball centre 0 and radius ϵ. Further $\rho_\epsilon \geqslant 0$ and

$$\int_{\mathbb{R}^n} \rho_\epsilon(x)\, dx = 1. \tag{1.2.6}$$

For,

$$\int_{\mathbb{R}^n} \rho_\epsilon(x) dx = \frac{k}{\epsilon^n} \int_{|x| \leqslant \epsilon} (-\epsilon^2/(\epsilon^2 - |x|^2))\, dx$$

$$= k \int_{|x| \leqslant 1} \exp(-1/(1 - |x|^2))\, dx = 1.$$

Thus the functions ρ_ϵ, $\epsilon \to 0$, have smaller supports, but preserve the volume contained under the graph. As $\epsilon \to 0$, these functions are concentrated at the origin. They will be used repeatedly in the sequel and are called **mollifiers.** ∎

A family of sets $\{E_i\}_{i \in I}$ in $\mathbb{R}^n$ is said to be *locally finite* if for every point x, there exists a neighbourhood of x which intersects only a finite number of the E_i. We now quote, without proof, a very important theorem. For complete details, see Appendix 1.

Theorem 1.2.1 (*Locally Finite C^∞ Partition of Unity*). Let Ω be an open set in $\mathbb{R}^n$ and let $\Omega = \bigcup_{i \in I} \Omega_i$, Ω_i, open. Then there exist C^∞ functions ϕ_i defined on Ω such that

(i) supp $(\phi_i) \subset \Omega_i$

(ii) $\{\text{supp}\,(\phi_i)\}_{i \in I}$ is locally finite

(iii) $0 \leqslant \phi_i(x) \leqslant 1$, for all $i \in I$, and

(iv) $\sum_{i \in I} \phi_i \equiv 1$. ∎

Remark 1.2.1 Since given any $x \in \Omega$, there exists a neighbourhood which will intersect only a finite number of the sets $\{\text{supp}\,(\phi_i)\}$, it follows that $\phi_i(x) = 0$ for all but finitely many i. Thus the sum in (iv) above is in fact a finite sum and is thus well defined. The name *partition of unity* is self-explanatory: the constant function 1 is partitioned into C^∞ functions whose support can be controlled. ∎

Corollary Let K be a compact set in $\mathbb{R}^n$. Then there exists a $\phi \in \mathscr{D}(\mathbb{R}^n)$ such that $\phi \equiv 1$ on K.

Proof: We consider a relatively compact open set U containing K. Now consider the covering of $\mathbb{R}^n$ consisting of $\{U, \mathbb{R}^n \setminus K\}$ and the partition of unity subordinate to this cover. Let ϕ and ψ be non-negative C^∞ functions with $\phi + \psi \equiv 1$, supp $(\phi) \subset U$ and supp $(\psi) \subset \mathbb{R}^n \setminus K$. Thus $\psi \equiv 0$ on K and hence $\phi \equiv 1$ on K. Also supp $(\phi) \subset U \subset \bar{U}$ which is compact. Thus $\phi \in \mathscr{D}\,(\mathbb{R}^n)$. ∎

The function ϕ constructed above is called a **cut-off function** with respect to the compact set K.

We have thus established that the class $\mathcal{D}$, also called the space of **test-functions,** is well endowed with functions. If Ω is any open set in $\mathbb{R}^n$, we can still talk of the space of C^∞ functions with compact support, the support being contained in Ω. This space will be denoted by $\mathcal{D}(\Omega)$.

We will now provide $\mathcal{D}(\Omega)$ with a topology which will make it a topological vector space. In fact, for the development of the theory, we will not need a complete description of the topology; we will only need to know what are convergent sequences in $\mathcal{D}(\Omega)$. Hence we will abstain, for the moment, from describing this topology and defer this task to Appendix 2. We will just define convergent sequences in $\mathcal{D}(\Omega)$.

Definition 1.2.1 A sequence of functions $\{\phi_m\}$ in $\mathcal{D}(\Omega)$ is said to converge to 0 if there exists a *fixed* compact set $K \subset \Omega$ such that supp $(\phi_m) \subset K$ for all m and ϕ_m and all its derivatives converge *uniformly* to zero on K. ∎

As indicated in Section 1.1, we will generalize the notion of a function by considering linear functionals on $\mathcal{D}$ (Ω) which are continuous with respect to the above mentioned topology.

Definition 1.2.2 A linear functional T on $\mathcal{D}(\Omega)$ is said to be a **distribution** on Ω if whenever $\phi_m \to 0$ in $\mathcal{D}(\Omega)$, we have $T(\phi_m) \to 0$. ∎

The space of distributions, which is the dual of the space of test-functions, is denoted by $\mathcal{D}'(\Omega)$. In case $\Omega = \mathbb{R}^n$, the symbol $\mathcal{D}'$ will also be used. We now proceed to give several examples of distributions.

Example 1.2.3 A function $f: \Omega \to \mathbb{R}$ (or $\mathbb{C}$) is said to be **locally integrable** if for every compact set $K \subset \Omega$,

$$\int_K |f| < +\infty. \tag{1.2.7}$$

For instance any continuous function is locally integrable. Another example of a locally integrable function (on $\mathbb{R}^2$) is r^{-1} where $r = |x|$. If B is the ball of radius ϵ centred at the origin, then

$$\int_B \frac{1}{r} = \int_0^\epsilon \int_0^{2\pi} \frac{1}{r}\, r\, d\theta\, dr,$$

which is finite.

Given a locally integrable function f, define $T_f : \mathcal{D}(\Omega) \to \mathbb{R}$ (or $\mathbb{C}$) by

$$T_f(\phi) = \int_\Omega f\phi. \tag{1.2.8}$$

Clearly T_f is a linear functional on $\mathcal{D}$ and it is easy to verify that it is a distribution.

If f and g are two locally integrable functions such that $f = g$ a.e. then it is obvious that $T_f = T_g$. In particular if $f = 0$ a.e., it defines the zero

distribution. In fact, the converse is also true. If $T_f = 0$ then $f = 0$ a.e., given that f is locally integrable. We omit a proof of this.

In future we will not distinguish between a locally integrable function and the distribution it generates. By saying that "a distribution T is a function" we mean that there exists a locally integrable function f such that $T = T_f$.

It is clear that any $f \in L^p(\Omega)$, $p \geqslant 1$ generates a distribution via (1.2.8). ∎

Example 1.2.4 **The Dirac Distribution**

Let $x \in \mathbf{R}^n$. Define δ_x by

$$\delta_x(\phi) = \phi(x), \; \phi \in \mathscr{D}. \tag{1.2.9}$$

Then it is easy to see that δ_x defines a distribution. In particular if $x = 0$, we write δ and this is the well-known "Dirac δ-function", introduced by P. A. M. Dirac early in this century. His symbolic calculus of the δ-function made no mathematical sense until correctly formulated in the framework of the theory of distributions developed essentially by L. Schwartz.

Note that this example is *not* contained in the preceding one, i.e. the Dirac distribution cannot be generated by a locally integrable function. Indeed assume that f were a locally integrable function such that

$$T_f = \delta.$$

For every $\epsilon > 0$ let $\phi_\epsilon \in \mathscr{D}$ with support in $B(0, \epsilon)$, $0 \leqslant \phi_\epsilon \leqslant 1$ and $\phi_\epsilon \equiv 1$ on $B(0; \epsilon/2)$. Now,

$$\delta(\phi_\epsilon) = 1 \text{ for all } \epsilon > 0.$$

However, on the other hand,

$$\delta(\phi_\epsilon) = \int_{\mathbf{R}^n} f\phi_\epsilon = \int_{B(0;\epsilon)} f\phi_\epsilon \leqslant \int_{B(0;\,\epsilon)} |f|$$

and so $\delta(\phi_\epsilon) \to 0$ as $\epsilon \to 0$ by the local integrability of f, which is a contradiction. ∎

Example 1.2.5 (Measures as distributions)

Let μ be either a complex Borel measure (see Rudin [1]) or a positive measure such that $\mu(K) < +\infty$ for every compact set K. Then the operation

$$\phi \to \int_{\mathbf{R}^n} \phi \, d\mu, \; \phi \in \mathscr{D} \tag{1.2.10}$$

defines a distribution on $\mathbf{R}^n$. It is easily seen that the above example generalizes all the preceding ones. ∎

Example 1.2.6 Consider the following linear functional defined on $\mathscr{D}(\mathbf{R})$:

$$\phi \to \phi'(0), \tag{1.2.11}$$

where ϕ' stands for the derivative of ϕ. Again it is easy to check that (1.2.11) defines a distribution. It is known as the **doublet** or **dipole** distribution in the literature of mathematical physics.

More generally, for any integer n, the following operation defines an element of $\mathcal{D}'(\mathbb{R})$:

$$\phi \to \phi^{(n)}(0), \phi \in \mathcal{D} \tag{1.2.12}$$

where $\phi^{(n)}$ stands for the nth derivative of ϕ.

The distributions cited in the above example are not covered by those described before. For example let us show that the doublet distribution cannot be generated by a measure. To see this let ϕ be a function in $\mathcal{D}(\mathbb{R})$ with support in $[-1, +1]$ and which is such that $0 \leqslant \phi \leqslant 1$ and $\phi \equiv 1$ on $[-1/2, 1/2]$. Define

$$\psi_k(x) = (\sin kx)\, \phi(x).$$

Then $\operatorname{supp}(\psi_k) \subset [-1, +1]$ and $|\psi_k| \leqslant 1$, $\psi_k'(0) = k$ for all $k > 0$. We denote the doublet distribution by $\delta^{(1)}$, then

$$\delta^{(1)}(\psi_k) = k, \quad \text{for all } k > 0 \tag{1.2.13}$$

If $\delta^{(1)}$ were generated by a measure μ, then

$$k = |\delta^{(1)}(\psi_k)| = \left| \int_{-1}^{+1} \phi(x) \sin(kx)\, d\mu \right| \leqslant |\mu|\,(B(0; 1)).$$

Hence

$$|\mu|\,(B(0; 1)) \geqslant k \quad \text{for all } k > 0,$$

which is impossible since $|\mu|\,(B(0; 1))$ is finite. ∎

Thus the space of distributions is indeed larger than the space of measures. The notation $\delta^{(1)}$ for the doublet is not accidental. It is related to the Dirac distribution via the process of derivation. If we imitate the procedure outlined at the end of the previous section, we would define the derivative of δ by a relation analogous to (1.1.9), i.e.,

$$\delta'(\phi) = -\delta(\phi') = -\phi'(0) = -\delta^{(1)}(\phi) \tag{1.2.14}$$

Thus by differentiating a measure (in the sense of distributions) we produce a new distribution which may no longer be a measure. The above process of differentiation can be iterated and distributions can now be infinitely derivable. We will make these notions precise in the next section.

We conclude this section with the notion of the *order* of a distribution.

Theorem 1.2.2 Let Ω be an open subset of $\mathbb{R}^n$. The following are equivalent:

(i) $T \in \mathcal{D}'(\Omega)$.

(ii) For every compact set $K \subset \Omega$ there exists a constant $C = C(K) > 0$ and an integer $N = N(K)$ such that

$$|T(\phi)| \leqslant C\|\phi\|_N \tag{1.2.15}$$

for all $\phi \in \mathcal{D}(\Omega)$ with $\operatorname{supp}(\phi) \subset K$, where $\|\phi\|_N$ is the maximum absolute value over Ω of ϕ and all its derivatives upto order N.

Proof: Assume (ii) is true. If $\phi_m \to 0$ in $\mathcal{D}(\Omega)$, then $\|\phi_m\|_N \to 0$ for all N and hence by virtue of (1.2.15), $T(\phi_m) \to 0$, and thus $T \in \mathcal{D}'(\Omega)$.

Conversely, if $T \in \mathcal{D}'(\Omega)$, let $\mathcal{D}_K$ stand for the subspace of functions in $\mathcal{D}(\Omega)$ with support in K and let $\phi_m \to 0$ in $\mathcal{D}_K$. Then $\phi_m \to 0$ in $\mathcal{D}(\Omega)$ and $T(\phi_m) \to 0$. Now it will be seen in Appendix 2 that the induced topology on $\mathcal{D}_K$ is metrizable and is generated by the semi-norms $\|\cdot\|_N$. Thus T restricted to $\mathcal{D}_K$ is continuous for any compact set $K \subset \Omega$ and (1.2.15) just expresses this fact. ∎

If it turns out that a single integer N serves for all K then T is said to be of **finite order** and the least such N is called the **order** of T. If not, T is said to be of **infinite order.** The distributions in Examples (1.2.3)–(1.2.5) are of order zero and the doublet has order 1.

1.3 SOME OPERATIONS WITH DISTRIBUTIONS

In this section we will study some familiar operations of the Calculus applied to distributions. We start with a description of the very useful multi-index notation of L. Schwartz.

Let $x \in \mathbb{R}^n$ with coordinates $(x_1, \ldots, x_n)$. A *multi-index* is an n-tuple

$$\alpha = (\alpha_1, \ldots, \alpha_n),\ \alpha_i \geqslant 0,\ \alpha_i \text{ integers.}$$

Associated to a multi-index α, we have the following symbols

$$\left.\begin{aligned} |\alpha| &= \alpha_1 + \ldots + \alpha_n \\ \alpha! &= \alpha_1! \ldots \alpha_n! \\ x^\alpha &= x_1^{\alpha_1} \ldots x_n^{\alpha_n}, \quad x \in \mathbb{R}^n \end{aligned}\right\} \tag{1.3.1}$$

We say that two multi-indices α and β are related by $\alpha \leqslant \beta$ if $\alpha_i \leqslant \beta_i$ for all $1 \leqslant i \leqslant n$. Finally we set

$$D^\alpha = \frac{\partial^{|\alpha|}}{\partial x_1^{\alpha_1} \ldots \partial x_n^{\alpha_n}}$$

Thus, for example, if $n = 2$ and $\alpha = (2, 1)$, then

$$D^\alpha = \frac{\partial^3}{\partial x_1^2 \partial x_2}$$

The first important operation on distributions is that of differentiation. We already have indicated in the preceding sections as to how we propose to do this. Let $T \in \mathcal{D}'(\mathbb{R})$. If $T = T_f$, f a C^1 function, then f' is locally integrable. Thus for $\phi \in \mathcal{D}(\mathbb{R})$

$$T_{f'}(\phi) = \int_{\mathbb{R}} f'\phi = -\int_{\mathbb{R}} f\phi' = -T_f(\phi').$$

Generalizing this, we define for any $T \in \mathcal{D}'(\mathbb{R})$,

$$T'(\phi) = -T(\phi'), \quad \phi \in \mathcal{D}(\mathbb{R}) \tag{1.3.3}$$

It is clear that $T' \in \mathcal{D}'(\mathbb{R})$. For if $\phi_n \to 0$ in $\mathcal{D}(\mathbb{R})$ then $\{\phi_n'\}$ is also a sequence in $\mathcal{D}(\mathbb{R})$ converging to zero. Hence $T'(\phi_n) = -T(\phi_n')$ converges to

zero. We can now iterate this:

$$T''(\phi) = -T'(\phi') = T(\phi'') \tag{1.3.4}$$

and more generally

$$T^{(k)}(\phi) = (-1)^k T(\phi^{(k)}). \tag{1.3.5}$$

In general if $T \in \mathscr{D}'(\Omega)$, $\Omega \subset \mathbb{R}^n$ an open set, then we define, for any multi-index α, the distribution $D^\alpha T$ by

$$(D^\alpha T)(\phi) = (-1)^{|\alpha|} T(D^\alpha \phi),\ \phi \in \mathscr{D}(\Omega) \tag{1.3.6}$$

Example 1.3.1 Consider the Dirac distribution δ on $\mathbb{R}$

$$\frac{d\delta}{dx}(\phi) = -\phi'(0) \tag{1.3.7}$$

which is, upto a sign, the doublet distribution. ∎

Example 1.3.2 Consider the **Heaviside function** on $\mathbb{R}$

$$H(x) = \begin{cases} 1, & x \geqslant 0 \\ 0, & x < 0 \end{cases} \tag{1.3.8}$$

This is locally integrable and hence defines a distribution. Let us denote, for the moment, the distribution by T_H. Let $\phi \in \mathscr{D}(\mathbb{R})$. Then

$$\frac{dT_H}{dx}(\phi) = -T_H\left(\frac{d\phi}{dx}\right) = -\int_0^\infty \frac{d\phi}{dx} = \phi(0) = \delta(\phi)$$

Thus we have

$$\frac{dT_H}{dx} = \delta. \quad ∎ \tag{1.3.9}$$

The above example raises an important and interesting point: if a locally integrable function has a classical derivative a.e. which is also locally integrable, then what is the relation between the distribution derivative and the distribution generated by the classical derivative? The preceding example shows that they are *not* necessarily the same. The Heaviside function is differentiable a.e. with derivative zero which generates the trivial distribution while the distribution derivative is the Dirac distribution. Of course in the case of a C^∞ function, they are the same by virtue of integration by parts. Another case where the two coincide is when the function is absolutely continuous. Let f be absolutely continuous on $\mathbb{R}$. Then it is differentiable a.e. and f' is integrable. Let $\phi \in \mathscr{D}(\mathbb{R})$ and let $\operatorname{supp}(\phi) \subset [-a, a]$. Consider the set

$$A = \{(x, y) \in \mathbb{R}^2 \mid -a \leqslant x < y \leqslant a\}$$

and let us evaluate the integral

$$I = \iint_A \phi'(x) f'(y)\, dx\, dy$$

in two ways using Fubini's theorem. On one hand

$$I = \int_{-a}^{a} f'(y) \left(\int_{-a}^{y} \phi'(x)\, dx \right) dy = \int_{-a}^{a} f'(y)\phi(y)\, dy = \int_{\mathbf{R}} f'(y)\phi(y)\, dy$$

and, on the other hand,

$$I = \int_{-a}^{a} \phi'(x) \left(\int_{x}^{a} f'(y)\, dy \right) dx = \int_{-a}^{a} \phi'(x)(f(a) - f(x))\, dx$$

$$= -\int_{-a}^{a} \phi'(x) f(x)\, dx = -\int_{\mathbf{R}} \phi'(x) f(x)\, dx.$$

Thus we have for all $\phi \in \mathscr{D}(\mathbf{R})$,

$$T_{f'}(\phi) = \int_{\mathbf{R}} f'\phi = -\int_{\mathbf{R}} f\phi' = -T_f(\phi') = (T_f)'(\phi).$$

Of course, there exist several examples where functions are not smooth and yet the classical and distribution derivatives coincide. We shall see some of them later.

Another important operation on distributions is the multiplication by C^∞ functions. Let $\Omega \subset \mathbf{R}^n$ be an open set and let $\{\phi_m\}$ be a sequence in $\mathscr{D}(\Omega)$ which converges to zero. Then by the classical Leibniz' formula for smooth functions, it is easy to see that $\psi\phi_m \to 0$ in $\mathscr{D}(\Omega)$. Hence for any distribution T, we have $T(\psi\phi_m) \to 0$. Thus the operation defined by

$$\phi \to T(\psi\phi),\ \phi \in \mathscr{D}(\Omega)$$

for a fixed $\psi \in C^\infty(\Omega)$ and $T \in \mathscr{D}'(\Omega)$, defines a distribution. We denote this by ψT. Thus

$$(\psi T)(\phi) = T(\psi\phi),\ \phi \in \mathscr{D}(\Omega) \tag{1.3.10}$$

If T were a function, i.e. $T = T_f$, then $\psi T = T_{\psi f}$. Thus the multiplication by a C^∞ function of a distribution extends the usual notion of multiplication of two functions. The familiar rules of calculus hold. For instance let $\psi \in C^\infty(\mathbf{R})$ and $T \in \mathscr{D}'(\mathbf{R})$. Let $\phi \in \mathscr{D}(\mathbf{R})$. Then

$$\frac{d}{dx}(\psi T)(\phi) = -(\psi T)\left(\frac{d\phi}{dx}\right) = -T\left(\psi \frac{d\phi}{dx}\right)$$

$$= -T\left(\frac{d}{dx}(\psi\phi)\right) + T\left(\frac{d\psi}{dx}\phi\right)$$

$$= \left(\psi \frac{dT}{dx} + \frac{d\psi}{dx} T\right)(\phi)$$

Thus we have the product rule

$$\frac{d}{dx}(\psi T) = \psi \frac{dT}{dx} + \frac{d\psi}{dx} T \tag{1.3.11}$$

This can be easily generalized as follows:

Theorem 1.3.1 (Leibniz' Formula) Let $\Omega \subset \mathbb{R}^n$ be an open set, $\psi \in C^\infty(\Omega)$, $T \in \mathscr{D}'(\Omega)$. Then for any multi-index α,

$$D^\alpha(\psi T) = \sum_{\beta \leqslant \alpha} \frac{\alpha!}{\beta!(\alpha - \beta)!} D^\beta \psi D^{\alpha-\beta} T$$

Proof: Exercise! (Hint: Use induction on $|\alpha|$). ∎

Finally we consider sequences of distributions. On the space of distributions $\mathscr{D}'(\Omega)$, which is a dual space, we consider the weak* topology; i.e. we say that a sequence of distributions $\{T_m\}$ converges to a distribution T if for every $\phi \in \mathscr{D}'(\Omega)$,

$$T_m(\phi) \to T(\phi)$$

Theorem 1.3.2 Let $T_m \to T$ in $\mathscr{D}'(\Omega)$. Then for any multi-index α, we have $D^\alpha T_m \to D^\alpha T$ in $\mathscr{D}'(\Omega)$.

Proof: Let $\phi \in \mathscr{D}(\Omega)$. Then

$$(D^\alpha T_m)(\phi) = (-1)^{|\alpha|} T_m(D^\alpha \phi)$$
$$\to (-1)^{|\alpha|} T(D^\alpha \phi) = D^\alpha T(\phi). \blacksquare$$

Example 1.3.3 Let $\{\rho_\epsilon\}$ be the mollifiers defined in Example 1.2.2. Then $\rho_\epsilon \to \delta$, the Dirac distribution, in $\mathscr{D}'(\mathbb{R})$. For, let $\phi \in \mathscr{D}(\mathbb{R}^n)$. Then

$$\int_{\mathbb{R}^n} \rho_\epsilon(x)\phi(x)\,dx = k\epsilon^{-n} \int_{\mathbb{R}^n} \phi(x) \exp\,(-\epsilon^2/(\epsilon^2 - |x|^2))\,dx$$

$$= k \int_{\mathbb{R}^n} \phi(\epsilon y) \exp\,(-1/(1 - |y|^2))\,dy$$

$$= \phi(0) + k \int_{\mathbb{R}^n} (\phi(\epsilon y) - \phi(0)) \exp\,(-1/(1 - |y|^2))\,dy$$

Now $\phi(\epsilon y) - \phi(0) \to 0$ pointwise and by the Dominated Convergence theorem, the integral converges to zero as $\epsilon \to 0$. ∎

This feature of the mollifiers is to be expected. These functions, as $\epsilon \to 0$, have small support but preserve the volume under the graph. Thus the curves must get steeper at the origin and eventually 'converge to the δ-function' which is zero outside the origin, infinitely large at 0 and has integral, unity.

1.4 SUPPORTS AND SINGULAR SUPPORTS OF DISTRIBUTIONS

In preceding sections we came across the notion of the support of a function, which was defined as the closure of the set on which the function does not vanish. Equivalently, the support can be said to be the complement of the largest open set on which the function vanishes. This latter version of the definition of the support of a function will help us to extend this notion to distributions as well. However we first have to make sense of

the words "*an open set on which a distribution is zero*" and then go on to examine if it is reasonable to talk of the *largest* such open set.

Let $\Omega \subset \mathbb{R}^n$ be an open set and $\Omega_0 \subset \Omega$ another open subset. If $\phi \in \mathcal{D}(\Omega_0)$, then by extending ϕ outside Ω_0 by zero, we produce a function $\tilde{\phi} \in \mathcal{D}(\Omega)$. Thus we have a canonical (conitinuous) injection

$$\mathcal{D}(\Omega_0) \to \mathcal{D}(\Omega)$$

Hence, given $T \in \mathcal{D}'(\Omega)$, we can define its 'restriction' to Ω_0, denoted $T|_{\Omega_0}$, as the following linear functional:

$$(T|_{\Omega_0})(\phi) = T(\tilde{\phi}), \quad \text{for all } \phi \in \mathcal{D}(\Omega_0) \tag{1.4.1}$$

It is elementary to verify that this defines a distribution on Ω_0. Thus by T vanishing on Ω_0 we mean that for every $\phi \in \mathcal{D}(\Omega_0)$, $T(\tilde{\phi}) = 0$, $\tilde{\phi}$ being the extension of ϕ by zero outside Ω_0.

Now let $\{\Omega_i, i \in I\}$ be a family of open sets in Ω such that $T|_{\Omega_i} = 0$ for all $i \in I$. Let $\{\phi_i\}$ be a locally finite C^∞-partition of unity subordinate to this family of open sets (cf. Theorem 1.2.1). Let $\tilde{\Omega} = \bigcup_{i \in I} \Omega_i$. If $\phi \in \mathcal{D}(\tilde{\Omega})$ then supp (ϕ), being compact, will only intersect finitely many of the Ω_i. Hence we can write

$$\phi = \sum_{i \in I} \phi\phi_i \tag{1.4.2}$$

and $\phi\phi_i \in \mathcal{D}(\Omega_i)$ for each $i \in I$. Thus $T(\phi\phi_i) = 0$ and so

$$T(\phi) = \sum_{i \in I} T(\phi\phi_i) = 0.$$

Since $\phi \in \mathcal{D}(\tilde{\Omega})$ was arbitrarily chosen, it follows that $T|_{\tilde{\Omega}} = 0$.

We conclude from the preceding arguments that the following definition is meaningful.

Definition 1.4.1 The **support** of a distribution is the complement of the largest open set on which the distribution vanishes. ∎

Example 1.4.1 If $T = T_f$, the distribution generated by a continuous function, then

$$\text{supp } T_f = \text{supp } (f). \quad ∎ \tag{1.4.3}$$

Example 1.4.2 The support of the Dirac distribution and all its derivatives is the set $\{0\}$. ∎

Distributions whose supports are compact (as in the case of Example 1.4.2 above) are called—naturally!—*distributions with compact support*. Such distributions are special.

Let f be a continuous function with compact support. Then if ψ is *any* C^∞ function on the given open set Ω, the integral

$$\int_\Omega f\psi$$

makes sense. This space of C^∞ functions on Ω—denoted henceforth by $\mathcal{E}(\Omega)$—is provided with a topology where a sequence $\{\psi_n\}$ converges to zero if and only if ψ_n and all its derivatives converge to zero uniformly on all compact subsets of Ω. With this topology the operation $\psi \to \int_\Omega f\psi$ is a continuous linear functional on $\mathcal{E}(\Omega)$. Thus the distribution T_f extends to a continuous linear functional on $\mathcal{E}(\Omega)$ when f has compact support. The same is true of any distribution with compact support.

To see this, let $T \in \mathcal{D}'(\Omega)$ with supp $(T) = K$, compact. Let $\phi \in \mathcal{D}(\Omega)$ with $\phi \equiv 1$ in a neighbourhood of K. Then for any $\tilde{\phi} \in \mathcal{D}(\Omega)$,

$$T(\tilde{\phi}) - T(\phi\tilde{\phi}) = T((1 - \phi)\tilde{\phi}) = 0$$

since $1 - \phi \equiv 0$ on K and hence supp $((1 - \phi)\tilde{\phi}) \subset \tilde{\Omega}\setminus K$. Thus we conclude that

$$T = \phi T \tag{1.4.4}$$

for any such ϕ, i.e. for all $\tilde{\phi} \in \mathcal{D}(\Omega)$

$$T(\tilde{\phi}) = T(\phi\tilde{\phi}). \tag{1.4.5}$$

We use (1.4.5) to extend the definition of T to all of $\mathcal{E}(\Omega)$. If $\psi \in \mathcal{E}(\Omega)$ then $\phi\psi \in \mathcal{D}(\Omega)$ and we define

$$T(\psi) = T(\phi\psi). \tag{1.4.6}$$

It is easy to check that if $\psi_m \to 0$ in $\mathcal{E}(\Omega)$, then $\phi\psi_m \to 0$ in $\mathcal{D}(\Omega)$ and so T is continuous on $\mathcal{E}(\Omega)$.

Reciprocally, given a continuous linear functional $\tilde{T}$ on $\mathcal{E}(\Omega)$, it defines a distribution when restricted to $\mathcal{D}(\Omega)$. For if $\phi_m \to 0$ in $\mathcal{D}(\Omega)$, trivially, $\phi_m \to 0$ in $\mathcal{E}(\Omega)$ as well. In other words with the two topologies described, we have the continuous inclusion

$$\mathcal{D}(\Omega) \to \mathcal{E}(\Omega) \tag{1.4.7}$$

and so $\tilde{T}$ restricted to $\mathcal{D}(\Omega)$—denoted T—is a distribution. *This distribution has compact support*. If not, given any m, the support of T must intersect $\Omega\setminus K_m$ where $\{K_m\}$ is an increasing family of compact sets covering Ω, i.e. $K_m \subset K_{m+1}$ and $\Omega = \bigcup_m K_m$. Such a family always exists for any open set Ω. Now let $\tilde{\phi}_m \in \mathcal{D}(\Omega)$ with supp $(\tilde{\phi}_m) \subset \Omega\setminus K_m$, and such that $T(\tilde{\phi}_m) \neq 0$. Normalizing, let $\{\phi_m\}$ be such that $T(\phi_m) = 1$ and supp $(\phi_m) \subset \Omega\setminus K_m$. Clearly given any compact set K, $K \subset K_m$ for m large enough and so $\phi_m \to 0$ in $\mathcal{E}(\Omega)$ and hence

$$T(\phi_m) = \tilde{T}(\phi_m) \to 0 \quad \text{as } m \to \infty,$$

which contradicts $T(\phi_m) = 1$ for all m.

Thus the class of distributions with compact support is identical with the restrictions of the dual of $\mathcal{E}(\Omega)$ *to* $\mathcal{D}(\Omega)$. For this reason, we will denote the class of distributions with compact support by $\mathcal{E}'(\Omega)$.

Theorem 1.4.1 Let $T \in \mathcal{E}'(\Omega)$. Then T has finite order.

Proof: Let $K = \text{supp}\,(T)$, K compact and let $\phi \in \mathcal{D}(\Omega)$ with $\phi \equiv 1$ in a neighbourhood of K. Then we saw that $T = \phi T$. Let $K_1 = \text{supp}\,(\phi)$. Then if $\tilde{\phi} \in \mathcal{D}_{\tilde{K}}$, $\tilde{K}$ a compact subset of Ω, $\phi\tilde{\phi}$ has support in K_1 and so

$$|T(\tilde{\phi})| = |T(\phi\tilde{\phi})| \leqslant C(K_1)\|\phi\tilde{\phi}\|_{N(K_1)} \tag{1.4.8}$$

By Leibniz formula,

$$\|\phi\tilde{\phi}\|_{N(K_1)} \leqslant C\|\tilde{\phi}\|_{N(K_1)}. \tag{1.4.9}$$

Combining (1.4.8) and (1.4.9) we see that $N(K_1)$ suffices for all compact sets $\tilde{K}$ and so T has finite order. ∎

We saw in Example 1.4.2 that the Dirac distribution and its derivatives span a subspace of $\mathcal{E}'(\Omega)$. In fact all elements in this subspace have support $\{0\}$. We will show that these are the only distributions with point support $\{0\}$. Before giving a proof of this, we recall a simple fact on linear functionals.

Lemma 1.4.1 Let E be a topological vector space and let $\Lambda, \Lambda_1, \ldots, \Lambda_n$ be linear functionals on E such that

$$\text{Ker}\,\Lambda \supset \bigcap_{i=1}^{n} \text{Ker}\,(\Lambda_i) \tag{1.4.10}$$

Then there exist scalars $\alpha_1, \ldots, \alpha_n$, such that

$$\Lambda = \sum_{i=1}^{n} \alpha_i \Lambda_i. \tag{1.4.11}$$

Proof: For simplicity, let the scalar field be $\mathbb{R}$. Consider the map

$$\Phi : E \to \mathbb{R}^{n+1}$$

defined by

$$\Phi(x) = (\Lambda x, \Lambda_1 x, \ldots, \Lambda_n x).$$

By hypothesis if $\Lambda_i x = 0$ for all i, then $\Lambda x = 0$. Thus

$$(1, 0, \ldots, 0) \notin \text{Range of } \Phi.$$

Now Φ being linear, its range is a linear subspace of $\mathbb{R}^{n+1}$, and it is automatically closed since it is finite dimensional. Thus by the Hahn-Banach Theorem there exists a linear functional on $\mathbb{R}^{n+1}$, i.e. $(n+1)$ scalars $\beta, \beta_1, \ldots, \beta_n$ such that

$$\begin{cases} \beta\Lambda x + \sum_{i=1}^{n} \beta_i \Lambda_i x = 0 & \text{for all } x \in E \\ \beta \neq 0 \end{cases} \tag{1.4.12}$$

Thus (1.4.11) follows with $\alpha_i = -\beta_i/\beta$. ∎

Theorem 1.4.2 Let $T \in \mathcal{D}'(\mathbb{R}^n)$ with $\text{supp}\,\{T\} = \{0\}$. Then there exists an integer $k \geqslant 0$ and scalars C_α, for every multi-index α with $|\alpha| \leqslant k$ such

that

$$T = \sum_{|\alpha| \leqslant k} C_\alpha D^\alpha \delta, \tag{1.4.13}$$

δ being the Dirac distribution.

Proof: The proof will follow directly from Lemma 1.4.1 if we can show that $T(\phi) = 0$ for all $\phi \in \mathcal{D}(\mathbb{R}^n)$ such that $D^\alpha \phi(0) = 0$ for all $|\alpha| \leqslant k$, where k is the order of T, which exists by Theorem 1.4.1.

Let $\epsilon > 0$ be given. Then for a sufficiently small compact neighbourhood of 0, we have

$$|D^\alpha \phi(x)| \leqslant \epsilon, \text{ for all } |\alpha| = k, \tag{1.4.14}$$

for all x belonging to this neighbourhood. Let x belong to this neighbourhood and set $g(t) = \phi(tx)$. Then by the Mean Value Theorem it follows that

$$|\phi(x)| \leqslant C\epsilon |x|^k \tag{1.4.15}$$

where C depends on n and k. Applying the same idea to $D^\alpha \phi$, $|\alpha| < k$, we deduce that

$$|D^\alpha \phi(x)| \leqslant C\epsilon |x|^{k-|\alpha|} \tag{1.4.16}$$

for all $|\alpha| \leqslant k$.

Now let $\eta \downarrow 0$. Let $\phi_1 \in \mathcal{D}(\mathbb{R}^n)$ with $\phi_1 \equiv 1$ in a neighbourhood of 0 and supp $(\phi_1) \subset B(0; 1)$. If we set

$$\phi_\eta(x) = \phi\left(\frac{x}{\eta}\right)$$

then $\phi_\eta \equiv 1$ in a neighbourhood of 0 and supp $(\phi_\eta) \subset B(0; \eta)$. By the Leibniz Formula,

$$D^\alpha(\phi\phi_\eta)(x) = \sum_{\beta \leqslant \alpha} C_{\alpha\beta} D^{\alpha-\beta}\phi_\eta(x) D^\beta \phi(x)$$

$$= \sum_{\beta \leqslant \alpha} C_{\alpha\beta} D^{\alpha-\beta}\phi_1\left(\frac{x}{\eta}\right) D^\beta \phi(x) \eta^{|\beta|-|\alpha|}$$

and hence for $|x| < \eta$,

$$\|\phi\phi_\eta\|_k \leqslant C\epsilon \|\phi_1\|_k, \tag{1.4.17}$$

using (1.4.16). Since $\phi_\eta \equiv 1$ in a neighbourhood of supp (T) it follows that

$$|T(\phi)| = |T(\phi_\eta \phi)| \leqslant C\|\phi_\eta \phi\|_k \leqslant C\epsilon \|\phi_1\|_k$$

and hence $T(\phi) = 0$ since $\epsilon > 0$ was arbitrarily chosen. ∎

In the beginning of this section we described the local behaviour of a given global distribution, i.e. given $T \in \mathcal{D}'(\Omega)$ we explained what was meant by $T|_{\Omega_0}$ $\Omega_0 \subset \Omega$. Conversely, given several locally defined distributions which are "compatible" in a sense to be made precise below, we can "patch them up" to define a global distribution. More precisely we prove the following theorem.

Theorem 1.4.3 Let $\Omega \subset \mathbb{R}^n$ be an open set and let $\{\Omega_i\}$, $i \in I$ constitute an open cover of Ω. Let $T_i \in \mathcal{D}'(\Omega_i)$ such that whenever $\Omega_i \cap \Omega_j \neq \emptyset$, $i \neq j$, then

$$T_i|_{\Omega_i \cap \Omega_j} = T_j|_{\Omega_i \cap \Omega_j}. \tag{1.4.18}$$

Then there exists a unique distribution $T \in \mathcal{D}'(\Omega)$ such that

$$T|_{\Omega_i} = T_i \text{ for all } i \in I. \tag{1.4.19}$$

Proof: Let $\{\phi_i\}$, $i \in I$ be a locally finite C^∞ partition of unity subordinate to the cover $\{\Omega_i\}$. Let $\phi \in \mathcal{D}(\Omega)$. Then the support of ϕ intersects only finitely many open sets Ω_i and $\phi\phi_i$ has support in Ω_i. We define

$$T(\phi) = \sum_{i \in I} T_i(\phi\phi_i) \tag{1.4.20}$$

which makes sense since the right-hand-side is essentially a finite sum. Let $\tilde{\phi}_m \to 0$ in $\mathcal{D}(\Omega)$. Let K be a compact set containing supp $(\tilde{\phi}_m)$ for all m. Let $i_1, i_2, \ldots, i_l$ be the indices such that $K \cap$ supp (ϕ_{i_j}) is non-empty for $1 \leqslant j \leqslant l$ and $K \cap$ supp $(\phi_i) = \emptyset$ for all other i. Thus

$$T(\tilde{\phi}_m) = \sum_{j=1}^{l} T_{i_j}(\tilde{\phi}_m \phi_{i_j})$$

Note that $\tilde{\phi}_m\phi_{i_j} \to 0$ in $\mathcal{D}(\Omega_{i_j})$. Thus $T(\tilde{\phi}_m) \to 0$ and so $T \in \mathcal{D}'(\Omega)$.

We now show that $T|_{\Omega_i} = T_i$. Let $\phi \in \mathcal{D}(\Omega_i)$. For any j,

$$\phi\phi_j \in \mathcal{D}(\Omega_i \cap \Omega_j).$$

Then by (1.4.18),

$$T_i(\phi\phi_j) = T_j(\phi\phi_j).$$

Hence

$$T(\phi) = \sum_j T_j(\phi\phi_j) = \sum_j T_i(\phi\phi_j) = T_i(\sum_j \phi\phi_j) = T_i(\phi).$$

The uniqueness of T is obvious. ∎

Let $\Omega \subset \mathbb{R}^n$ be an open set and let $T \in \mathcal{D}'(\Omega)$. If there exists an $f \in C^\infty(\Omega)$ such that $T = T_f$, we say that T is C^∞ on Ω. Let $\{\Omega_i\}$, $i \in I$ be a family of open sets in Ω such that $T|_{\Omega_i}$ is C^∞ on Ω_i. Then by virtue of the preceding theorem, T is C^∞ on $\bigcup_{i \in I} \Omega_i$. Thus, a given a distribution T on Ω, it makes sense to talk of the largest open subset on which T is C^∞. Thus we can define the following set:

Definition 1.4.2 Let $T \in \mathcal{D}'(\Omega)$. The **singular support** of T is the complement of the largest open set on which T is C^∞. ∎

Clearly the singular support of T, denoted by sing. supp (T), is closed in Ω and

$$\text{sing supp } (T) \subset \text{supp } (T). \tag{1.4.21}$$

1.5 CONVOLUTION OF FUNCTIONS

We will now study an important and useful operation on functions which we will later extend to certain classes of distributions. Let $f, g \in L^1(\mathbb{R}^n)$. For $x \in \mathbb{R}^n$, consider the integral,

$$h(x) = \int_{\mathbb{R}^n} f(x-y)g(y)\,dy \tag{1.5.1}$$

This integral is well-defined since $F(x, y) = f(x - y)g(y)$ is measurable in the product space $\mathbb{R}_x^n \times \mathbb{R}_y^n$ and by Fubini's theorem and the property of translation invariance of the Lebesgue measure,

$$\int_{\mathbb{R}^n \times \mathbb{R}^n} |F(x, y)| \, dx \, dy = \int_{\mathbb{R}^n} |g(y)| \, dy \int_{\mathbb{R}^n} |f(x - y)| \, dx$$
$$= \|g\|_{L^1(\mathbb{R}^n)} \|f\|_{L^1(\mathbb{R}^n)} < +\infty .$$

Hence, again by Fubini's Theorem, the x-section of F, viz.

$$F_x^*(y) = f(x - y)g(y)$$

is in $L^1(\mathbb{R}^n)$ and $h(x)$ is well-defined. Further $h \in L^1(\mathbb{R}^n)$ as a function of x and by the preceding computation,

$$\|h\|_{L^1(\mathbb{R}^n)} \leqslant \|f\|_{L^1(\mathbb{R}^n)} \|g\|_{L^1(\mathbb{R}^n)} . \tag{1.5.2}$$

Definition 1.5.1 Let $f, g \in L^1(\mathbb{R}^n)$. Then the function $h \in L^1(\mathbb{R}^n)$ defined via the equation (1.5.1) is called the **convolution** of f and g and is denoted by

$$h = f * g. \quad \blacksquare \tag{1.5.3}$$

Theorem 1.5.1 The convolution is a commutative and associative binary operation on $L^1(\mathbb{R}^n)$.

Proof: Let $f, g \in L^1(\mathbb{R}^n)$. Then

$$(f * g)(x) = \int_{\mathbb{R}^n} f(x - y)g(y) \, dy = \int_{\mathbb{R}^n} f(z)g(x - z) \, dz = (g * f)(x)$$

using the change of variable $z = x - y$. This proves the commutativity.

If $f, g, h \in L^1(\mathbb{R}^n)$, then by use of the change of variable $z = t - y$ and by Fubini's theorem,

$$((f * g) * h)(x) = \int_{\mathbb{R}_y^n} (f * g)\,(x - y)h(y) \, dy$$
$$= \int_{\mathbb{R}_y^n} \int_{\mathbb{R}_z^n} f(x - y - z)g(z)h(y) \, dz \, dy$$
$$= \int_{\mathbb{R}_t^n} f(x - t) \int_{\mathbb{R}_y^n} g(t - y)h(y) \, dy \, dt$$
$$= \int_{\mathbb{R}_t^n} f(x - t)(g * h)(t) \, dt$$
$$= (f * (g * h))(x)$$

which proves the associativity. $\blacksquare$

Theorem 1.5.2 Let $1 < p < \infty$ and $f \in L^1(\mathbb{R}^n)$, $g \in L^p(\mathbb{R}^n)$. The $f * g$ is well-defined and further, $f * g \in L^p(\mathbb{R}^n)$ with

$$\|f * g\|_{L^p(\mathbb{R}^n)} \leqslant \|f\|_{L^1(\mathbb{R}^n)} \|g\|_{L^p(\mathbb{R}^n)} \tag{1.5.4}$$

Proof. Let q be the conjugate exponent of p, i.e. $1/p + 1/q = 1$. Let $h \in L^q(\mathbb{R}^n)$. Then $(x, y) \to f(x - y)g(y)h(x)$ is measurable and

$$\int_{\mathbb{R}^n_x}\int_{\mathbb{R}^n_y} |f(x-y)g(y)h(x)|\,dx\,dy = \int_{\mathbb{R}^n_x} |h(x)| \int_{\mathbb{R}^n_y} |f(x-y)g(y)|\,dy\,dx$$

$$= \int_{\mathbb{R}^n_x} |h(x)| \int_{\mathbb{R}^n_t} |f(t)g(x-t)|\,dt\,dx$$

$$= \int_{\mathbb{R}^n_t} |f(t)|\,dt \int_{\mathbb{R}^n_x} |h(x)|\,|g(x-t)|\,dx$$

$$\leqslant \|h\|_{L^q(\mathbb{R}^n)}\|g\|_{L^p(\mathbb{R}^n)}\|f\|_{L^1(\mathbb{R}^n)} < +\infty$$

where we have used Holder's inequality and the fact that by the translation invariance of the Lebesgue measure $g(x)$ and $g(x-t)$ have the same L^p norm. Thus by Fubini's theorem

$$\int_{\mathbb{R}^n_y} h(x)f(x-y)g(y)\,dy$$

exists for almost for all x and since we can choose $h(x) \neq 0$ for all x (e.g. $h(x) = \exp(-|x|^2)$, $|x|^2 = \sum_{i=1}^{n} x_i^2$) we deduce that $f * g$ defined by (1.5.1) is well defined. Also

$$h \to \int (f * g)h$$

is a continuous linear functional on $L^q(\mathbb{R}^n)$ with norm bounded by $\|g\|_{L^p(\mathbb{R}^n)}\|f\|_{L^1(\mathbb{R}^n)}$ which shows, by the Riesz Representation Theorem, that $f * g \in L^p(\mathbb{R}^n)$ and that (1.5.4) holds. ∎

Remark 1.5.1 The inequality (1.5.4) is a particular case of *Young's inequality*. Let $1 \leqslant p, q, r < \infty$ such that

$$(1/p) + (1/q) = 1 + (1/r) \tag{1.5.5}$$

If $f \in L^p(\mathbb{R}^n)$, $g \in L^q(\mathbb{R}^n)$ then $f * g \in L^r(\mathbb{R}^n)$ and

$$\|f * g\|_{L^r(\mathbb{R}^n)} \leqslant \|f\|_{L^p(\mathbb{R}^n)}\,\|g\|_{L^q(\mathbb{R}^n)}. \quad ∎ \tag{1.5.6}$$

If f is a continuous function on $\mathbb{R}^n$ and g continuous with *compact support* then again the integral in (1.5.1) makes sense. Thus in this case also we use (1.5.1) to define the convolution $f * g$. More generally, if $f_1, \ldots, f_k$ are continuous functions on $\mathbb{R}^n$ such that all but at most one of them has compact support, then again we can define $f_1 * \ldots * f_k$ by considering the functions as follows (for instance):

$$f_1 * (f_2 * \ldots (f_{k-1} * f_k)).$$

The actual order will be unimportant by commutativity and associativity. That between each parenthesis, at least one element has compact support is a consequence of the following result.

Theorem 1.5.3 Let f be continuous and g continuous with compact support. Then

$$\operatorname{supp}(f * g) \subset \operatorname{supp}(f) + \operatorname{supp}(g) \tag{1.5.7}$$

where for sets A and B in $\mathbb{R}^n$,

$$A+B=\{x+y \mid x \in A, y \in B\}. \tag{1.5.8}$$

Proof: Set $A=\operatorname{supp}(f)$, $B=\operatorname{supp}(g)$, B compact. Then $A+B$ is closed, for if $x_k+y_k \in A+B$, $x_k+y_k \to z$, then (for a subsequence) $y_k \to y \in B$. Hence $x_k \to z-y$ and the limit must be in A. Thus $z-y=x \in A$ and so $z \in A+B$.

In order that $f*g(x) \neq 0$, we need clearly only consider the integral in (1.5.1) over the set $B=\operatorname{supp}(g)$. Then necessarily, $x-y \in A=\operatorname{supp}(f)$ for all y in a subset (of non-zero measure) of B. Hence $x \in \operatorname{supp}(f)+\operatorname{supp}(g)$ and so

$$\operatorname{supp}(f*g)=\overline{\{x \mid (f*g)(x) \neq 0\}} \subset \operatorname{supp}(f)+\operatorname{supp}(g). \quad ∎$$

One of the important properties of the convolution is that it has a smoothing effect on functions. More precisely, we have the following result.

Theorem 1.5.4 Let f be continuous and g continuous with compact support. If one of them is C^∞ then $f*g$ is C^∞.

Proof: Let f be C^∞. It suffices to show that

$$\frac{\partial}{\partial x_i}(f*g)=\left(\frac{\partial f}{\partial x_i}\right)*g, \; 1 \leqslant i \leqslant n. \tag{1.5.9}$$

(i) To show $f*g$ is continuous: Consider $h \in \mathbb{R}^n$, and

$$|(f*g)(x+h)-(f*g)(x)| \leqslant \int |(f(x+h-y)-f(x-y)| \, |g(y)| \, dy.$$

It suffices to consider the above integral over $K=\operatorname{supp}(g)$, which is compact. Hence for fixed x, the set

$$x-K=\{x-y \mid y \in K\}$$

is compact and hence f is uniformly continuous over it. Hence given $\epsilon>0$, there exists $\eta>0$ such that if $|h|<\eta$ then $|f(x-y+h)-f(x-y)| < \epsilon$ and so the integrand is bounded by $\epsilon|g(y)|$ which is integrable and tends to zero as $h \to 0$. Thus $(f*g)(x+h) \to (f*g)(x)$ as $h \to 0$ and so $f*g$ is continuous.

(ii) To prove (1.5.9): Let $e_i=(0,\ldots,1,\ldots,0)$, where 1 occurs in the ith place; then

$$\frac{\partial}{\partial x_i}(f*g)(x)=\lim_{h \to 0}\frac{(f*g)(x+he_i)-(f*g)(x)}{h}$$

Again if $K=\operatorname{supp}(g)$, we have

$$\begin{aligned}\frac{1}{h}[(f*g)(x+he_i)-(f*g)(x)] &= \frac{1}{h}\int_K (f(x+he_i-y)-f(x-y))g(y)\,dy \\ &= \int_K \frac{\partial f}{\partial x_i}((x-y)+\theta he_i)g(y)\,dy\end{aligned}$$

with $0<\theta<1$. Again $\dfrac{\partial f}{\partial x_i}$ is continuous and hence bounded on the com-

pact set K and as $h \to 0$ the integrand converges to $\frac{\partial f}{\partial x_i}(x - y)g(y)$. The result now follows, once more, from the Dominated Convergence Theorem. ∎

Remark 1.5.2 If α is any multi-index, g continuous with compact support, f a C^∞ function, then

$$D^\alpha(f * g)(x) = ((D^\alpha f) * g)(x). \quad ∎ \tag{1.5.10}$$

Thus convolution with a smooth function produces a smooth function out of 'any' function. This fact used together with the sequence of mollifiers considered in Example 1.2.2 provides us with a powerful technique to prove a variety of density theorems. We illustrate this now.

Theorem 1.5.5 Let f be a continuous function with compact support in $\mathbb{R}^n$. Then there exist a compact set K containing the support of f and a family $\{\phi_\epsilon\}$ of C^∞ functions with compact support contained in K such that $\phi_\epsilon \to f$ uniformly in K as $\epsilon \to 0$.

Proof: For $0 < \epsilon < 1$, let ρ_ϵ be the mollifiers constructed previously (cf. Example 1.2.2). Define

$$\phi_\epsilon = \rho_\epsilon * f$$

Then ϕ_ϵ is C^∞ and

$$\begin{aligned} \text{supp}\,(\phi_\epsilon) &\subset \text{supp}\,(f) + B(0; \epsilon) \\ &\subset \text{supp}\,(f) + B(0; 1) = K. \end{aligned}$$

Thus K is compact since it is the sum of two compact sets and contains the supports of f and all the ϕ_ϵ. Let $\eta > 0$ be given. Since f is uniformly continuous on K, there exists $\epsilon > 0$ such that $|y| < \epsilon$ implies $|f(x - y) - f(x)| < \eta$ for all $x \in K$. Hence

$$\begin{aligned} |\phi_\epsilon(x) - f(x)| &\leqslant \int_{|y| \leqslant \epsilon} |f(x - y) - f(x)|\rho_\epsilon(y)\, dy \\ &\leqslant \eta \int_{|y| \leqslant \epsilon} \rho_\epsilon(y)\, dy = \eta \end{aligned}$$

Hence $\phi_\epsilon \to f$ uniformly on K. ∎

Theorem 1.5.6 Let $1 \leqslant p < \infty$. Then $\mathscr{D}(\mathbb{R}^n)$ is dense in $L^p(\mathbb{R}^n)$.

Proof: *Step 1* Let S be the class of measurable simple functions ϕ such that

$$\mu\{x \mid \phi(x) \neq 0\} < +\infty,$$

μ being the Lebesgue measure. If $\phi \in S$ then ϕ is zero outside a set of finite measure and hence $\phi \in L^p(\mathbb{R}^n)$. Let $f \in L^p(\mathbb{R}^n)$ and $f \geqslant 0$. Then there exist simple functions ϕ_m such that $0 \leqslant \phi_m \leqslant f$, $\phi_m \uparrow f$. Since $f \in L^p(\mathbb{R}^n)$, it follows that $\phi_m \in S$. Also $|\phi_m - f|^p \leqslant 2^p|f|^p$ and since $|f|^p$ is integrable, by the Dominated Convergence Theorem it follows that

$\phi_m \to f$ in $L^p(\mathbb{R}^n)$. If f is any element in $L^p(\mathbb{R}^n)$, write $f = f^+ - f^-$ with $f^+ \geqslant 0, f^- \geqslant 0$ and so there exist $\phi_m, \psi_m \in S$ with $\phi_m \to f^+, \psi_m \to f^-$ in $L^p(\mathbb{R}^n)$ and so $\phi_m - \psi_m \to f$ in $L^p(\mathbb{R}^n)$. Thus S is dense in $L^p(\mathbb{R}^n)$.

Step 2 Let $\phi \in S$. Given $\epsilon > 0$, there exists a continuous function g with compact support such that $g = \phi$ except possibly on a set of measure ϵ and such that

$$|g| \leqslant \|\phi\|_{L^\infty(\mathbb{R}^n)}$$

(This is a consequence of Lusin's Theorem.) Hence

$$\|g - \phi\|_{L^p(\mathbb{R}^n)} \leqslant 2\epsilon^{(1/p)}\|\phi\|_{L^\infty(\mathbb{R}^n)}$$

and it follows from Step 1 above that continuous functions with compact support are dense in $L^p(\mathbb{R}^n)$.

Step 3 Let f be a continuous function with compact support and let $\phi_m \in \mathcal{D}(\mathbb{R}^n)$ with supp (ϕ_m), supp $(f) \subset K$, K a fixed compact set, such that $\phi_m \to f$ uniformly on K. Then

$$\int_{\mathbb{R}^n} |f - \phi_m|^p = \int_K |f - \phi_m|^p \leqslant (\sup_K |f - \phi_m|)^p \mu(K) \to 0$$

as $m \to \infty$. Thus ϕ_m approximates f in the L^p-norm as well and the result now follows on combining this fact with the conclusion of Step 2. ∎

1.6 CONVOLUTION OF DISTRIBUTIONS

We will now generalize the notion of convolution to certain classes of distributions. There are notably two ways of doing this: one is via the tensor product of distributions as described in L. Schwartz [1]; the other—the one which we will follow—is described in Rudin [2] and in Hormander [1]. In this latter approach, generalizing from the notion of convolution of two functions, we will first define the convolution of a distribution and a test function. Then we will extend this to the convolution of two distributions at least one of them having compact support.

We first introduce some useful notations. Let u be a function on $\mathbb{R}^n$ and let $x \in \mathbb{R}^n$. We define

$$(\tau_x u)(y) = u(y - x) \tag{1.6.1}$$

the translation operator, and

$$\check{u}(y) = u(-y) \tag{1.6.2}$$

Starting from these relations, the following are easy to verify:

$$(\tau_x u)^\vee = \tau_{-x}(\check{u}) \tag{1.6.3}$$

$$\tau_x \tau_y = \tau_{x+y} \tag{1.6.4}$$

Given two functions u and v on $\mathbb{R}^n$ we have

$$\int_{\mathbb{R}^n} (\tau_x u)v = \int_{\mathbb{R}^n} u(y-x)v(y)\,dy = \int_{\mathbb{R}^n} u(z)v(z+x)\,dz = \int_{\mathbb{R}^n} u(\tau_{-x}v). \tag{1.6.5}$$

In the same vein, if $T \in \mathscr{D}'(\mathbb{R}^n)$ we define $\tau_x T \in \mathscr{D}'(\mathbb{R}^n)$ by

$$(\tau_x T)(\phi) = T(\tau_{-x}\phi) \quad \text{for all } \phi \in \mathscr{D}(\mathbb{R}^n) \tag{1.6.6}$$

which is exactly (1.6.5) if T is a function. It is easy to see that $\tau_x T$ is indeed a distribution.

If u and v are functions whose convolution can be defined, then we have

$$(u*v)(x) = \int_{\mathbb{R}^n} u(y)v(x-y)\,dy = \int_{\mathbb{R}^n} u(y)(\tau_x \check{v})(y)\,dy \tag{1.6.7}$$

Thus we are led to

Definition 1.6.1 Let $T \in \mathscr{D}'(\mathbb{R}^n)$ and $\phi \in \mathscr{D}(\mathbb{R}^n)$. Then the convolution of T and ϕ is a *function* $T*\phi$ defined by

$$(T*\phi)(x) = T(\tau_x\check{\phi}). \quad \blacksquare$$

Note that if $\phi \in \mathscr{D}(\mathbb{R}^n)$, so does $\tau_x\check{\phi}$ so that the above equation has a meaning and if T is a function, this is exactly (1.6.7).

We now prove some of the basic properties of the convolution defined above.

Theorem 1.6.1 Let $T \in \mathscr{D}'(\mathbb{R}^n)$ and $\phi \in \mathscr{D}(\mathbb{R}^n)$. Then:

(i) for any $x \in \mathbb{R}^n$,

$$\tau_x(T*\phi) = (\tau_x T)*\phi = T*(\tau_x\phi) \tag{1.6.8}$$

(ii) if α is any multi-index,

$$D^\alpha(T*\phi) = (D^\alpha T)*\phi = T*(D^\alpha\phi) \tag{1.6.9}$$

In particular, $T*\phi \in \mathscr{E}(\mathbb{R}^n)$.

(iii) If $\psi \in \mathscr{D}(\mathbb{R}^n)$, then

$$T*(\phi*\psi) = (T*\phi)*\psi \tag{1.6.10}$$

(iv) If $T \in \mathscr{D}'(\mathbb{R}^n)$ such that $T*\phi = 0$ for all $\phi \in \mathscr{D}(\mathbb{R}^n)$ then $T = 0$.

Proof: (i) The proof of (1.6.8) is by a straightforward computation using relations (1.6.3)–(1.6.6).

$$\begin{aligned}
\tau_x(T*\phi)(y) &= (T*\phi)(y-x) = T(\tau_{y-x}\check{\phi}) \\
&= T(\tau_{-x}\tau_y\check{\phi}) = (\tau_x T)(\tau_y\check{\phi}) = (\tau_x T*\phi)(y) \\
&= T(\tau_y(\tau_x\phi)^\vee) = (T*\tau_x\phi)(y).
\end{aligned}$$

(ii) It suffices to prove (1.6.9) when $\alpha = (0, 0, \ldots, 1, \ldots, 0)$ with 1 at the ith place. By iterating this suitably, we can deduce the general case.

Let e_i be the ith (standard) basis vector of $\mathbb{R}^n$. Then

$$\frac{\partial}{\partial x_i}(x)(T*\phi) = \lim_{h\to 0}\frac{(T*\phi)(x)-(T*\phi)(x-he_i)}{h}$$

$$= \lim_{h\to 0}\frac{1}{h}[(T*\phi)(x)-\tau_{he_i}(T*\phi)(x)]$$

$$= \lim_{h\to 0}\frac{1}{h}[(T*\phi)(x)-(T*(\tau_{he_i}\phi))(x)]$$

$$= \lim_{h\to 0}\frac{1}{h}(T*(\phi-\tau_{he_i}\phi))(x)$$

$$= \lim_{h\to 0} T\left(\tau_x\frac{(\phi-\tau_{he_i}\phi)^{\vee}}{h}\right)$$

$$= T\left(\tau_x\left(\frac{\partial\phi}{\partial x_i}\right)^{\vee}\right) = \left(T*\frac{\partial\phi}{\partial x_i}\right)(x).$$

since $\dfrac{\phi-\tau_{he_i}\phi}{h}\to\dfrac{\partial\phi}{\partial x_i}$ in $\mathcal{D}(\mathbb{R}^n)$.

Iterating this we get one part of (1.6.9). To prove the other, we have

$$\frac{\partial}{\partial x_i}(T*\phi)(x) = \lim_{h\to 0}\left(\left(\frac{T-\tau_{he_i}T}{h}\right)*\phi\right)(x)$$

$$= \lim_{h\to 0}\left(\frac{T-\tau_{he_i}T}{h}\right)(\tau_x\check{\phi})$$

$$= \frac{\partial T}{\partial x_i}(\tau_x\check{\phi}) = \left(\frac{\partial T}{\partial x_i}*\phi\right)(x)$$

Here we use the fact that $\dfrac{T-\tau_{he_i}T}{h}$ converges to $\dfrac{\partial T}{\partial x_i}$ in $\mathcal{D}'(\Omega)$. (Proof: Exercise!)

(iii) Let $\phi, \psi \in \mathcal{D}(\mathbb{R}^n)$. Now it suffices to prove that

$$(T*(\phi*\psi))(0) = ((T*\phi)*\psi)(0) \tag{1.6.11}$$

Then (1.6.10) will follow since for any $x \in \mathbb{R}^n$ we can apply (1.6.11) with $\tau_{-x}\psi$ instead of ψ and then use (1.6.8). But

$$(T*(\phi*\psi))(0) = T((\phi*\psi)^{\vee})$$

Now

$$(\phi*\psi)^{\vee}(x) = (\phi*\psi)(-x) = \int\phi(-x-y)\psi(y)\,dy$$

$$= \int\tau_y\check{\phi}(x)\check{\psi}(y)\,dy$$

and it suffices to consider this integral over the compact set, supp $(\check{\psi})$. This integral could be considered as the limit, as $\epsilon\to 0$, of the Riemann sum

$$\sum_p \tau_{\epsilon p}\check{\phi}(x)\check{\psi}(\epsilon p)$$

where the sum extends over all integral lattice points p in $\mathbb{R}^n$. Again this is essentially only a finite sum since for any fixed number $\epsilon \to 0$, only a finite number of ϵp will lie in the set supp $(\check{\psi})$, which is compact. Thus

$$(\phi * \psi)^\vee(\cdot) = \lim_{\epsilon \to 0} \sum_p \tau_{\epsilon p}\check{\phi}(\cdot)\check{\psi}(\epsilon p)$$

in the topology of $\mathscr{D}(\mathbb{R}^n)$ and so

$$\begin{aligned}(T * (\phi * \psi))(0) &= T((\phi * \psi)^\vee) = \lim_{\to 0} \sum_p T(\tau_{\epsilon p}\check{\phi})\check{\psi}(\epsilon p) \\ &= \int (T * \phi)(y)\psi(-y)\, dy = ((T * \phi) * \psi)(0).\end{aligned}$$

This completes the proof of (1.6.10).

(iv) Finally, if $T \in \mathscr{D}'(\mathbb{R}^n)$ such that $T * \phi = 0$ for all $\phi \in \mathscr{D}(\mathbb{R}^n)$, then as $\check{\phi} \in \mathscr{D}(\mathbb{R}^n)$ for any $\phi \in \mathscr{D}(\mathbb{R}^n)$,

$$0 = (T * \check{\phi})(0) = T(\phi)$$

Thus $T = 0$. ∎

The definition of convolution can be extended to $\psi \in \mathscr{E}(\mathbb{R}^n)$ provided T has compact support. Thus if $T \in \mathscr{E}'(\mathbb{R}^n)$ and $\psi \in \mathscr{E}(\mathbb{R}^n)$ we can again define the function $T * \psi$ by

$$(T * \psi)(x) = T(\tau_x\check{\psi}). \tag{1.6.12}$$

Once again we can prove properties analogous to those described in Theorem 1.6.1.

Theorem 1.6.2 Let $T \in \mathscr{E}'(\mathbb{R}^n)$, $\psi \in \mathscr{E}(\mathbb{R}^n)$. Then

$$\tau_x(T * \psi) = T * (\tau_x\psi) = (\tau_x T) * \psi, \tag{1.6.13}$$

$$D^\alpha(T * \psi) = (D^\alpha T) * \psi = T * D^\alpha\psi \tag{1.6.14}$$

for any multi-index α. In particular $T * \psi \in \mathscr{E}(\mathbb{R}^n)$. Further, if $\phi \in \mathscr{D}(\mathbb{R}^n)$, then $T * \phi \in \mathscr{D}(\mathbb{R}^n)$ and

$$T * (\psi * \phi) = (T * \psi) * \phi = (T * \phi) * \psi. \tag{1.6.15}$$

Proof: The proofs of relations (1.6.13) and (1.6.14) follows the same lines as those for (1.6.8), (1.6.9). Let $K =$ supp (T) and $H =$ supp (ϕ), K, H compact. Now

$$(T * \phi)(x) = T(\tau_x\check{\phi}) \tag{1.6.16}$$

The support of $\tau_x\check{\phi}$ is the set $x - H$ and so the right-hand-side of (1.6.16) will vanish unless $(x - H) \cap K \neq \emptyset$, i.e., $x \in K + H$. Thus

$$\text{supp } (T * \phi) \subset \text{supp } (T) + \text{supp } (\phi)$$

and the set of the right-hand-side, being a sum of compact sets, is compact. Hence

$$T * \phi \in \mathscr{D}(\mathbb{R}^n) \text{ for } T \in \mathscr{E}'(\mathbb{R}^n) \text{ and } \phi \in \mathscr{D}(\mathbb{R}^n).$$

Finally we prove (1.6.15). Again we only show that these functions agree at the origin. The rest follows, as in the previous theorem, using translations.

Let $\psi_0 \in \mathcal{D}(\mathbb{R}^n)$ such that $\check{\psi}_0 = \check{\psi}$ in a neighbourhood of $K + H$, which is compact. Now supp $(T * \phi) \subset K + H$ and so

$$\begin{aligned}((T * \phi) * \psi)(0) &= \int_{K+H} (T * \phi)(y)\psi(-y)\, dy \\ &= \int_{K+H} (T * \phi)(y)\psi_0(-y)\, dy \\ &= ((T * \phi) * \psi_0)(0) \\ &= ((T * \psi_0) * \phi)(0)\end{aligned}$$

using Theorem 1.6.1, as both $\phi, \psi_0 \in \mathcal{D}(\mathbb{R}^n)$. Now if $-s \in H$, then $\tau_s\check{\psi} = \tau_s\check{\psi}_0$ in a neighbourhood of K and hence $T * \psi = T * \psi_0$ in $-H$. Thus

$$((T * \psi) * \phi)(0) = ((T * \psi_0) * \phi)(0).$$

It follows then that if $\phi \in \mathcal{D}(\mathbb{R}^n)$, $\psi \in \mathcal{E}(\mathbb{R}^n)$, we have

$$(T * \psi) * \phi = (T * \phi) * \psi \tag{1.6.17}$$

To complete the proof of (1.6.15), let $\tilde{\phi} \in \mathcal{D}(\mathbb{R}^n)$ be arbitrary. Then using the convolution of functions,

$$\begin{aligned}((T * \phi) * \psi) * \tilde{\phi} &= ((T * \phi) * \tilde{\phi}) * \psi \\ &= (T * \tilde{\phi}) * (\phi * \psi)\end{aligned}$$

Now $\tilde{\phi} \in \mathcal{D}(\mathbb{R}^n)$ and $\phi * \psi \in \mathcal{E}(\mathbb{R}^n)$. Thus by (1.6.17), we get

$$((T * \phi) * \psi) * \tilde{\phi} = (T * (\phi * \psi)) * \tilde{\phi}$$

Since $\tilde{\phi}$ is arbitrary, it follows from part (iv) of Theorem 1.6.1 that

$$(T * \phi) * \psi = T * (\phi * \psi).$$

This completes the proof. ∎

Now let S and T be two distributions, at least one of which has compact support. Consider the linear map $L\colon \mathcal{D}(\mathbb{R}^n) \to \mathcal{E}(\mathbb{R}^n)$ defined by

$$L(\phi) = T * (S * \phi) \tag{1.6.18}$$

If $S \in \mathcal{E}'(\mathbb{R}^n)$ then $S * \phi \in \mathcal{D}(\mathbb{R}^n)$ and so L is defined. If $T \in \mathcal{E}'(\mathbb{R}^n)$, then $S * \phi \in \mathcal{E}(\mathbb{R}^n)$ and again L is defined. Also by (1.6.8) and (1.6.13), for any $x \in \mathbb{R}^n$, we have

$$\tau_x L = L\tau_x. \tag{1.6.19}$$

Now consider the functional

$$\phi \to (L(\check{\phi}))(0),\ \phi \in \mathcal{D}(\mathbb{R}^n). \tag{1.6.20}$$

This is linear and if $\phi_m \to 0$ in $\mathcal{D}(\mathbb{R}^n)$ then $S * \check{\phi}_m \to 0$ in $\mathcal{E}$ (and in $\mathcal{D}$ if $S \in \mathcal{E}'$). Thus it follows that (1.6.20) defines a distribution which we call $T * S$. Further if $x \in R^n$,

$$\begin{aligned}(T * (S * \phi))(x) &= (L\phi)(x) = (\tau_{-x}L\phi)(0) \\ &= (L(\tau_{-x}\phi))(0) = (T * S)((\tau_{-x}\phi)^{\vee}) \\ &= (T * S)(\tau_x\check{\phi}) = ((T * S) * \phi)(x).\end{aligned}$$

Taking into account part (iv) Theorem 1.6.1, it follows that:

Definition 1.6.2 Let $T, S \in \mathcal{D}'(\mathbb{R}^n)$ with at least one of compact support. The **convolution** of T and S, denoted $T * S$, is the unique distribution characterized by the following equivalent conditions;

$$(T * S) * \phi = T * (S * \phi) \quad \text{for all } \phi \in \mathcal{D}(\mathbb{R}^n) \tag{1.6.21}$$

$$(T * S)(\phi) = (T * (S * \check{\phi}))(0). \tag{1.6.22}$$

We now prove the basic properties of the convolution product defined on distributions.

Theorem 1.6.3 Let T_i, $i = 1, 2, 3$, be distributions on $\mathbb{R}^n$.

(i) If at least one of T_1 and T_2 has compact support, then

$$T_1 * T_2 = T_2 * T_1. \tag{1.6.23}$$

(ii) If S_1 and S_2 are the supports of T_1 and T_2 respectively, at least one of them compact, then

$$\text{supp}\,(T_1 * T_2) \subset \text{supp}\,(T_1) + \text{supp}\,(T_2). \tag{1.6.24}$$

(iii) If at least two of T_i, $i = 1, 2, 3$, have compact support, then

$$T_1 * (T_2 * T_3) = (T_1 * T_2) * T_3. \tag{1.6.25}$$

(iv) If α is any multi-index,

$$D^\alpha(T_1 * T_2) = D^\alpha T_1 * T_2 = T_1 * D^\alpha T_2. \tag{1.6.26}$$

Proof: (i) Let ϕ_1 and ϕ_2 be functions in $\mathcal{D}(\mathbb{R}^n)$. Then

$$\begin{aligned}(T_1 * T_2) * (\phi_1 * \phi_2) &= T_1 * (T_2 * (\phi_1 * \phi_2)) \\ &= T_1 * ((T_2 * \phi_1) * \phi_2) \\ &= T_1 * (\phi_2 * (T_2 * \phi_1))\end{aligned}$$

If $T_1 \in \mathcal{E}'(\mathbb{R}^n)$ then we can apply (1.6.15); if $T_2 \in \mathcal{E}'(\mathbb{R}^n)$ then $T_2 * \phi_1$ is also in $\mathcal{D}(\mathbb{R}^n)$ and we apply (1.6.10). In either case we deduce that

$$(T_1 * T_2) * (\phi_1 * \phi_2) = (T_1 * \phi_2) * (T_2 * \phi_1) \tag{1.6.27}$$

Similarly

$$(T_2 * T_1) * (\phi_2 * \phi_1) = (T_2 * \phi_1) * (T_1 * \phi_2) \tag{1.6.28}$$

But convolution of functions is commutative and so the right-hand sides of the above relations are equal. Hence, as $\phi_1 * \phi_2 = \phi_2 * \phi_1$, we conclude that

$$((T_1 * T_2) * \phi_1) * \phi_2 = ((T_2 * T_1) * \phi_1) * \phi_2. \tag{1.6.29}$$

As ϕ_2 was arbitrary, Part (iv) of Theorem 1.6.1 implies that

$$(T_1 * T_2) * \phi_1 = (T_2 * T_1) * \phi_1$$

for all $\phi_1 \in \mathcal{D}(\mathbb{R}^n)$ and again by the same result, we deduce (1.6.23).

(ii) By virtue of the commutativity of the convolution, there is no loss

of generality in assuming that S_2 is compact. By virtue of (1.6.22),

$$(T_1 * T_2)(\phi) = (T_1 * (T_2 * \check{\phi}))(0) = T_1((T_2 * \check{\phi})^{\vee})$$

But,

$$\text{supp}\,(T_2 * \check{\phi}) \subset \text{supp}\,(T_2) - \text{supp}\,(\phi)$$

Hence $(T_1 * T_2)(\phi) = 0$ unless supp $(T_1) \cap (\text{supp}\,(\phi) - S_2)$ is non-empty, i.e. supp $(\phi) \cap (S_1 + S_2) \neq \emptyset$ which proves (1.6.24).

(iii) By (ii) above if two distributions have compact support, so does their convolution product. Thus if atleast two of T_i, $i = 1, 2, 3$ have compact support, both $(T_1 * T_2) * T_3$ and $T_1 * (T_2 * T_3)$ are well-defined. Assume that $T_3 \in \mathcal{E}'(\mathbb{R}^n)$. Let $\phi \in \mathcal{D}(\mathbb{R}^n)$. Then

$$(T_1 * (T_2 * T_3)) * \phi = T_1 * ((T_2 * T_3) * \phi) = T_1 * (T_2 * (T_3 * \phi))$$

$$((T_1 * T_2) * T_3) * \phi = (T_1 * T_2) * (T_3 * \phi) = T_1 * (T_2 * (T_3 * \phi))$$

since $T_3 * \phi \in \mathcal{D}(\mathbb{R}^n)$ as well. Since ϕ is arbitrary, the associative law follows from above by the usual application of Part (iv) of Theorem 1.6.1. If $T_3 \notin \mathcal{E}'(\mathbb{R}^n)$, then both T_1 and T_2 are in $\mathcal{E}'(\mathbb{R}^n)$ by hypothesis. Hence

$$\begin{aligned} T_1 * (T_2 * T_3) &= T_1 * (T_3 * T_2) = (T_3 * T_2) * T_1 \\ &= T_3 * (T_2 * T_1) \text{ (by preceding argument)} \\ &= (T_2 * T_1) * T_3 = (T_1 * T_2) * T_3. \end{aligned}$$

(iv) Let α be any multi-index. Let $\phi \in \mathcal{D}(\mathbb{R}^n)$. Then

$$\begin{aligned} D^\alpha(T_1 * T_2) * \phi &= (T_1 * T_2) * D^\alpha\phi \quad \text{(by (1.6.9))} \\ &= T_1 * (T_2 * D^\alpha\phi) \quad \text{(by (1.6.21))} \\ &= T_1 * (D^\alpha T_2 * \phi) \quad \text{(by (1.6.9))} \\ &= (T_1 * D^\alpha T_2) * \phi \quad \text{(by (1.6.21))} \end{aligned}$$

Hence, as usual, we deduce that

$$D^\alpha(T_1 * T_2) = T_1 * D^\alpha T_2.$$

The other relation follows similarly. ∎

The dirac distribution enjoys a special status vis-a-vis the convolution product.

Theorem 1.6.4 Let $T \in \mathcal{D}'(\mathbb{R}^n)$. Then

$$T = \delta * T = T * \delta \tag{1.6.30}$$

More generally, for any multi-index α,

$$D^\alpha T = (D^\alpha\delta) * T \tag{1.6.31}$$

Proof Let $\phi \in \mathcal{D}(\mathbb{R}^n)$. Then

$$(\delta * \phi)(x) = \delta(\tau_x\check{\phi}) = \tau_x\check{\phi}(0) = \check{\phi}(-x) = \phi(x)$$

Hence $\delta * \phi = \phi$. Now

$$(\delta * T) * \phi = (T * \delta) * \phi = T * (\delta * \phi) = T * \phi$$

and (1.6.30) follows. If α is any multi-index, then by (1.6.26) and (1.6.30),

$$D^\alpha T = D^\alpha T * \delta = T * D^\alpha \delta = D^\alpha \delta * T. \quad \blacksquare$$

Remark 1.6.1. If $T_1, \ldots, T_n$ are distributions such that all but at most one have compact support, it is possible to define the convolution $T_1 * \ldots * T_n$. The parentheses are unnecessary since we can group them two-by-two as we please by virtue of the associative law. ▮

Example 1.6.1. Given three distributions with more than one having non-compact support, it is at times still possible to define the convolution product two-by-two. However, now the order in which we do this is important as the associative law is no longer valid in general. For instance consider the following distributions: $T_1 = 1$, the distribution generated by the constant function 1; $T_2 = \delta'$; $T_3 = H$, the Heaviside function (Cf. Example 1.3.2). Now

$$\delta' * H = \delta * H' = \delta * \delta = \delta$$

$$1 * \delta' = 1' * \delta = 0.$$

Thus we have

$$1 * (\delta' * H) = 1 * \delta = 1$$

$$(1 * \delta') * H = 0 * H = 0. \quad \blacksquare$$

1.7 FUNDAMENTAL SOLUTIONS

Let us consider a differential operator

$$L = \sum_{|\alpha| \leq m} a_\alpha D^\alpha. \tag{1.7.1}$$

Let S be a given distribution on $\mathbb{R}^n$. If $T \in \mathcal{D}'(\mathbb{R}^n)$ such that

$$L(T) = S, \text{ i.e. } \sum_{|\alpha| \leq m} a_\alpha D^\alpha T = S, \tag{1.7.2}$$

we call T a distribution solution. If S is a function then we have the familiar notion of a partial differential equation

$$Lu = f. \tag{1.7.3}$$

While we may not have a smooth function u satisfying (1.7.3), we may look for distribution solutions of that equation.

Of particular interest is the special distribution, δ. Let us assume that there exists a distribution E such that

$$L(E) = \delta. \tag{1.7.4}$$

Let S be any distribution with compact support. Consider the distribution $S * E$ which is now well-defined. Then

$$L(S * E) = \sum_{|\alpha| \leq m} a_\alpha D^\alpha (S * E)$$

$$= \sum_{|\alpha| \leq m} a_\alpha (S * D^\alpha E) = S * \left(\sum_{|\alpha| \leq m} a_\alpha D^\alpha E \right)$$

$$= S * \delta = S.$$

Thus $T = S * E$ is a distribution solution of (1.7.2). Thus the importance of distributions E satisfying (1.7.4) is clear, at least in the context of existence of solutions of partial differential equations. Such distributions have other uses as well (see comments at the end of this chapter).

Definition 1.7.1 Let L be a differential operator. A **fundamental solution** of L is a distribution E such that $L(E)=\delta$ (in the sense of distributions). ∎

Example 1.7.1 Consider the operator d/dx on $\mathbb{R}$. The Heaviside function H (cf. Example 1.3.2) is a fundamental solution of this operator. If C is any constant, then the function $H(x) + C$ is also a fundamental solution, for

$$\frac{d}{dx}(H + C) = \frac{dH}{dx} = \delta. \quad ∎$$

As we see from the above example, the fundamental solution is not uniquely defined. We can add to a fundamental solution any distribution solution of the homogeneous equation, $L(T) = 0$.

We will now compute the fundamental solution of the Laplace operator

$$\Delta = \sum_{i=1}^{n} \frac{\partial^2}{\partial x_i^2}.$$

The exercise will also help us to illustrate computations involving distribution derivatives. We will also come across fundamental solutions of other operators in the sequel. In what follows, given a vector $x=(x_1 \ldots, x_n)$ in $\mathbb{R}^n$, we will use the notation

$$r = |x| = \left(\sum_{i=1}^{n} |x_i|^2 \right)^{1/2} \tag{1.7.5}$$

Theorem 1.7.1 Let $n = 2$. Then the function $u(x) = \frac{1}{2\pi} \log r$ is a fundamental solution for Δ.

Proof: *Step 1* u defines a distribution, since it is locally integrable. If K is any compact set not containing 0, then as u is smooth on K, it is summable over K. If K contains a neighbourhood of 0, say $B(0; a)$ then

$$\int_{B(0;\, a)} |\log r| = -\int_0^{2\pi} d\theta \int_0^a r \log r \, dr \quad (a < 1)$$

and as $r \log r \to 0$ as $r \to 0$, the above integral is finite.

Step 2 u is harmonic on $\mathbb{R}^2 \setminus \{0\}$, i.e. $\Delta u = 0$. Since u is a smooth function on $\mathbb{R}^2 \setminus \{0\}$, its distribution derivatives are just the usual partial

derivatives. Now the proof is just a straightforward computation on the function

$$u(x_1, x_2) = \frac{1}{4\pi} \log (x_1^2 + x_2^2).$$

Step 3 Let $\phi \in \mathcal{D}(\mathbb{R}^2)$. We must prove that

$$\int_{\mathbb{R}^2} u \, \Delta\phi = \phi(0).$$

Let supp $(\phi) \subset B(0; R)$, $R > 0$. Let $0 < r < R$ and set Ω_r equal to the annulus between the circles of radius R and r, i.e.

$$\Omega_r = B(0; R) \diagdown B(0; r).$$

Let S_R be the circle of radius R and S_r that of radius r, with centre origin (fig. 3).

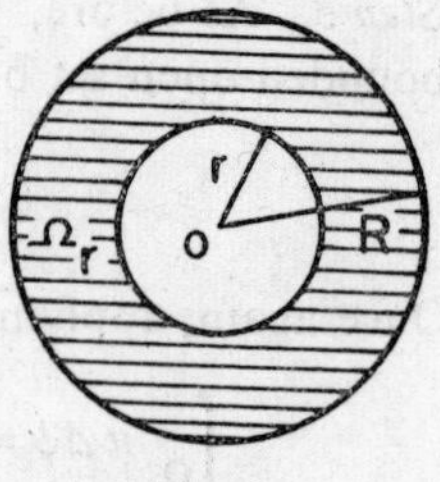

Fig. 3

Now,

$$\int_{\mathbb{R}^2} u \, \Delta\phi = \int_{B(0;\, R)} u \, \Delta\phi = \lim_{r \to 0} \int_{\Omega_r} u \, \Delta\phi$$

$$\int_{\Omega_r} u \, \Delta\phi = \frac{1}{2\pi} \int_{\Omega_r} \log r \cdot \Delta\phi$$

$$= \frac{1}{2\pi} \int_{\Omega_r} \Delta(\log r)\phi + \int_{S_R} \left(u \frac{\partial \phi}{\partial r} - \phi \frac{\partial u}{\partial r} \right) - \int_{S_r} \left(u \frac{\partial \phi}{\partial r} - \phi \frac{\partial u}{\partial r} \right)$$

using Green's identity in plane domains. But as ϕ vanishes outside a compact set which is contained in $B(0; R)$ the integral on S_R vanishes. Also by step (ii), the integral on Ω_r is zero. Hence

$$\int_{\Omega_r} u \, \Delta\phi = -\frac{1}{2\pi} \int_0^{2\pi} r \log r \frac{\partial \phi}{\partial r} \, d\theta + \frac{1}{2\pi} \int_0^{2\pi} \phi(r, \theta) \frac{\partial}{\partial r} (\log r) r \, d\theta.$$

But ϕ is a smooth function with compact support in $\mathbb{R}^2$ and so $\left| \frac{\partial \phi}{\partial r} \right| \leqslant M$ everywhere, and $r \log r \to 0$ as $r \to 0$. Hence the first integral on the right-hand side tends to zero as $r \to 0$.

The second integral is given by

$$\frac{1}{2\pi} \int_0^{2\pi} \phi(r, \theta) \, d\theta = \frac{1}{2\pi} \int_0^{2\pi} (\phi(r, \theta) - \phi(0)) \, d\theta + \phi(0).$$

But ϕ is uniformly continuous and so the integral on the right-hand side above vanishes as $r \to 0$. Thus

$$\lim_{r \to 0} \int_{\Omega_r} \Delta \, u\phi = \phi(0)$$

which proves the theorem. ∎

Theorem 1.7.2 Let $n \geqslant 3$ and ω_n the surface measure of the unit sphere in $\mathbb{R}^n$. Then $u(x) = -1/((n-2)\omega_n r^{n-2})$ is a fundamental solution to Δ.

Proof: *Step 1* u is locally integrable and thus defines a distribution. As before, we only need to verify this in a neighbourhood of the origin.

$$\int_{B(0;\,a)} r^{2-n} = \int_0^a r^{2-n} r^{n-1} \omega_n \, dr = \frac{\omega_n a^2}{2} < +\infty.$$

Step 2 r^{2-n} is harmonic in $\mathbb{R}^n \setminus \{0\}$. Given a function v depending only on r, we have

$$\Delta v = \left(\frac{\partial^2}{\partial r^2} + \frac{(n-1)}{r} \frac{\partial}{\partial r} \right) v.$$

Set $v = r^{2-n}$.

Step 3 As before, let $\phi \in \mathscr{D}(\mathbb{R}^n)$ with supp $(\phi) \subset B(0; R)$. If Ω_r is the bounded open set between $B(0; R)$ and $B(0; r)$; then

$$(\Delta u)(\phi) = \int_{B(0;\,R)} \Delta u \phi = \lim_{r \to 0} \int_{\Omega_r} u \Delta \phi.$$

Once again, applying Green's identity, and using step 2,

$$\int_{\Omega_r} u \, \Delta \phi = \frac{1}{\omega_n (n-2)} \int_{S_r} \left(r^{2-n} \frac{\partial \phi}{\partial r} - \phi \frac{\partial}{\partial r} (r^{2-n}) \right) dS_r,$$

S_r being the sphere of radius r, centre 0. Since ϕ is continuous with compact support, $\left| \frac{\partial \phi}{\partial r} \right| \leqslant M$ everywhere for some constant $M > 0$ and $\int_{S_r} r^{2-n} \, dS_r = \int_{S_1} r \, dS_1$ and thus the first term tends to zero as $r \to 0$. Hence

$$\lim_{r \to 0} \int_{\Omega_r} u \, \Delta \phi = \lim_{r \to 0} \frac{(n-2)}{(n-2)\omega_n} \int_{S_1} \phi(r, y) \frac{1}{r^{n-1}} r^{n-1} \, dS_1(y)$$

$$= \lim_{r \to 0} \frac{1}{\omega_n} \int_{S_1} (\phi(r, y) - \phi(0)) \, dS_1(y) + \phi(0)$$

$$= \phi(0).$$

Since ϕ is uniformly continuous. ∎

Let f be a locally integrable function defined on a bounded open set Ω. Extending it by zero outside Ω, f generates a distribution with compact support (contained in $\bar{\Omega}$) on $\mathbb{R}^n$. Then by virtue of the discussion at the beginning of this section, the convolution with the fundamental solution gives a solution to the partial differential equation with right hand side f. Thus

$$u(x) = \frac{1}{2\pi} \int_{\mathbb{R}^2} \log |x - y| f(y) \, dy \tag{1.7.6}$$

and

$$u(x) = \frac{-1}{4\pi} \int_{\mathbb{R}^3} \frac{f(y)}{|x - y|} \, dy \tag{1.7.7}$$

solve the equation

$$\Delta u = f \tag{1.7.8}$$

in $\mathbb{R}^2$ and $\mathbb{R}^3$ respectively. Such expressions are familiar from classical texts on partial differential equations. Of course it is not strictly necessary for f to have compact support; it is enough if f has suitable decay properties for the convolutions to be defined.

We have exhibited here fundamental solutions of some operators. We will encounter some others later. An important result regarding fundamental solutions is the *Malgrange-Ehrenpreis Theorem* which states that if L is a constant coefficient operator, i.e. the a_α in (1.7.1) are constants, then a fundamental solution always exists. The proof, which we do not present here, depends on the Hahn-Banach Theorem and can be found, for instance, in Rudin [2].

1.8 THE FOURIER TRANSFORM

The Fourier Transform together with the convolution product provides us with a useful and powerful tool in the study of partial differential equations. We will first define the Fourier transform for functions in $L^1(\mathbb{R}^n)$ and then extend it to a class of distributions.

Definition 1.8.1 Let $f \in L^1(\mathbb{R}^n)$. The **Fourier Transform** of f, denoted by $\hat{f}$, is a function defined on $\mathbb{R}^n$ by the formula

$$\hat{f}(\xi) = \int_{\mathbb{R}^n} \exp(-2\pi i x \cdot \xi) f(x)\, dx, \; i = \sqrt{-1}, \tag{1.8.1}$$

where $x \cdot \xi = \sum_{j=1}^{n} x_j \xi_j$ is the usual Euclidean inner-product in $\mathbb{R}^n$. ∎

Since $f \in L^1(\mathbb{R}^n)$, it is immediate to see that $\hat{f}(\xi)$ is well-defined for every $\xi \in \mathbb{R}^n$. In fact we have

$$|\hat{f}(\xi)| \leqslant \|f\|_{L^1(\mathbb{R}^n)}. \tag{1.8.2}$$

A simple application of the Dominated Convergence Theorem shows that $\hat{f}$ is a continuous function; but we have more:

Theorem 1.8.1 The function $\hat{f}$ is a uniformly continuous function.

Proof Let $\epsilon > 0$. Since $f \in L^1(\mathbb{R}^n)$ then there exists an $R > 0$ such that

$$\int_{\mathbb{R}^n \setminus B(0;\, R)} |f| < \epsilon/4. \tag{1.8.3}$$

Let $\eta > 0$ be chosen such that

$$4\pi R \eta \int_{B(0;\, R)} |f| < \epsilon. \tag{1.8.4}$$

Now let $|h| < \eta$, $h \in \mathbb{R}^n$. Then for any $y \in \mathbb{R}^n$,

$$|\hat{f}(y+h) - \hat{f}(y)| = \left| \int_{\mathbb{R}^n} f(x) \exp(-2\pi i x \cdot y)(\exp(-2\pi i x \cdot h) - 1)\, dx \right|$$

$$\leqslant \int_{\mathbb{R}^n} |f(x)|\,|\exp(-2\pi i x\cdot h) - 1|\,dx$$

$$= \int_{\mathbb{R}^n} |f(x)| 2 |\sin(\pi x\cdot h)|\,dx$$

$$\leqslant 2\int_{\mathbb{R}^n \setminus B(0;\,R)} |f(x)|\,dx + 2\pi \int_{B(0;\,R)} |f(x)|\,|x\cdot h|\,dx$$

$$< \frac{\epsilon}{2} + 2\pi R\eta \int_{B(0;\,R)} |f(x)|\,dx$$

$$< \frac{\epsilon}{2} + \frac{\epsilon}{2} = \epsilon.$$

Thus $\hat{f}$ is uniformly continuous. ∎

The following properties of the Fourier transform are easy to establish and are very useful.

Theorem 1.8.2 Let $f \in L^1(\mathbb{R}^n)$, and let $h \in \mathbb{R}^n$. Then

$$(\tau_h \hat{f})(\xi) = \exp(-2\pi i h\cdot\xi)\hat{f}(\xi), \tag{1.8.5}$$

$$\tau_h(\hat{f}) = (\exp(2\pi i h\cdot(\cdot))f(\cdot))\hat{\ }. \tag{1.8.6}$$

If $\lambda > 0$ and $g(x) = f(x/\lambda)$ for $x \in \mathbb{R}^n$, then

$$\hat{g}(\xi) = \lambda^n \hat{f}(\lambda\xi). \tag{1.8.7}$$

Finally if $f, g \in L^1(\mathbb{R}^n)$, then

$$\widehat{f*g} = \hat{f}\hat{g}. \tag{1.8.8.}$$

Proof Relations (1.8.5)–(1.8.7) follow by straightforward computations starting from the definition of the Fourier transform. Let $h = f*g$, $f, g \in L^1(\mathbb{R}^n)$. Then we know by (1.5.2) that $h \in L^1(\mathbb{R}^n)$, so its Fourier transform is defined.

$$\begin{aligned}
\hat{h}(\xi) &= \int_{\mathbb{R}^n} \exp(-2\pi i x\cdot\xi)\left(\int_{\mathbb{R}^n} f(x-y)g(y)\,dy\right)dx \\
&= \int_{\mathbb{R}^n} g(y)\left(\int_{\mathbb{R}^n} \exp(-2\pi i x\cdot\xi) f(x-y)\,dx\right)dy \\
&= \int_{\mathbb{R}^n} \exp(-2\pi i y\cdot\xi)g(y)\left(\int_{\mathbb{R}^n} f(x-y)\exp[-2\pi i(x-y)\cdot\xi]\,dx\right)dy \\
&= \int_{\mathbb{R}^n} \exp(-2\pi i y\cdot\xi)g(y)\,dy \int_{\mathbb{R}^n} f(z)\exp(-2\pi i z\cdot\xi)\,dz \\
&= \hat{g}(\xi)\hat{f}(\xi),
\end{aligned}$$

by repeated use of Fubini's theorem and the translation invariance of the Lebesgue measure. ∎

This last property of the Fourier transform of a convolution being the algebraic product of the individual Fourier transforms is one which renders this concept very useful. The other important property is its

behaviour with respect to derivatives of functions which we will see in the next section.

We conclude this section with the computation of the Fourier transform of a standard function.

Example 1.8.1 Let $f(x) = e^{-|x|^2}$, $x \in \mathbb{R}^n$. Then it is easy to see that $f \in L^1(\mathbb{R}^n)$. We first compute the Fourier transform when $n = 1$. Thus for $\xi \in \mathbb{R}$,

$$\hat{f}(\xi) = \int_{-\infty}^{\infty} \exp(-2\pi i x\xi) f(x)\, dx$$

$$= \int_{-\infty}^{\infty} \exp(-\pi^2\xi^2) \exp[-(x + \pi i\xi)^2]\, dx.$$

We can evaluate this integral using Cauchy's theorem in the complex plane. Consider the contour $\Gamma = \bigcup_{i=1}^{4} \Gamma_i$ shown in Fig. 4.

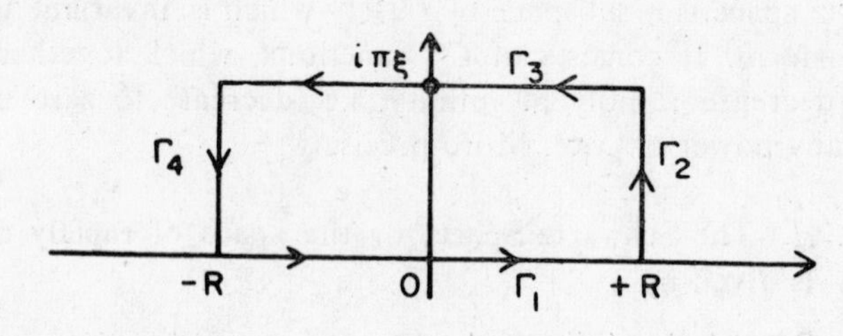

Fig. 4

By Cauchy's theorem, $\int_\Gamma \exp(-z^2)\, dz = 0$. Further

$$\left|\int_{\Gamma_2} \exp(-z^2)\, dz\right| = \left|\int_0^{\pi\xi} \exp[-(R + iy)^2]\, dy\right|$$

$$= \left|\int_0^{\pi\xi} \exp(-R^2) \exp(-2iRy) \exp(y^2)\, dy\right|$$

$$\leqslant C \exp(-R^2)$$

and so this integral tends to zero as $R \to +\infty$.

Similarly

$$\lim_{R\to+\infty} \int_{\Gamma_4} \exp(-z^2)\, dz = 0.$$

Thus

$$\int_{-\infty}^{\infty} \exp[-(x + i\pi\xi)^2]\, dx = -\lim_{R\to\infty} \int_{\Gamma_3} \exp(-z^2)\, dz$$

$$= \int_{-\infty}^{\infty} \exp(-x^2)\, dx = \sqrt{\pi},$$

as is well-known. Hence,

$$\hat{f}(\xi) = \sqrt{\pi} \exp(-\pi^2\xi^2). \tag{1.8.9}$$

Now for any general n,

$$\begin{aligned}\hat{f}(\xi) &= \int_{\mathbb{R}^n} \exp(-2\pi i x\cdot\xi)\exp(-|x|^2)\,dx \\ &= \int_{\mathbb{R}^n} \exp\left(-\sum_{j=1}^{n}(x_j^2 + 2\pi i x_j\xi_j)\right)dx \\ &= (\sqrt{\pi})^n \exp\left(-\sum_{j=1}^{n}\pi^2\xi_j^2\right)\end{aligned}$$

using (1.8.9). Thus

$$\hat{f}(\xi) = (\pi)^{n/2}\exp(-\pi^2|\xi|^2). \tag{1.8.10}$$

Upto a normalizing factor, f is preserved by the Fourier transform. A more sleek proof of this can be found in Rudin [2]. ∎

1.9 THE SCHWARTZ SPACE, $\mathcal{S}$

The Schwartz space is a subspace of $L^1(\mathbb{R}^n)$ which is invariant under the Fourier transform. It consists of C^∞ functions, which together with all derivatives, decrease rapidly at infinity, i.e. decrease to zero at infinity faster than any power of $|x|^{-1}$. More precisely

Definition 1.9.1 The **Schwartz Space,** or the space of **rapidly decreasing functions,** $\mathcal{S}$, is given by

$$\mathcal{S} = \{f \in \mathcal{E}(\mathbb{R}^n) \mid \lim_{|x|\to\infty} |x^\beta D^\alpha f(x)| = 0 \quad \text{for all multi-indices } \alpha \text{ and } \beta\}. \tag{1.9.1}$$ ∎

The following statements are easy to verify for $f \in \mathcal{E}(\mathbb{R}^n)$:

(i) $f \in \mathcal{S}$ if, and only if, for every polynomial $P(x)$ and for every differential operator L with constant coefficients, the function

$$P(x)\,Lf(x)$$

is bounded in $\mathbb{R}^n$.

(ii) $f \in \mathcal{S}$ if, and only if, for any integer $k \geqslant 0$, and any multi-index α, the function

$$(1 + |x|^2)^k D^\alpha f(x)$$

is bounded in $\mathbb{R}^n$.

Example 1.9.1 Any function $\phi \in \mathcal{D}$ is trivially in $\mathcal{S}$. ∎

Example 1.9.2 Let $n = 1$ and $f(x) = \exp(-x^2)$. Then $f \in \mathcal{S}$. Indeed it suffices in this case to check that $|x|^k \exp(-x^2)$ tends to zero as $|x| \to \infty$ and this was done in Lemma 1.2.1. ∎

As was done in case of spaces $\mathcal{D}$ and $\mathcal{E}$, we describe the topology on $\mathcal{S}$ via its convergent sequences. We say that a sequence $f_k \to 0$ in $\mathcal{S}$ if for every polynomial $P(x)$ and every differential operator L with constant coefficients, the functions

$$P(x)\,Lf_k(x)$$

converge to zero uniformly in $\mathbb{R}^n$. A linear mapping from $\mathcal{S}$ to a topological vector space will be called continuous if it is sequentially continuous. In view of this, we have the following continuous inclusions:

$$\mathcal{D} \to \mathcal{S} \to \mathcal{E} \tag{1.9.2}$$

Theorem 1.9.1 Let $f, g \in \mathcal{S}$, α a multi-index, L a differential operator with constant coefficients. Then

$$D^\alpha f \in \mathcal{S},\ P(\cdot)f(\cdot) \in \mathcal{S},\ L(P(\cdot)f(\cdot)) \in \mathcal{S},\ P(\cdot)Lf(\cdot) \in \mathcal{S},$$

and $fg \in \mathcal{S}$. Also the map

$$f \to L(P(\cdot)f(\cdot))$$

is continuous from $\mathcal{S}$ into itself.

Proof This is an immediate consequence of the definition and its equivalent forms. The details are left as an exercise to the reader. ∎

Theorem 1.9.2 $\mathcal{S} \subset L^1(\mathbb{R}^n)$ and the inclusion is continuous.

Proof Let $f \in \mathcal{S}$. Then for any integer $k \geqslant 0$, there exists a constant $M_k > 0$ such that

$$\sup_{x \in \mathbb{R}^n} |f(x)(1 + |x|^2)^k| \leqslant M_k. \tag{1.9.3}$$

Now, for $k > n/2$, it is well known that $(1 + |x|^2)^{-k} \in L^1(\mathbb{R}^n)$ (use polar coordinates). Hence

$$\begin{aligned} \int_{\mathbb{R}^n} |f(x)|\, dx &= \int_{\mathbb{R}^n} |f(x)|(1 + |x|^2)^k (1 + |x|^2)^{-k}\, dx \\ &\leqslant M_k \int_{\mathbb{R}^n} (1 + |x|^2)^{-k}\, dx < +\infty. \end{aligned}$$

Thus $f \in L^1(\mathbb{R}^n)$. Also if $C = \int_{\mathbb{R}^n} (1 + |x|^2)^{-k}\, dx$, then

$$\|f\|_{L^1(\mathbb{R}^n)} \leqslant C \sup_{x \in \mathbb{R}^n} (|f(x)|(1 + |x|^2)^k). \tag{1.9.4}$$

Hence if $f_m \to 0$ in $\mathcal{S}$ it follows that $\|f_m\|_{L^1(\mathbb{R}^n)} \to 0$ and the continuity of the inclusion follows. ∎

Remark 1.9.1 The same proof can be modified to show that for any p such that $1 \leqslant p \leqslant +\infty$,

$$\mathcal{S} \subset L^p(\mathbb{R}^n). \tag{1.9.5}$$

In particular for $1 \leqslant p < +\infty$, we know that (cf. Theorem 1.5.6) $\mathcal{D}$ is dense in $L^p(\mathbb{R}^n)$. Hence it follows from (1.9.2) and (1.9.5) that $\mathcal{S}$ is dense in $L^p(\mathbb{R}^n)$, $1 \leqslant p < +\infty$. ∎

The utility of $\mathcal{S}$ stems from its congenial behaviour with respect to the Fourier transform.

Theorem 1.9.3 Let $f \in \mathcal{S}$. Then $\hat{f} \in \mathcal{S}$ and the map $f \to \hat{f}$ of $\mathcal{S}$ into itself is a continuous linear map.

Proof Since $f \in \mathcal{S}$, $f \in L^1(\mathbb{R}^n)$ and the Fourier transform is defined.

Step 1 $f \in \mathcal{S}$ implies $\hat{f} \in \mathcal{E}$.
We have

$$\hat{f}(\xi) = \int_{\mathbb{R}^n} \exp(-2\pi i x \cdot \xi) f(x)\, dx.$$

Now $x_j f(x) \in \mathcal{S}$ for any $1 \leqslant j \leqslant n$ and so is integrable. Hence by the Dominated Convergence Theorem we can justify the differentiation under the integral sign to get

$$\frac{\partial \hat{f}}{\partial \xi_j}(\xi) = -2\pi i \int_{\mathbb{R}^n} \exp(-2\pi i x \cdot \xi) x_j f(x)\, dx. \tag{1.9.6}$$

More generally, it follows by iteration that, for any multi-index α,

$$D^\alpha \hat{f}(\xi) = (-2\pi i)^{|\alpha|} (x^\alpha f(x))\hat{\ }(\xi). \tag{1.9.7}$$

which being the Fourier transform of a L^1-function, is continuous (Theorem 1.8.1). Thus $\hat{f} \in \mathcal{E}$.

Step 2 We will now show that

$$2\pi i \xi_j \hat{f}(\xi) = \widehat{\frac{\partial f}{\partial x_j}}(\xi) = \int_{\mathbb{R}^n} \exp(-2\pi i x \cdot \xi) \frac{\partial f}{\partial x_j}(x)\, dx. \tag{1.9.8}$$

This is essentially a consequence of Green's identity in space. We have

$$\int_{\mathbb{R}^n} \exp(-2\pi i x \cdot \xi) \frac{\partial f}{\partial x_j}(x)\, dx = \lim_{R \to \infty} \int_{B(0;\, R)} \exp(-2\pi i x \cdot \xi) \frac{\partial f}{\partial x_j}(x)\, dx. \tag{1.9.9}$$

Integrating the integral on the ball $B(0;\, R)$ by parts, we get

$$\int_{B(0;\, R)} \exp(-2\pi i x \cdot \xi) \frac{\partial f}{\partial x_j}(x)\, dx = \int_{B(0;\, R)} 2\pi i \xi_j \exp(-2\pi i x \cdot \xi) f(x)\, dx + \frac{1}{R} \int_{|x| = R} \exp(-2\pi i x \cdot \xi) f(x) x_j\, dS_R \tag{1.9.10}$$

But

$$\left| \frac{1}{R} \int_{|x|=R} \exp(-2\pi i x \cdot \xi) f(x) x_j\, dS_R \right| \leqslant \frac{1}{R} \int_{|x|=R} |x_j|\, |f(x)|\, dS_R$$
$$\leqslant \frac{M_k}{R} \int_{|x|=R} |x_j| (1 + |x|^2)^{-k}\, dS_R$$

where $k \geqslant 0$ is any integer (of our choice) and M_k as in (1.9.3). Now

$$\frac{M_k}{R}\int_{|x|=R} |x_j|(1+|x|^2)^{-k}\, dS_R = \frac{M_k R^{n-1}}{(1+R^2)^k}\int_{|y|=1} |y_j|\, dS_1$$

which tends to zero as $R \to \infty$ provided k is chosen large enough. Thus combining (1.9.9) and (1.9.10) we get (1.9.8).

Step 3 Iterating (1.9.8) we deduce, for any multi-index β,

$$(2\pi i\xi)^\beta \hat{f}(\xi) = (D^\beta f)\hat{\ }(\xi). \tag{1.9.11}$$

Combining (1.9.7) and (1.9.11) we get, for any two multi-indices α and β,

$$\xi^\beta D^\alpha \hat{f}(\xi) = \frac{(-2\pi i)^{|\alpha|}}{(2\pi i)^{|\beta|}}\,(D^\beta(x^\alpha f(x)))\hat{\ }(\xi) \tag{1.9.12}$$

which, since $D^\beta(x^\alpha f(x)) \in \mathcal{S} \subset L^1(\mathbb{R}^n)$, is bounded. Hence $\hat{f} \in \mathcal{S}$.

Step 4 Again by (1.9.12),

$$|\xi^\beta D^\alpha \hat{f}(\xi)| \leqslant C\int_{\mathbb{R}^n} (1+|x|^2)^k |D^\beta(x^\alpha f(x))|(1+|x|^2)^{-k}\, dx$$

Hence

$$\sup_{\xi\in\mathbb{R}^n} |\xi^\beta D^\alpha \hat{f}(\xi)| \leqslant C_1 \sup_{\mathbb{R}^n} ((1+|x|^2)^k |D^\beta(x^\alpha f(x))|)$$

which shows that if $f_m \to 0$ in $\mathcal{S}$ then $\hat{f}_m \to 0$ in $\mathcal{S}$ thus proving the continuity of the Fourier transform as a linear operator on $\mathcal{S}$. ∎

Remark 1.9.2 The relations (1.9.7) and (1.9.11) which are 'dual' to each other is one of the important properties of the Fourier transform. It converts differentiation into algebraic products with polynomials and vice-versa. ∎

Corollary **(Riemann-Lebesgue Lemma)** Let $f \in L^1(\mathbb{R}^n)$. Then $\hat{f}$ is a uniformly continuous function which vanishes at infinity.

Proof If $f \in \mathcal{S}$, then $\hat{f} \in \mathcal{S}$ and hence vanishes at infinity. Let $f \in L^1(\mathbb{R}^n)$ and let $f_m \in \mathcal{S}$ with $f_m \to f$ in $L^1(\mathbb{R}^n)$ (cf. Remark 1.9.1). Then for any $\xi \in \mathbb{R}^n$, by (1.8.2),

$$|\hat{f}(\xi) - \hat{f}_m(\xi)| \leqslant \|f - f_m\|_{L^1(\mathbb{R}^n)}$$

and thus $\hat{f}_m \to \hat{f}$ uniformly on $\mathbb{R}^n$. But the space of continuous functions which vanish at infinity is a Banach space with the uniform norm and so $\hat{f}$ also vanishes at infinity. ∎

We conclude this section with an application of the Fourier transform to partial differential equations.

Example 1.9.3 Consider the following initial-value problem for the heat equation:

$$\left.\begin{aligned} \frac{\partial u}{\partial t} - \Delta u = 0 \qquad \text{in } \mathbb{R}^n \times [0, \infty) \\ u(x, 0) = u_0(x) \text{ in } \mathbb{R}^n. \end{aligned}\right\} \tag{1.9.13}$$

Hence $u_0(x)$ represents the initial temperature distribution and $u(x, t)$ is the temperature at time 't'. We assume that we can form the Fourier transform for the function $u(\cdot, t)$ and u_0. Then by (1.9.11) it follows that

$$\left.\begin{aligned} \frac{\partial \hat{u}}{\partial t}(\xi, t) + 4\pi^2|\xi|^2\hat{u}(\xi, t) = 0 \\ \hat{u}(\xi, 0) = \hat{u}_0(\xi). \end{aligned}\right\} \tag{1.9.14}$$

Treating ξ as a parameter, we solve the resulting ordinary differential equation in t to get

$$\hat{u}(\xi, t) = \hat{u}_0(\xi) \exp(-4\pi^2|\xi|^2 t). \tag{1.9.15}$$

Now,

$$\begin{aligned} \exp(-4\pi^2|\xi|^2 t) &= \exp(-\pi^2|2\sqrt{t}\xi|^2) \\ &= (\pi)^{-n/2}(2\sqrt{t})^n \cdot \frac{\pi^{n/2}}{(2\sqrt{t})^n} \exp(-\pi^2|2\sqrt{t}\xi|^2) \\ &= (4\pi t)^{-n/2}\hat{g}(\xi) \end{aligned}$$

where $g(x) = \exp(-|x|^2/4t)$, using (1.8.7) and Example 1.8.1. Hence by (1.8.8) it follows that

$$u(x, t) = (4\pi t)^{-n/2}(u_0 * g)(x),$$

provided the convolution is defined. In particular, if u_0 has compact support or has good decay properties at infinity, we can write the solution to the initial-value problem (1.9.13) as

$$u(x, t) = \frac{1}{(4\pi t)^{n/2}} \int_{\mathbb{R}^n} \exp(-|x - y|^2/4t) u_0(y)\, dy, \; t > 0 \tag{1.9.16}$$

Keeping in mind the relation between the fundamental solution and the solution of a given problem, we may conjecture that

$$E(x, t) = (4\pi t)^{-n/2} \exp(-|x|^2/4t) H(t) \tag{1.9.17}$$

be a fundamental solution to the heat operator. Indeed this is so and a rigorous proof may be found in Treves [4].

We wish to record here one interesting property of the solution of the heat equation. If $u_0\ (\neq 0)$ has compact support (so that formula (1.9.16) is valid) and if $u_0 \geqslant 0$ then the formula shows that for all $x \in \mathbb{R}^n$, $u(x, t) > 0$ whenever $t > 0$. Thus even a local disturbance spreads over the entire space with immediate effect. This property is known in the literature as the *infinite speed of propagation of signals.*

1.10 THE FOURIER INVERSION FORMULA

In the previous section we saw that the Fourier transform mapped the Schwartz space $\mathcal{S}$ into itself. We now show that it is invertible on $\mathcal{S}$.

Lemma 1.10.1 (*The Weak Parseval Relation*) Let $f, g \in \mathcal{S}$. Then

$$\int_{\mathbb{R}^n} \hat{f}(\xi)g(\xi)\, d\xi = \int_{\mathbb{R}^n} f(x)\hat{g}(x)\, dx. \tag{1.10.1}$$

Proof This is an immediate consequence of Fubini's theorem applied to the measurable function $\exp(-2\pi i x\cdot\xi)f(x)g(\xi)$ (in the (x, ξ)-space). Thus

$$\begin{aligned}\int_{\mathbb{R}^n} \hat{f}(\xi)g(\xi)\, d\xi &= \int_{\mathbb{R}^n} g(\xi)\int_{\mathbb{R}^n} \exp(-2\pi i x\cdot\xi)f(x)\, dx\, d\xi \\ &= \int_{\mathbb{R}^n} f(x)\int_{\mathbb{R}^n} \exp(-2\pi i x\cdot\xi)g(\xi)\, d\xi\, dx \\ &= \int_{\mathbb{R}^n} f(x)\hat{g}(x)\, dx. \quad \blacksquare\end{aligned}$$

Theorem 1.10.1 (**Fourier Inversion Formula**) Let $g \in \mathcal{S}$. Then

$$g(x) = \int_{\mathbb{R}^n} \exp(2\pi i x.\xi)\hat{g}(\xi)\, d\xi. \tag{1.10.2}$$

Proof: Let $\phi \in \mathcal{S}$ be arbitrary and let $\lambda > 0$. Define

$$f(x) = \phi(x/\lambda).$$

Then $f \in \mathcal{S}$ and by (1.8.7),

$$\hat{f}(\xi) = \lambda^n\hat{\phi}(\lambda\xi).$$

Applying (1.10.1) to f and g, we get

$$\int_{\mathbb{R}^n} g(\xi)\lambda^n\hat{\phi}(\lambda\xi)\, d\xi = \int_{\mathbb{R}^n} \hat{g}(x)\phi(x/\lambda)\, dx.$$

By an obvious change of variable on the left-hand side,

$$\int_{\mathbb{R}^n} g(\xi/\lambda)\hat{\phi}(\xi)\, d\xi = \int_{\mathbb{R}^n} \hat{g}(x)\phi(x/\lambda)\, dx.$$

As $\lambda \to \infty$, $g(\xi/\lambda) \to g(0)$ and $\phi(x/\lambda) \to \phi(0)$. Also as $\hat{\phi}, \hat{g} \in \mathcal{S}$, we can pass to the limit by the Dominated Convergence Theorem and hence

$$g(0)\int_{\mathbb{R}^n} \hat{\phi}(\xi)\, d\xi = \phi(0)\int_{\mathbb{R}^n} \hat{g}(x)\, dx.$$

We now set $\phi(x) = \exp(-|x|^2)$. Then (cf. Example 1.8.1)

$$\hat{\phi}(\xi) = (\pi)^{n/2}\exp(-\pi^2|\xi|^2),$$

and

$$\int_{\mathbb{R}^n} \hat{\phi}(\xi)\, d\xi = 1, \ \phi(0) = 1.$$

Thus we deduce that

$$g(0) = \int_{\mathbb{R}^n} \hat{g}(\xi)\, d\xi$$

which proves (1.10.2) at $x = 0$. For any other x apply this result with g replaced by $\tau_{-x}g$ and use relation (1.8.5). ∎

Corollary The Fourier transform is a (topological) isomorphism of $\mathcal{S}$ onto itself.

Proof We have two maps $\mathcal{F}: \mathcal{S} \to \mathcal{S}$ and $\overline{\mathcal{F}}: \mathcal{S} \to \mathcal{S}$ such that

$$\mathcal{F}(f)(\xi) = \int_{\mathbb{R}^n} \exp(-2\pi i x \cdot \xi) f(x)\, dx,$$

and,

$$\overline{\mathcal{F}}(f)(x) = \int_{\mathbb{R}^n} \exp(2\pi i x)\xi \cdot f(\xi)\, d\xi.$$

We can show that $\overline{\mathcal{F}}$ is continuous from $\mathcal{S}$ into itself just as we did for $\mathcal{F}$. Thus $\mathcal{F}$, $\overline{\mathcal{F}}$ are continuous and by the above theorem

$$\mathcal{F} \circ \overline{\mathcal{F}} = \overline{\mathcal{F}} \circ \mathcal{F} = Ids$$

This proves the result. ∎

This corollary will be strengthened further.

Lemma 1.10.2 (*Strong Parseval Relation*) Let $f, g \in \mathcal{S}$. Then

$$\int_{\mathbb{R}^n} f\bar{g} = \int_{\mathbb{R}^n} \hat{f}\,\overline{\hat{g}}. \tag{1.10.3}$$

Proof Since $\bar{g} \in \mathcal{S}$, we have, by the Fourier inversion formula, a function $h \in \mathcal{S}$ such that $\hat{h} = \bar{g}$. Then by (1.10.1),

$$\int_{\mathbb{R}^n} f\bar{g} = \int_{\mathbb{R}^n} f\hat{h} = \int_{\mathbb{R}^n} \hat{f}h$$

But, again by the Fourier inversion formula,

$$h(x) = \int_{\mathbb{R}^n} \exp(2\pi i x \cdot \xi)\bar{g}(\xi)\, d\xi = \overline{\hat{g}(x)}.$$

Hence the result. ∎

Corollary: Let $f \in \mathcal{S}$. Then

$$\|f\|_{L^2(\mathbb{R}^n)} = \|\hat{f}\|_{L^2(\mathbb{R}^n)} \quad ∎ \tag{1.10.4}$$

This corollary leads us to the first extension of the Fourier transform to a larger class of functions.

Theorem 1.10.2 (*Plancherel*) There exists a unique isometry

$$\mathcal{P}: L^2(\mathbb{R}^n) \to L^2(\mathbb{R}^n)$$

which is onto and such that

$$\mathcal{P}(f) = \hat{f}, \text{ for every } f \in \mathcal{S}. \tag{1.10.5}$$

Proof The Fourier transform $\mathcal{F} : \mathcal{S} \to \mathcal{S}$ is an $L^2(\mathbb{R}^n)$-isometry by the preceding corollary. But $\mathcal{S}$ is dense in $L^2(\mathbb{R}^n)$ and so $\mathcal{F}$ extends uniquely to an operator of $L^2(\mathbb{R}^n)$ into itself. If $f \in L^2(\mathbb{R}^n)$, let $f_k \in \mathcal{S}$ such that $f_k \to f$ in $L^2(\mathbb{R}^n)$. Then

$$\|\mathcal{P}(f)\|_{L^2(\mathbb{R}^n)} = \lim_{k\to\infty} \|\mathcal{P}(f_k)\|_{L^2(\mathbb{R}^n)} = \lim_{k\to\infty} \|\hat{f}_k\|_{L^2(\mathbb{R})^n}$$

$$= \lim_{k\to\infty} \|f_k\|_{L^2(\mathbb{R}^n)} = \|f\|_{L^2(\mathbb{R}^n)}$$

Thus $\mathcal{P}$ is an isometry and hence has a closed range. But as $\mathcal{F}$ is onto, Range $(\mathcal{P}) \supset \mathcal{S}$ which is dense in $L^2(\mathbb{R}^n)$. Thus $\mathcal{P}$ is onto as well. ∎

1.11 TEMPERED DISTRIBUTIONS

Let $\phi \in \mathcal{D}(\mathbb{R}^n)$ with $\phi \equiv 1$ on the unit ball in $\mathbb{R}^n$. Set $\phi_m(x) = \phi(x/m)$. It can now be checked that $\phi_m f \to f$ in $\mathcal{S}$. Thus we have that $\mathcal{D} \subset \mathcal{S}$, and the inclusion is *dense*.

Hence the dual of $\mathcal{S}$, denoted $\mathcal{S}'$, can be identified with a subspace of $\mathcal{D}'(\mathbb{R}^n)$.

Definition 1.11.1 The space of distributions $\mathcal{S}'$ is called the space of **tempered distributions.** ∎

Example 1.11.1 Any distribution with compact support is tempered. Since $\mathcal{D} \subset \mathcal{S} \subset \mathcal{E}$ and all inclusions are dense, it follows that $\mathcal{E}' \subset \mathcal{S}'$. ∎

Example 1.11.2 Let μ be a measure on $\mathbb{R}^n$. It is said to be *slowly increasing* if for some integer $k \geqslant 0$,

$$\int_{\mathbb{R}^n} \frac{d\mu(x)}{(1 + |x|^2)^k} < +\infty. \tag{1.11.1}$$

In particular, any bounded measure is slowly increasing. Define now

$$T_\mu(f) = \int_{\mathbb{R}^n} f(x)\, d\mu(x),\ f \in \mathcal{S}. \tag{1.11.2}$$

Then T_μ is a linear functional on $\mathcal{S}$ and

$$|T_\mu(f)| \leqslant (\sup_{x\in\mathbb{R}^n} |f(x)(1 + |x|^2)^k|) \int_{\mathbb{R}^n} \frac{d\mu(x)}{(1 + |x|^2)^k}$$

and it follows that if $f_m \to 0$ in $\mathcal{S}$ then $T_\mu(f_m) \to 0$. Thus T_μ is a tempered distribution. ∎

Example 1.11.3 If $1 \leqslant p \leqslant +\infty$, then $L^p(\mathbb{R}^n) \subset \mathcal{S}'$. If $f \in L^p(\mathbb{R}^n)$, define

$$T_f(\phi) = \int_{\mathbb{R}^n} f\phi,\ \phi \in \mathcal{S}.$$

If $1 < p < \infty$, let q be such that $(1/p) + (1/q) = 1$. Then for $k > n/2q$, $g(x) = (1 + |x|^2)^{-k} \in L^q(\mathbb{R}^n)$. Hence

$$|T_f(\phi)| \leqslant (\sup_{x \in \mathbb{R}^n} |(1 + |x|^2)^k \phi(x)|) \, \|f\|_{L^p(\mathbb{R}^n)} \|g\|_{L^q(\mathbb{R}^n)}$$

by Hölder's inequality which proves, on one hand, that T_f is tempered, and, on the other, that $f \to T_f$ is continuous on $L^p(\mathbb{R}^n)$.

If $p = 1$,

$$|T_f(\phi)| \leqslant \|\phi\|_{L^\infty(\mathbb{R}^n)} \|f\|_{L^1(\mathbb{R}^n)}$$

and if $p = \infty$,

$$|T_f(\phi)| < \sup_{x \in \mathbb{R}^n} |(1 + |x|^2)^k \phi(x)| \, \|f\|_{L^\infty(\mathbb{R}^n)} \|g\|_{L^1(\mathbb{R}^n)}$$

where g is as above with $k > n/2$. ∎

We can now define the Fourier transform on $\mathcal{S}'$ by duality. Let $T \in \mathcal{S}'$. Then define $\hat{T} \in \mathcal{S}'$ by

$$\hat{T}(f) = T(\hat{f}), f \in \mathcal{S}. \tag{1.11.3}$$

Since $f \to \hat{f}$ is continuous on $\mathcal{S}$, it follows that $\hat{T} \in \mathcal{S}'$.

$\mathcal{S} \subset L^1(\mathbb{R}^n) \subset \mathcal{S}'$ and so *a priori* there seem to be two definitions of the Fourier transform on $\mathcal{S}$—one inherited from $L^1(\mathbb{R}^n)$ and the other from $\mathcal{S}'$. But if $f \in \mathcal{S}$, for any $g \in \mathcal{S}$, we have

$$\hat{T}_f(g) = \int_{\mathbb{R}^n} f\hat{g} = \int_{\mathbb{R}^n} \hat{f}g = T_{\hat{f}}(g)$$

by the weak Parseval relation. Hence $\hat{T}_f = T_{\hat{f}}$ and we are assured that both definitions of the Fourier transform coincide on $\mathcal{S}$. Since $\mathcal{F} : \mathcal{S} \to \mathcal{S}$ is an isomorphism on $\mathcal{S}$, the Fourier transform is a bijection on $\mathcal{S}'$ as well. By Theorem 1.9.1, it follows that if P is a polynomial and if α is any multi-index, then for any $T \in \mathcal{S}'$, we have $PT \in \mathcal{S}'$ and $D^\alpha T \in \mathcal{S}'$. We now prove the relation between the Fourier transform and the process of differentiation.

Theorem 1.11.1 Let $T \in \mathcal{S}'$ and let α be a multi-index. Then

$$D^\alpha \hat{T} = (-2\pi i)^{|\alpha|} (x^\alpha T)\hat{} \tag{1.11.4}$$

$$(D^\alpha T)\hat{} = (2\pi i)^{|\alpha|} \xi^\alpha \hat{T}. \tag{1.11.5}$$

Proof Let $f \in \mathcal{S}$. Then

$$\begin{aligned}(x^\alpha T)\hat{}(f) &= (x^\alpha T)(\hat{f}) = T(x^\alpha \hat{f}(x)) \\ &= \frac{1}{(2\pi i)^{|\alpha|}} T(\widehat{D^\alpha f}) = \frac{1}{(2\pi i)^{|\alpha|}} \hat{T}(D^\alpha f) \\ &= \frac{1}{(-2\pi i)^{|\alpha|}} D^\alpha \hat{T}(f),\end{aligned}$$

using (1.9.11). This proves (1.11.4). The proof of (1.11.5) is similar. ∎

Our final result in this section is to study the effect of the Fourier transform on the Dirac distribution.

Theorem 1.11.2 We have

$$\hat{\delta} = 1 \tag{1.11.6}$$

$$\left(\frac{\partial \delta}{\partial x_k}\right)^{\wedge} = 2\pi i \xi_k \tag{1.11.7}$$

$$\hat{1} = \delta. \tag{1.11.8}$$

Proof Let $\phi \in \mathcal{S}$. Then

$$\hat{\delta}(\phi) = \delta(\hat{\phi}) = \hat{\phi}(0) = \int_{\mathbb{R}^n} \phi(x)\, dx.$$

Thus $\hat{\delta} = 1$, i.e., the distribution induced by the constant function 1. Now,

$$\left(\frac{\partial \delta}{\partial x_k}\right)^{\wedge} = 2\pi i \xi_k \hat{\delta} = 2\pi i \xi_k.$$

The relation (1.11.8) follows by the Fourier inversion formula. ∎

We conclude by settling another question of consistency. Since $\mathcal{S} \subset L^2(\mathbb{R}^n) \subset \mathcal{S}'$, there are two definitions of the Fourier transform on $L^2(\mathbb{R}^n)$—one inherited from $\mathcal{S}'$ and the other, the operator $\mathcal{P}$ defined in the Plancherel theorem. We show that these are the same.

We consider once again the (weak $*$) convergence of tempered distributions, i.e., $T_m \to T$ (in the weak $*$ sense) if for every $\phi \in \mathcal{S}$, $T_m(\phi) \to T(\phi)$. If $T_m \to T$ in $\mathcal{S}'$ then $\hat{T}_m \to \hat{T}$ in $\mathcal{S}'$, for,

$$\hat{T}_m(\phi) = T_m(\hat{\phi}) \to T(\hat{\phi}) = \hat{T}(\phi).$$

Let $f \in L^2(\mathbb{R}^n)$ and let $f_k \in \mathcal{S}$ such that $f_k \to f$ in $L^2(\mathbb{R}^n)$. Then as the inclusion of $L^2(\mathbb{R}^n)$ in $\mathcal{S}'$ is continuous, $f_k \to f$ in $\mathcal{S}'$ as well. Hence $\hat{f}_k \to \hat{f}$ in $\mathcal{S}'$. On the other hand $\mathcal{P}(f_k) \to \mathcal{P}(f)$ in $L^2(\mathbb{R}^n)$ and hence in $\mathcal{S}'$. But as $f_k \in \mathcal{S}$, we already have seen that $\mathcal{P}(f_k) = \hat{f}_k$. Now it follows from the uniqueness of the weak $*$ limit that $\hat{f} = \mathcal{P}(f)$.

COMMENTS

1. *Historical Remarks* It was as early as the end of the nineteenth century that the engineer Heaviside [1, 2] introduced his rules of symbolic calculus in papers where computations which were hardly justified were used to solve physical problems. Later, in the beginning of this century, the physicist Dirac [1] introduced the "δ-function" and its derivatives and the formulae of symbolic calculus become even more inacceptable for the mathematicians of the day. Attempts were made to justify these manipulations but they did not satisfy the engineers or the physicists. Around 1940, H. Cartan and N. Bourbaki considered Radon measures as linear functionals on suitable function spaces and such objects included the δ-function but not its derivatives. It was in 1944 that L. Schwartz discovered the theory of distributions which laid the proper mathematical foundation

of this calculus and it covered the derivatives of the δ-function which were not measures (cf. Example 1.2.6). His discovery did not stem from a view of justifying the symbolic calculus but from an attempt to solve, in higher dimensions, a problem on polyharmonic functions which had been solved in 2-dimensions by Choquet and Deny [1]. Since the discovery of this theory of distributions, fresh life has been infused into several branches of Analysis and Partial Differential Equations.

2. *Distributions* The theory of distributions has provided a natural class of objects which generalize functions and where the usual rules of the calculus hold. However, we have seen that for a function to define a distribution it must be locally integrable. Nevertheless it is possible, by means of suitable limit processes, to produce distributions of functions which are not locally integrable. One such example is the function $f(x) = x^{-1}$ on $\mathbb{R}$. The distribution $PV\left(\frac{1}{x}\right)$ is defined by

$$PV\left(\frac{1}{x}\right)(\phi) = \lim_{\epsilon\to 0} \int_{|x|>\epsilon} \frac{\phi(x)}{x}\, dx.$$

The limit on the right-hand side is the *principal value* of the integral $\int_{-\infty}^{\infty} x^{-1}\phi(x)\, dx$ and hence the letters '*PV*' in the notation for the distribution. A simple computation by integration by parts reveals that

$$\lim_{\epsilon\to 0} \int_{|x|>\epsilon} x^{-1}\phi(x)\, dx = -\int_{\mathbb{R}} \log|x|\, \phi'(x)\, dx$$

and hence, as $\log|x|$ is locally integrable, we say that

$$\frac{d}{dx}(\log|x|) = PV\left(\frac{1}{x}\right)$$

in the dense of distributions.

There also exist functions, which, whatever procedure one employs, do not give distributions. One example is the function $\exp(x^{-1})$. There is no distribution on $\mathbb{R}$ which equals this function on $\mathbb{R}\setminus\{0\}$ (cf. Treves [2]).

The spaces $\mathscr{D}$ and $\mathscr{D}'$ are topological vector spaces and the generalization of classical results on normed spaces are valid here. These are useful in proving results on distributions. For instance, the generalization of the Hahn-Banach Theorem to $\mathscr{D}'(\mathbb{R}^n)$ helps to prove (in the usual way) that the inclusion

$$\mathscr{D}(\mathbb{R}^n) \subset \mathscr{D}'(\mathbb{R}^n)$$

is dense. Similarly by the generalization of the Banach-Steinhaus Theorem to $\mathscr{D}(\mathbb{R}^n)$ we can prove the following result: "If $\{T_m\}$ is a sequence of distributions such that for every $\phi \in \mathscr{D}(\mathbb{R}^n)$, $\{T_m(\phi)\}$ is convergent, then the mapping

$$\phi \to T(\phi) = \lim_{m\to\infty} T_m(\phi)$$

is a distribution" (cf. Rudin [2]).

3. *Fundamental Solutions* The importance of the fundamental solution of a differential operator regarding existence of solutions to linear partial differential equations was seen in this chapter. Also formulae like (1.7.7) give us information regarding the behaviour of the solution at infinity. Another very important use of the fundamental solution is regarding the regularity of solutions. When we solve a differential equation in the sense of distributions, we wish to know whether the distribution solution is a classical solution, i.e. if the data were smooth, if the solution is smooth. We say that a differential operator, L, is **hypoelliptic** if the solution is smooth whenever the data is smooth, i.e. T is C^∞ whenever $L(T)$ is C^∞. In other words, if

$$\text{sing}\cdot\text{supp}\,(T) \subset \text{sing}\cdot\text{supp}\,(L(T)).$$

Obviously if L is hypoelliptic then since δ is C^∞ on $\mathbb{R}^n\diagdown\{0\}$, it follows that every fundamental solution of L has to be C^∞ on $\mathbb{R}^n\diagdown\{0\}$. If L has *constant coefficients*, the converse is true. If L has a fundamental solution which is C^∞ on $\mathbb{R}^n\diagdown\{0\}$, then L is hypoelliptic and every fundamental solution will be C^∞ on $\mathbb{R}^n\diagdown\{0\}$. A partial proof of this fact (i.e. restricted to $L(T) \in \mathcal{E}(\mathbb{R}^n)$) is given in Exercise 1.15. For a full proof see Treves [1].

The operator $\partial^2/\partial x\,\partial y$ is not hypoelliptic. Any function of the form $f(x) + g(y)$ is a solution of

$$\frac{\partial^2 T}{\partial x\,\partial y} = 0$$

and it is enough to have f and g at most C^2! By virtue of Theorems 1.7.1 and 1.7.2, the Laplace operator is hypoelliptic. An interesting consequence is an elegant proof of the following fact: If $T \in \mathcal{D}'(\mathbb{R}^n)$ such that $\dfrac{\partial T}{\partial x_i} = 0$ for all $1 \leqslant i \leqslant n$, then $T = T_C$, the distribution generated by a constant function. (*Proof*: By hypothesis, we can deduce that $\dfrac{\partial^2 T}{\partial x_i^2} = 0$ and hence $\Delta T = 0$; by hypoellipticity, T is a C^∞ function and so must be a constant by classical analysis!) The operator d/dx on $\mathbb{R}$ is hypoelliptic and a similar proof holds for the case of a distribution whose derivative is zero. For a more direct proof, see Exercise 1.2.

4. *The Fourier Transform* If f is an integrable function then by the weak Parseval relation (1.10.1) we have that

$$T_{\hat{f}}(\phi) = T_f(\hat{\phi}).$$

We are tempted to define the Fourier transform $\hat{T}$ of a distribution T by the relation

$$\hat{T}(\phi) = T(\hat{\phi}).$$

Unfortunately, if $\phi \in \mathcal{D}(\mathbb{R}^n)$ then $\hat{\phi} \notin \mathcal{D}(\mathbb{R}^n)$ unless $\phi \equiv 0$. This is a consequence of the *Paley-Wiener Theorem* (Cf. Rudin [2]) which characterizes

entire functions (of n complex variables) with a certain exponential growth as being exactly the Fourier transforms (defined on $\mathbb{C}^n$) of functions in $\mathcal{D}(\mathbb{R}^n)$. Thus ϕ will be the restriction to $\mathbb{R}^n$ of an entire function on $\mathbb{C}^n$ and so cannot have compact support. Thus we have to resort to the Schwartz space $\mathcal{S}$ and its dual $\mathcal{S}'$, the space of tempered distributions, and define the Fourier transform on $\mathcal{S}'$ by the duality relation above. The Fourier transform of a distribution with compact support (which is hence tempered) is the restriction to $\mathbb{R}^n$ of an entire function on $\mathbb{C}^n$ by an extension of the Paley-Wiener theorem. It can then be shown that if S has compact support and T is tempered, then $S * T$ (which is defined) has an analogous property of functions:

$$(S * T)\hat{} = \hat{S}\hat{T}$$

which makes sense since $\hat{S}$ is a function.

5. *Bibliography* For a thorough study and for an excellent introduction to the theory of distributions, see L. Schwartz [1]. Other important references are Treves [2], Hormander [1] and Rudin [2]. The latter reference has also a rich store of exercises. The section on convolution of distributions closely follows the treatment given in this reference. A very nice survey and introduction to the theory of distributions can be found in Horvath [1] and for some further applications to the theory of partial differential equations, see Treves [3].

EXERCISES

1.1 Let f be a continuous function on $\mathbb{R}^n$ such that

$$\int_{\mathbb{R}^n} f\phi = 0, \quad \text{for all } \phi \in \mathcal{D}(\mathbb{R}^n).$$

Show that $f \equiv 0$.

1.2 (a) Let $\phi \in \mathcal{D}(\mathbb{R})$. Show that there exists a $\psi \in \mathcal{D}(\mathbb{R})$ such that $\psi' = \phi$ if, and only if,

$$\int_{\mathbb{R}} \phi(x)\, dx = 0.$$

(b) Deduce that if $T \in \mathcal{D}'(\mathbb{R})$ such that $\dfrac{dT}{dx} = 0$, then $T = T_C$, i.e., the distribution generated by a constant function.

(c) Deduce that if $\dfrac{dT}{dx} \in \mathcal{E}(\mathbb{R})$ for $T \in \mathcal{D}'(\mathbb{R})$, then $T \in \mathcal{E}(\mathbb{R})$.

1.3 Find all distribntion solutions to the equation

$$\frac{dT}{dx} + aT = 0, \quad a \in \mathbb{R}.$$

1.4 Let e_i be the ith standard basis vector of $\mathbb{R}^n$. Show that

$$\frac{1}{h}(T - \tau_{he_i}T) \to \frac{\partial T}{\partial x_i} \quad \text{in } \mathscr{D}'(\mathbb{R}^n)$$

as $h \to 0$.

1.5 Prove Theorem 1.3.1.

1.6 Find all functions $f \in \mathscr{E}(\mathbb{R})$ such that $f\delta' = 0$.

1.7 Let $T \in \mathscr{D}'(\mathbb{R}^n)$ and let α be a multi-index. Show that

$$\text{supp}\,(D^\alpha T) \subset \text{supp}\,(T).$$

1.8 Let $T \in \mathscr{D}'(\mathbb{R}^n)$ and $\phi \in \mathscr{D}(\mathbb{R}^n)$. Show that $\phi T \in \mathscr{E}'(\mathbb{R}^n)$.

1.9 Find the set sing. supp (H).

1.10 Show that the inclusion $\mathscr{D}(\mathbb{R}^n) \subset \mathscr{E}(\mathbb{R}^n)$ is dense.

1.11 Let $T \in \mathscr{D}'(\mathbb{R})$ and let $x \in \mathbb{R}$. If $\delta_x(\phi) = \phi(x)$ for $\phi \in \mathscr{D}(\mathbb{R})$, compute $T * \delta_x$.

1.12 If $T \in \mathscr{D}'(\mathbb{R}^n)$ and $\{\rho_\epsilon\}$ the family of mollifiers, show that $T * \rho_\epsilon \to T$ in $\mathscr{D}'(\mathbb{R}^n)$.

1.13 Show that the function

$$E(x) = \begin{cases} e^{-\lambda x}, & x \geqslant 0 \\ 0, & x < 0 \end{cases}$$

is a fundamental solution to the operator $\left(\frac{d}{dx} + \lambda\right)$ on $\mathbb{R}$.

1.14 Consider the differential operator $\left(\frac{d^2}{dx^2} + a\frac{d}{dx} + b\right) = L$, a, b constants. Let f and g satisfy $Lf=0$, $Lg=0$, $f(0)=g(0)$, $f'(0)-g'(0)=1$. Consider the function

$$F(x) = \begin{cases} f(x), & x \leqslant 0 \\ g(x), & x > 0 \end{cases}.$$

Show that $-F$ is a fundamental solution for L.

1.15 (a) Let $L = \sum_{|\alpha| \leqslant m} a_\alpha D^\alpha$ be a differential operator with constant coefficients such that L has a fundamental solution E which is C^∞ on $\mathbb{R}^n \setminus \{0\}$. Let $\phi \in \mathscr{D}(\mathbb{R}^n)$ with $\phi \equiv 1$ in a neighbourhood of 0. Show that $P = \phi E$ is a distribution with compact support and such that

$$L(P) = \delta + \zeta,$$

$\zeta \in \mathscr{D}(\mathbb{R}^n)$. (Such a distribution P is called a *Parametrix*.)

(b) Deduce that if $T \in \mathscr{D}'(\mathbb{R}^n)$ with $L(T) \in \mathscr{E}(\mathbb{R}^n)$ then $T \in \mathscr{E}(\mathbb{R}^n)$.

(c) Deduce that every fundamental solution of L is C^∞ on $\mathbb{R}^n \setminus \{0\}$.

1.16 Prove Theorem 1.8.2.

1.17 Prove Theorem 1.9.1.

1.18 Show that any polynomial on $\mathbb{R}^n$ defines a tempered distribution.

1.19 Let $h \in \mathbb{R}^n$ and $\delta_h(\phi) = \phi(h)$, $\phi \in \mathscr{D}(\mathbb{R}^n)$. Prove that

(i) $\hat{\delta}_h = \exp(-2\pi i h \cdot \xi)$,

(ii) $[\exp(2\pi i h \cdot x)]^\wedge = \delta_h$.

1.20 If T is a tempered distribution which is harmonic, i.e. $\Delta(T) = 0$, show that T is a polynomial.

TWO

Sobolev Spaces

2.1 DEFINITION AND BASIC PROPERTIES

In this chapter we will study some of the important properties of a class of function spaces—known as Sobolev spaces—which will provide the proper functional setting for the study of the partial differential equations of the following chapters.

Henceforth Ω will stand for an open set in $\mathbb{R}^n$ and $\partial\Omega$ for its boundary.

Definition 2.1.1 Let $m > 0$ be an integer and let $1 \leqslant p \leqslant \infty$. The **Sobolev space** $W^{m,p}(\Omega)$ is defined by

$$W^{m,p}(\Omega) = \{u \in L^p(\Omega) \mid D^\alpha u \in L^p(\Omega) \quad \text{for all } |\alpha| \leqslant m\}. \quad ∎ \tag{2.1.1}$$

In other words, $W^{m,p}(\Omega)$ is the collection of all functions in $L^p(\Omega)$ such that all distribution derivatives upto order m are also in $L^p(\Omega)$. Clearly $W^{m,p}(\Omega)$ is a vector space. (In all that follows we will consider functions with values in $\mathbb{R}$ and the corresponding function spaces as vector spaces over $\mathbb{R}$). We provide it with the norm:

$$\|u\|_{m,p,\Omega} = \sum_{|\alpha|\leqslant m} \|D^\alpha u\|_{L^p(\Omega)} \tag{2.1.2}$$

or, equivalently, for $1 < p < \infty$,

$$\|u\|_{m,p,\Omega} = \left(\sum_{|\alpha|\leqslant m} \int_\Omega |D^\alpha u|^p \right)^{1/p} = \left(\sum_{|\alpha|\leqslant m} \|D^\alpha u\|^p_{L^p(\Omega)} \right)^{1/p}. \tag{2.1.3}$$

It is a straightforward exercise to check that these define norms and that they are equivalent. (We will not distinguish in the future between these two norms though they are only equivalent and not equal. We will use the same notation for both and take care in any computation that we consistently use only one of the two formulae.)

Notation (i) The case $p = 2$ will play a special role in the sequel. These spaces will be denoted by $H^m(\Omega)$. Thus

$$H^m(\Omega) = W^{m,2}(\Omega) \tag{2.1.4}$$

and for $u \in H^m(\Omega)$, we denote its norm by $\|u\|_{m,\Omega}$. i.e.

$$\|u\|_{m,\Omega} = \|u\|_{m,2,\Omega}. \tag{2.1.5}$$

(ii) We will also often use the *semi-norms* which consist of the L^p-norms of the highest order derivatives. We denote these by $|\cdot|_{m, p, \Omega}$. Thus for $u \in W^{m,p}(\Omega)$.

$$|u|_{m, p, \Omega} = \sum_{|\alpha| = m} \|D^\alpha u\|_{L^p(\Omega)} \tag{2.1.6}$$

with the obvious modification (when $1 < p < \infty$) if we use formula (2.1.3) to define the norm. Consistent to case (i) above if $p = 2$ we only write $|\cdot|_{m, \Omega}$ instead of $|\cdot|_{m, 2, \Omega}$.

(iii) We can naturally consider the space $L^p(\Omega)$ as a special case of the Sobolev class, viz. when $m = 0$. i.e. we do not bother about derivatives. In particular we denote the L^p-norm of a function by $|\cdot|_{0, p, \Omega}$ (since in this case the semi-norm and norm are the same). Again the $L^2(\Omega)$-norm will be denoted by $|\cdot|_{0, \Omega}$. ∎

The spaces $H^m(\Omega)$ have a natural inner-product defined by

$$(u, v)_{m, \Omega} = \sum_{|\alpha| \leqslant m} \int_\Omega D^\alpha u D^\alpha v, \quad \text{for } u, v \in H^m(\Omega). \tag{2.1.7}$$

This inner-product yields the norm given by formula (2.1.3) when $p = 2$.

In case $\Omega = \mathbb{R}^n$ the space $H^m(\mathbb{R}^n)$ can also be defined via the Fourier transform. Let $u \in H^m(\mathbb{R}^n)$. Then, by definition, $D^\alpha u \in L^2(\mathbb{R}^n)$ for all $|\alpha| \leqslant m$. Hence the Fourier transform of $D^\alpha u$ is well-defined and we have (cf. (1.11.5))

$$(D^\alpha u)\hat{} = (2\pi i)^{|\alpha|} \xi^\alpha \hat{u},$$

and so $\xi^\alpha \hat{u}(\xi) \in L^2(\mathbb{R}^n)$ for all $|\alpha| \leqslant m$. Conversely if $u \in L^2(\mathbb{R}^n)$ such that $\xi^\alpha \hat{u}(\xi) \in L^2(\mathbb{R}^n)$ for all $|\alpha| \leqslant m$, we have $D^\alpha u \in L^2(\mathbb{R}^n)$ for all $|\alpha| \leqslant m$ and so $u \in H^m(\mathbb{R}^n)$. We can express this in a more compact form using the following algebraic lemma.

Lemma 2.1.1 There exist positive constants M_1 and M_2 depending only on m and n such that

$$M_1(1 + |\xi|^2)^m \leqslant \sum_{|\alpha| \leqslant m} |\xi^\alpha|^2 \leqslant M_2(1 + |\xi|^2)^m \tag{2.1.8}$$

for all $\xi \in \mathbb{R}^n$.

Proof Recall that $|\xi|^2 = \xi_1^2 + \ldots + \xi_n^2$ and $|\xi^\alpha| = |\xi_1|^{\alpha_1} \ldots |\xi_n|^{\alpha_n}$. By a simple induction argument on m we can see that the same powers of ξ occur in $(1 + |\xi|^2)^m$ and $\sum_{|\alpha| \leqslant m} |\xi^\alpha|^2$, albeit with different coefficients, which depend only on n and m. Since the number of terms is finite and depends again only on n and m, the inequalities (2.1.8) follow. ∎

By virtue of the preceding lemma we can define the space $H^m(\mathbb{R}^n)$ as follows:

$$H^m(\mathbb{R}^n) = \{u \in L^2(\mathbb{R}^n) \mid (1 + |\xi|^2)^{m/2} \hat{u}(\xi) \in L^2(\mathbb{R}^n)\}. \tag{2.1.9}$$

Again, by the Plancherel Theorem, it follows that the norm $\|\cdot\|_{m, \mathbb{R}^n}$ on

$H^m(\mathbb{R}^n)$ is equivalent to the norm:

$$\|u\|_{H^m(\mathbb{R}^n)} = \left(\int_{\mathbb{R}^n} (1 + |\xi|^2)^m |\hat{u}(\xi)|^2 \, d\xi\right)^{1/2}. \tag{2.1.10}$$

The advantage of this definition is that it is immediate to generalize to all $s \geqslant 0$. If $s \geqslant 0$ we define $H^s(\mathbb{R}^n)$ as

$$H^s(\mathbb{R}^n) = \{u \in L^2(\mathbb{R}^n) \mid (1 + |\xi|^2)^{s/2} \hat{u}(\xi) \in L^2(\mathbb{R}^n)\} \tag{2.1.11}$$

with associated norm

$$\|u\|_{H^s(\mathbb{R}^n)} = \left(\int_{\mathbb{R}^n} (1 + |\xi|^2)^s |\hat{u}(\xi)|^2 \, d\xi\right)^{1/2}. \tag{2.1.12}$$

We will investigate such spaces defined over other open sets of $\mathbb{R}^n$ later.

Let us return to the spaces $W^{m,p}(\Omega)$. The map

$$u \in W^{1,p}(\Omega) \to \left(u, \frac{\partial u}{\partial x_1}, \ldots, \frac{\partial u}{\partial x_n}\right) \in (L^p(\Omega))^{n+1} \tag{2.1.13}$$

is an isometry of $W^{1,p}(\Omega)$ into $(L^p(\Omega))^{n+1}$ if we provide the latter space with the norm

$$\|u\| = \sum_{i=1}^{n+1} |u_i|_{0,p,\Omega} \quad \text{or} \quad \|u\| = \left(\sum_{i=1}^{n+1} |u_i|_{0,p,\Omega}^p\right)^{1/p}$$

for $u = (u_i) \in (L^p(\Omega))^{n+1}$, depending on whether we use the formula (2.1.2) or (2.1.3) for the norm on $W^{1,p}(\Omega)$. This is a useful fact to remember and will be used in the proof of the following result.

Theorem 1.2.1 For every $1 \leqslant p \leqslant \infty$, the space $W^{1,p}(\Omega)$ is a Banach space. If $1 < p < \infty$, it is reflexive and if $1 \leqslant p < \infty$, it is separable. In particular $H^1(\Omega)$ is a separable Hilbert space.

Proof: Let $\{u_m\}$ be a Cauchy sequence in $W^{1,p}(\Omega)$. It follows from the definition of the norm that $\{u_m\}$ and $\left\{\frac{\partial u_m}{\partial x_i}\right\}$, $1 \leqslant i \leqslant n$ are all Cauchy sequences in $L^p(\Omega)$. Let $u_m \to u$ and $\frac{\partial u_m}{\partial x_i} \to v_i$, $1 \leqslant i \leqslant n$, in $L^p(\Omega)$. The completeness of the space $W^{1,p}(\Omega)$ will be proved if we show that $\frac{\partial u}{\partial x_i} = v_i$ in the sense of distributions so that, on one hand, $u \in W^{1,p}(\Omega)$ and, on the other $u_m \to u$ in $W^{1,p}(\Omega)$.

Let $\phi \in \mathscr{D}(\Omega)$. We need to show that

$$\int_\Omega u \frac{\partial \phi}{\partial x_i} = -\int_\Omega v_i \phi. \tag{2.1.14}$$

Now, since $u_m \in W^{1,p}(\Omega)$, we know that

$$\int_\Omega u_m \frac{\partial \phi}{\partial x_i} = -\int_\Omega \frac{\partial u_m}{\partial x_i} \phi \tag{2.1.15}$$

Since $\phi \in \mathscr{D}(\Omega)$, $\phi \in L^q(\Omega)$ for all $1 \leqslant q \leqslant \infty$ and so we pass to the limit

on both sides of (2.1.15) as $m \to \infty$ to obtain (2.1.14). Thus $W^{1,p}(\Omega)$ is complete and if $p = 2$, we get that $H^1(\Omega)$ is a Hilbert space.

Now $(L^p(\Omega))^{n+1}$ is reflexive for $1 < p < \infty$ and separable for $1 \leqslant p < \infty$. Since $W^{1,p}(\Omega)$ is complete, its image under the isometry (2.1.13) is a closed subspace of $(L^p(\Omega))^{n+1}$ which inherits the corresponding properties. ∎

Remark 2.1.1 The results of this theorem can be proved by the same way for any integer $m \geqslant 2$. In future, unless absolutely necessary, we will establish theorems only for the spaces $W^{1,p}(\Omega)$. The extensions to higher order spaces will often be obvious. ∎

Remark 2.1.2 In the course of the proof of the preceding theorem we have proved the following fact : "if $u_m \to u$ in $L^p(\Omega)$ and $\dfrac{\partial u_m}{\partial x_i} \to v_i$ in $L^p(\Omega)$ for each $1 \leqslant i \leqslant n$, then $u \in W^{1,p}(\Omega)$ and $\dfrac{\partial u}{\partial x_i} = v_i$". Indeed, we can weaken the hypotheses even further. What was needed to pass to the limit in (2.1.15) and obtain (2.1.14) was only the weak convergence of $\left\{\dfrac{\partial u_m}{\partial x_i}\right\}$ (weak $*$ if $p = \infty$). Since bounded sequences have weakly convergent (weak $*$ convergent when $p = \infty$) subsequences, it is enough to know that $u_m \to u$ in $L^p(\Omega)$ and $\left\{\dfrac{\partial u_m}{\partial x_i}\right\}$ are *bounded* $(1 < p \leqslant \infty)$ to deduce that $u \in W^{1,p}(\Omega)$. ∎

Let us recall that though we talk of L^p-functions, the space $L^p(\Omega)$ is really only made up of *equivalence classes* of functions (under the equivalence relation given by equality a.e.). Thus by saying that "u is a continuous function" in $L^p(\Omega)$ we mean that the corresponding equivalence class has a representative which is a continuous function. In this spirit we prove the following result characterizing the space $W^{1,p}(I)$ where $I \subset \mathbb{R}$ is an open interval.

Theorem 2.1.2 Let $I \subset \mathbb{R}$ be an open interval and let $u \in W^{1,p}(I)$. Then u is *absolutely continuous.*

Proof Let $x_0 \in I$ and define

$$\bar{u}(x) = \int_{x_0}^{x} u'(t)\, dt, \tag{2.1.16}$$

which, by definition, is absolutely continuous. Hence its classical derivative exists a.e. and is equal a.e. to u'; this is also its distribution derivative (cf. Section 1.3). Hence, in the sense of distributions,

$$(u - \bar{u})' = 0$$

and so $u - \bar{u} = c$, a constant a.e. (cf. Exercise 1.2). Thus $u = \bar{u} + c$ a.e. and the latter function is absolutely continuous. ∎

We can deduce an important property of $W^{1,p}(I)$ from the preceding theorem, when I is a *bounded* interval. Let, for instance, $I = (0, 1)$. Then if $u \in W^{1,p}(I)$, we can write

$$u(x) = u(0) + \int_0^x u'(t)\, dt. \qquad (2.1.17)$$

Hence by Holder's inequality, if q is the conjugate exponent of p, i.e., $p^{-1} + q^{-1} = 1$, we have

$$|u(0)| \leqslant |u(x)| + |u'|_{0,p,I}|x|^{1/q}.$$

Thus it follows (on integration) that

$$|u(0)| \leqslant C(|u|_{0,p,I} + |u'|_{0,p,I}) = C\|u\|_{1,p,I} \qquad (2.1.18)$$

where $C > 0$ is a constant not depending on u. Now using (2.1.17) and (2.1.18) we also deduce that for any $x \in I$

$$|u(x)| \leqslant C\|u\|_{1,p,I}, \quad C > 0, \text{ independent of } u. \qquad (2.1.19)$$

Let B be the unit ball in $W^{1,p}(I)$. Then

$$B = \{u \in W^{1,p}(I) \mid \|u\|_{1,p,I} \leqslant 1\} \qquad (2.1.20)$$

It follows that if $i: W^{1,p}(I) \to C(\bar{I})$ is the inclusion map (established in Theorem 2.1.2 and continuous by virtue of (2.1.19)) then $B = i(B)$ is a uniformly bounded set in $C(\bar{I})$. Again if $x, y \in I$, by (2.1.17), we have

$$|u(x) - u(y)| \leqslant |u'|_{0,p,I}|x - y|^{1/q} \leqslant \|u\|_{1,p,I}|x - y|^{1/q} \qquad (2.1.21)$$

from which it follows that B is equicontinuous in $C(\bar{I})$. It follows from the *Ascoli-Arzela Theorem* that B is relatively compact in $C(\bar{I})$; in other words, the map $i: W^{1,p}(I) \to C(\bar{I})$ is a *compact operator*. This is an important property of the Sobolev spaces and will be studied in detail later.

Finally, we introduce an important subspace of the space $w^{m,p}(\Omega)$. If $1 \leqslant p < \infty$, we know that $\mathcal{D}(\Omega)$ is dense in $L^p(\Omega)$. Also, if $\phi \in \mathcal{D}(\Omega)$, so does every derivative of ϕ and so $\mathcal{D}(\Omega) \subset W^{m,p}(\Omega)$, for any m and p. If $1 \leqslant p < \infty$, we define the space $W_0^{m,p}(\Omega)$ as the closure of $\mathcal{D}(\Omega)$ in $W^{m,p}(\Omega)$. Thus $W_0^{m,p}(\Omega)$ is a closed subspace of $W^{m,p}(\Omega)$ and its elements can be approximated in the $W^{m,p}(\Omega)$ norm by C^∞ functions with compact support. In general this is a strict subspace of $W^{m,p}(\Omega)$, except when $\Omega = \mathbb{R}^n$ as we show below.

Theorem 2.1.3 Let $1 \leqslant p < \infty$. Then for any integer $m \geqslant 0$,

$$W^{m,p}(\mathbb{R}^n) = W_0^{m,p}(\mathbb{R}^n). \qquad (2.1.22)$$

Proof: We will present the details for the case $m = 1$. We have to show that if $u \in W^{1,p}(\mathbb{R}^n)$, then there exists a sequence $\{\phi_k\}$ in $\mathcal{D}(\mathbb{R}^n)$ such that $\phi_k \to u$ in $W^{1,p}(\mathbb{R}^n)$.

Step 1. Let $(\rho_\epsilon\}$ be the family of mollifiers (cf. Example 1.2.2). Then if $u \in L^p(\mathbb{R}^n)$, we have $\rho_\epsilon * u \to u$ in $L^p(\mathbb{R}^n)$. First of all since $\rho_\epsilon \in L^1(\mathbb{R}^n)$, the

convolution is well-defined and is in $L^p(\mathbb{R}^n)$, by Theorem 1.5.2. Let ϕ be a continuous function with compact support such that $|\phi - u|_{0, p, \mathbb{R}^n} < \delta/3$ where $\delta > 0$ is a pre-assigned number. Now choose $\epsilon > 0$ small enough such that $|\phi * \rho_\epsilon - \phi|_{0, p, \mathbb{R}^n} < \delta/3$ (which is possible by Theorem 1.5.6). Hence

$$|u - u * \rho_\epsilon|_{0, p, \mathbb{R}^n} \leqslant |u - \phi|_{0, p, \mathbb{R}^n} + |\phi - \rho_\epsilon * \phi|_{0, p, \mathbb{R}^n} + |\phi * \rho_\epsilon - u * \rho_\epsilon|_{0, p, \mathbb{R}^n} < \delta$$

since, by (1.5.4),

$$|(\phi - u) * \rho_\epsilon|_{0, p, \mathbb{R}^n} \leqslant |\phi - u|_{0, p, \mathbb{R}^n} |\rho_\epsilon|_{0, 1, \mathbb{R}^n} < \delta/3$$

This proves the claim made in the beginning of this step.

Step 2 Now, if $u \in W^{1, p}(\mathbb{R}^n)$, then $u * \rho_\epsilon$ is a C^∞ function and $D^\alpha(u * \rho_\epsilon) = D^\alpha u * \rho_\epsilon = u * D^\alpha \rho_\epsilon$ for any multi-index α. By Step 1, $u * \rho_\epsilon \to u$ and $D^\alpha(u * \rho_\epsilon) = \rho_\epsilon * D^\alpha u \to D^\alpha u$ in $L^p(\mathbb{R}^n)$. Hence $u * \rho_\epsilon \to u$ in $W^{1, p}(\mathbb{R}^n)$.

Step 3 Let ζ be a function in $\mathscr{D}(\mathbb{R}^n)$ such that $0 \leqslant \zeta \leqslant 1$, $\zeta \equiv 1$ on $B(0; 1)$ and supp $(\zeta) \subset B(0; 2)$. We consider the sequence $\{\zeta_k\}$ in $\mathscr{D}(\mathbb{R}^n)$ defined by

$$\zeta_k(x) = \zeta(x/k) \tag{2.1.23}$$

Let $\epsilon_k \downarrow 0$. Set $u_k = \rho_{\epsilon_k} * u$. Then $u_k \in \mathscr{E}(\mathbb{R}^n)$ and $u_k \to u$ in $W^{1, p}(\mathbb{R}^n)$ by Step 2. Now define $\phi_k \in \mathscr{D}(\mathbb{R}^n)$ by

$$\phi_k(x) = \zeta_k(x) u_k(x). \tag{2.1.24}$$

We now show that $\phi_k \to u$ in $W^{1, p}(\mathbb{R}^n)$, which will complete the proof. Since $\zeta_k \equiv 1$ on $B(0; k)$ we have $u_k = \phi_k$ on $B(0; k)$. Hence

$$|u_k - \phi_k|_{0, p, \mathbb{R}^n} = \left(\int_{|x|>k} |u_k(x) - \phi_k(x)|^p \, dx\right)^{1/p} \leqslant 2\left(\int_{|x|>k} |u_k|^p\right)^{1/p}$$

since $|\phi_k| \leqslant |u_k|$. Now the latter integral tends to zero as $k \to \infty$ since $u_k \to u$ in $L^p(\mathbb{R}^n)$. Thus $\phi_k \to u$ in $L^p(\mathbb{R}^n)$. Similarly, since

$$\frac{\partial \phi_k}{\partial x_i} = \frac{\partial u_k}{\partial x_i}$$

on $B(0; k)$, we get $\dfrac{\partial \phi_k}{\partial x_i} \to \dfrac{\partial u}{\partial x_i}$ in $L^p(\mathbb{R}^n)$ by an analogous argument. Thus $\phi_k \to u$ in $W^{1, p}(\mathbb{R}^n)$. ∎

As usual when $p = 2$ we write $H_0^m(\Omega)$ for $W_0^{m, 2}(\Omega)$ and so

$$H_0^m(\mathbb{R}^n) = H^m(\mathbb{R}^n). \tag{2.1.25}$$

In the following sections we will study the following aspects of the Sobolev spaces: (i) Approximation theorems: Often many results are easy to prove for smooth functions and, by a density argument, we can generalize to an arbitrary function in the given space. So we study the approximation of elements of $W^{m, p}(\Omega)$ by smooth functions. (ii) Extension theorems: Results are usually easy to prove on the whole space $\mathbb{R}^n$. In

order to prove it for an open set Ω a usual technique is to extend a function on Ω to $\mathbb{R}^n$ use the result available there and restrict it to Ω again. To do this we need to know if on one hand an element of $W^{m,p}(\Omega)$ can be extended to $W^{m,p}(\mathbb{R}^n)$ and on the other hand, if such an extension is continuous on $W^{m,p}(\Omega)$. This property will depend on the nature of Ω as we shall see later. (iii) Inclusion theorems: We saw that when $n = 1$ the elements of $W^{1,p}(I)$ were in $C(\bar{I})$. By iteration, higher order spaces will be included in spaces of smoother functions. In higher dimensions, this may not be true unless we consider spaces of fairly high order. However elements of $W^{1,p}(\Omega)$ can be shown to be included in Lebesgue spaces $L_q(\Omega)$ with q larger than p. These properties are very useful. (iv) Compactness theorems: We saw that the inclusion $W^{1,p}(I) \to C(\bar{I})$ was compact. We will study the compactness of the various inclusions cited above. Again compactness of inclusions have interesting applications to partial differential equations, especially in the study of eigenvalue problems. (v) Trace theory: In the one-dimensional case by Theorem 2.1.2, it follows that for bounded domains, it makes sense to talk of the boundary values of elements of $W^{1,p}(I)$, I a bounded interval. In general as the n-dimensional measure of $\partial\Omega$, the boundary of an open set $\Omega \subset \mathbb{R}^n$, is zero, it is not meaningful *a priori* to talk of the value of u on $\partial\Omega$ when $u \in W^{1,p}(\Omega)$, unless, say, u is at least continuous. The object of the trace theory is to give a meaning to $u|_{\partial\Omega}$, called the **trace** of u. This will be essential in the next two chapters where boundary value problems for partial differential operators will be studied. To do this we will also need to define the spaces $W^{s,p}(\Omega)$ for all real numbers s.

An important reference for the results presented in this chapter is Adams [1]. In this reference, several generalizations of the results we present can be found. We will often only present proofs in the simplest of cases in order that the exposition is not encumbered by technical details. Another eminently readable version with several illuminating remarks can be found in Brezis [1].

2.2 APPROXIMATION BY SMOOTH FUNCTIONS

In this section we will consider a certain number of density results and prove some simple consequences of these.

Theorem 2.2.1 (*Friedrichs*) Let $1 \leqslant p < \infty$, and let $u \in W^{1,p}(\Omega)$. Then there exists a sequence $\{u_m\}$ in $\mathscr{D}(\mathbb{R}^n)$ such that $u_m \to u$ in $L^p(\Omega)$ and $\left.\frac{\partial u_m}{\partial x_i}\right|_{\Omega'} \to \left.\frac{\partial u}{\partial x_i}\right|_{\Omega'}$ in $L^p(\Omega')$ for every $1 \leqslant i \leqslant n$ and for every $\Omega' \subset\subset \Omega$ (i.e. Ω' relatively compact in Ω).

Proof: *Step 1* Consider the family of mollifers $\{\rho_\epsilon\}$ and let $\bar{u}$ be the extension of u by zero outside Ω. Then $\rho_\epsilon * \bar{u} \to \bar{u}$ in $L^p(\mathbb{R}^n)$ (cf. Step 1 of the proof of Theorem 2.1.3) and so $\rho_\epsilon * \bar{u} \to u$ in $L^p(\Omega)$. Let α be any multi-

index such that $|\alpha| = 1$. Let Ω' be relatively compact in Ω, i.e. $\bar{\Omega}'$ is compact and $\bar{\Omega}' \subset \Omega$. Then dist $(\Omega', \partial\Omega) > 0$. Let ϵ be smaller than this quantity. It is then clear that for $x \in \Omega'$, we have

$$\rho_\epsilon * \bar{u}(x) = \rho_\epsilon * u(x),$$

and so $D^\alpha(\rho_\epsilon * \bar{u}) = \rho_\epsilon * D^\alpha u$ on Ω' and hence is in $L^p(\Omega')$ since $D^\alpha u \in L^p(\Omega')$. Also it follows that $D^\alpha(\rho_\epsilon * \bar{u}) \to D^\alpha u$ in $L^p(\Omega')$ and so $\rho_\epsilon * \bar{u} \to u$ in $W^{1,p}(\Omega')$.

Step 2 By the preceding step we can construct (by choosing $\epsilon_m \downarrow 0$) a sequence of functions $v_m (= \rho_{\epsilon_m} * \bar{u})$ such that $v_m \to u$ in $L^p(\Omega)$ and $\left.\frac{\partial v_m}{\partial x_i}\right|_{\Omega'} \to \left.\frac{\partial u}{\partial x_i}\right|_{\Omega'}$ in $L^p(\Omega')$ for $\Omega' \subset\subset \Omega$. Now if $\{\zeta_m\}$ in the sequence of cut-off functions as defined in Step 3 of the proof of Theorem 2.1.3, then $u_m = \zeta_m v_m$ is in $\mathscr{D}(\mathbb{R}^n)$ and the sequence $\{u_m\}$ has the same convergence properties as $\{v_m\}$. ∎

In the above result, the derivatives converge only on relatively compact sets. There do exist examples of open sets where convergence of the derivatives cannot be achieved on the entire open set. We describe below a class of open sets for which it is possible to have the derivatives converge in $L^p(\Omega)$.

Definition 2.2.1 Let Ω be an open set in $\mathbb{R}^n$. An **extension operator** P for $W^{1,p}(\Omega)$ is a bounded linear operator

$$P : W^{1,p}(\Omega) \to W^{1,p}(\mathbb{R}^n)$$

such that $Pu|_\Omega = u$ for every $u \in W^{1,p}(\Omega)$. ∎

By virtue of the fact that P is a bounded linear operator, it follows that

$$\|Pu\|_{1,p,\Omega} \leqslant C\|u\|_{1,p,\Omega} \tag{2.2.1}$$

where $C > 0$ is a constant, which, in general, will only depend on Ω and p. Thus if Ω is such that an extension P exists then we can consider $W^{1,p}(\Omega)$ as the space of restrictions to Ω of functions of $W^{1,p}(\mathbb{R}^n)$. A sufficient condition for the existence of P is the smoothness of the boundary $\partial\Omega$ and these considerations will be taken up in the next section.

Theorem 2.2.2 If Ω is an open set in $\mathbb{R}^n$ such that an extension operator $P : W^{1,p}(\Omega) \to W^{1,p}(\mathbb{R}^n)$ exists then given $u \in W^{1,p}(\Omega)$ there exists a sequence $\{u_m\}$ in $\mathscr{D}(\mathbb{R}^n)$ such that $u_m|_\Omega \to u$ in $W^{1,p}(\Omega)$.

Proof Let P be an extension operator. Then choosing $\epsilon_m \downarrow 0$ we have

$$\zeta_m(\rho_{\epsilon_m} * P(u)) \to P(u) \quad \text{in } W^{1,p}(\mathbb{R}^n). \tag{2.2.2}$$

Then their restrictions to Ω converge in $W^{1,p}(\Omega)$. ∎

Thus the above theorem tells us that if Ω admits an extension operator, then elements of $W^{1,p}(\Omega)$ can be approximated by functions in $C^\infty(\Omega)$ *which are restrictions of functions of* $\mathcal{D}(\mathbb{R}^n)$. However, a more difficult theorem of Meyers and Serrin says that the set of all functions in $C^\infty(\Omega) \cap W^{1,p}(\Omega)$ is dense in $W^{1,p}(\Omega)$ when $1 \leqslant p < \infty$, *for all open sets* Ω (cf. Adams [1]).

None of these results extend to the case $p = \infty$. If $\Omega = \mathbb{R}^n$, it is known that the completion of continuous functions with compact support with respect to the L^∞-norm is the space of continuous functions which vanish at infinity. Thus while the function $u \equiv 1$ is in $W^{1,\infty}(\mathbb{R}^n)$, it cannot be approximated by elements of $\mathcal{D}(\mathbb{R}^n)$. We give below another simple example.

Example 2.2.1 Let Ω be an open interval, say $(-1, 1)$. Consider the function

$$u(x) = \begin{cases} 0, & x \leqslant 0 \\ x, & x \geqslant 0. \end{cases} \tag{2.2.3}$$

Then as u is absolutely continuous, its distribution derivative is given by

$$u'(x) = \begin{cases} 0, & x < 0 \\ 1, & x > 0. \end{cases}$$

Let $\phi \in C^\infty(\Omega)$ such that

$$|\phi' - u'|_{0,\infty,\Omega} < \epsilon.$$

Thus if $x < 0$, $|\phi'(x)| < \epsilon$ and if $x > 0$, $|\phi'(x) - 1| < \epsilon$ or $\phi'(x) > 1 - \epsilon$. Thus by continuity $\phi'(0) < \epsilon$ and $\geqslant 1 - \epsilon$ which is impossible if $\epsilon < 1/2$. Thus u cannot be approximated in $W^{1,\infty}(\Omega)$ by smooth functions. ∎

Let us present a few simple applications of these approximation results.

Theorem 2.2.3 (Chain Rule) Let $G \in C^1(\mathbb{R})$ such that $G(0) = 0$ and $|G'(s)| \leqslant M$ for all $s \in \mathbb{R}$. Let $u \in W^{1,p}(\Omega)$. Then the function $G \circ u \in W^{1,p}(\Omega)$ and

$$\frac{\partial}{\partial x_i}(G \circ u) = (G' \circ u)\frac{\partial u}{\partial x_i}, \quad 1 \leqslant i \leqslant n. \tag{2.2.4}$$

Proof Since the derivative of G is bounded by M and $G(0) = 0$, by the Mean Value Theorem, we have

$$|G(s)| \leqslant M|s|, \quad s \in \mathbb{R}. \tag{2.2.5}$$

Thus $|G \circ u(x)| = |G(u(x))| \leqslant M|u(x)|$ for every $x \in \Omega$ and so $G \circ u \in L^p(\Omega)$. Similarly $(G' \circ u)\dfrac{\partial u}{\partial x_i} \in L^p(\Omega)$ for $1 \leqslant i \leqslant n$. Assume now that $1 \leqslant p < \infty$. Then by Theorem 2.2.1, there exists a sequence $\{u_m\}$ in $\mathcal{D}(\mathbb{R}^n\}$ such that $u_m \to u$ in $L^p(\Omega)$ and $\dfrac{\partial u_m}{\partial x_i} \to \dfrac{\partial u}{\partial x_i}$ in $L^p(\Omega')$ for every $\Omega' \subset\subset \Omega$. Let $\phi \in \mathcal{D}(\Omega)$. Then choose $\Omega' \subset\subset \Omega$ such that supp $(\phi) \subset \Omega' \subset\subset \Omega$. Then

since u_m is smooth, we have by the usual chain rule,

$$\int_{\Omega} (G \circ u_m) \frac{\partial \phi}{\partial x_i} = \int_{\Omega'} (G \circ u_m) \frac{\partial \phi}{\partial x_i} = -\int_{\Omega'} (G' \circ u_m) \frac{\partial u_m}{\partial x_i} \phi \tag{2.2.6}$$

Now $G \circ u_m \to G \circ u$ in $L^p(\Omega)$ since

$$|G \circ u_m(x) - G \circ u(x)| \leqslant M|u_m(x) - u(x)|$$

Also $G' \circ u_m$ is uniformly bounded by M and so $(G' \circ u_m) \dfrac{\partial u_m}{\partial x_i} \to (G' \circ u) \dfrac{\partial u}{\partial x_i}$ in $L^p(\Omega')$ and so we can pass to the limit in (2.2.6) and thus prove (2.2.4). If $p = \infty$, fix $\phi \in \mathcal{D}(\Omega)$ and choose Ω' such that supp $(\phi) \subset \Omega' \subset\subset \Omega$. Then as Ω' is relatively compact, $u \in W^{1,\infty}(\Omega)$ implies that $u \in W^{1,p}(\Omega')$ for all $1 \leqslant p \leqslant \infty$ and so (2.2.4) is valid by the preceding method. ∎

This result can be generalized to Lipschitz continuous functions G. We will prove this in the context of the spaces $W_0^{1,p}(\Omega)$ shortly.

Theorem 2.2.4 Let $1 \leqslant p < \infty$, and let $u \in W^{1,p}(\Omega)$ such that u vanishes outside a compact set contained in Ω. Then $u \in W_0^{1,p}(\Omega)$.

Proof $K \subset \Omega$ be a compact set such that $u \equiv 0$ on Ω/K. Let Ω' be such that $K \subset \Omega' \subset\subset \Omega$. Let $\phi \in \mathcal{D}(\Omega')$ be a cut-off function such that $\phi \equiv 1$ on K. Then we know that $\phi u = u$. Now by Theorem 2.2.1, there exists a sequence $\{u_m\}$ in $\mathcal{D}(\mathbb{R}^n)$ such that $u_m \to u$ in $L^p(\Omega)$ and $\dfrac{\partial u_m}{\partial x_i} \to \dfrac{\partial u}{\partial x_i}$, $1 \leqslant i \leqslant n$ in $L^p(\Omega')$. Consequently $\phi u_m \to \phi u$ in $W^{1,p}(\Omega)$ and as $\phi u_m \in \mathcal{D}(\Omega)$ it follows that ϕu, i.e. u is in $W_0^{1,p}(\Omega)$. ∎

Theorem 2.2.5 (*Stampacchia*) Let G be a Lipschitz continuous function of $\mathbb{R}$ into itself such that $G(0) = 0$. Then if Ω is bounded, $1 < p < \infty$, and $u \in W_0^{1,p}(\Omega)$, we have $G \circ u \in W_0^{1,p}(\Omega)$.

Proof Let $u \in W_0^{1,p}(\Omega)$ and let $u_m \in \mathcal{D}(\Omega)$ such that $u_m \to u$ in $W^{1,p}(\Omega)$. Define

$$v_m = G \circ u_m.$$

Since u_m has compact support and $G(0) = 0$, v_m has compact support. Also v_m is Lipschitz continuous; for

$$\begin{aligned} |v_m(x) - v_m(y)| &= |G(u_m(x)) - G(u_m(y))| \\ &\leqslant K|u_m(x) - u_m(y)| \\ &\leqslant K_m|x-y| \end{aligned}$$

as u_m is a smooth function with compact support and G is Lipschitz continuous. Hence $v_m \in L^p(\Omega)$. Also it follows that $\left|\dfrac{\partial v_m}{\partial x_i}\right| \leqslant K_m$, $1 \leqslant i \leqslant n$ and as Ω is bounded, $\dfrac{\partial v_m}{\partial x_i} \in L^p(\Omega)$. Thus $v_m \in W^{1,p}(\Omega)$ and has compact support. Thus by Theorem 2.2.4, $v_m \in W_0^{1,p}(\Omega)$. Also from the relation

$$|v_m(x) - G(u(x))| \leqslant K|u_m(x) - u(x)| \tag{2.2.7}$$

it follows that $v_m \to G \circ u$ in $L^p(\Omega)$. Further if e_i is the standard ith basis vector in $\mathbb{R}^n$, we have

$$\frac{|v_m(x + he_i) - v_m(x)|}{|h|} \leqslant \frac{K|u_m(x + he_i) - u_m(x)|}{|h|}$$

and so

$$\limsup_{m\to\infty} \left|\frac{\partial v_m}{\partial x_i}\right|_{0,p,\Omega} \leqslant K \limsup_{m\to\infty} \left|\frac{\partial u_m}{\partial x_i}\right|_{0,p,\Omega} \tag{2.2.8}$$

But $\left\{\frac{\partial u_m}{\partial x_i}\right\}$ is a convergent sequence in $L^p(\Omega)$ and so from (2.2.8) it follows that $\left\{\frac{\partial v_m}{\partial x_i}\right\}$ is bounded for each $1 \leqslant i \leqslant n$. Hence by the Remark 2.1.2 it follows that $\{v_m\}$ converges in $W^{1,p}(\Omega)$ and that $G \circ u \in W_0^{1,p}(\Omega)$. ∎

Remark 2.2.1 If Ω is a bounded open set for which Theorem 2.2.2 is valid, then we can choose $u_m \in \mathcal{D}(\mathbb{R}^n)$ with $u_m \to u$ in $W^{1,p}(\Omega)$ for $u \in W^{1,p}(\Omega)$. The same proof will then go through and we will have the above theorem for $W^{1,p}(\Omega)$ as well. ∎

Corollary Let $u \in H_0^1(\Omega)$, Ω is a bounded open set in $\mathbb{R}^n$. Then $|u|$, u^+ and u^- belong to $H_0^1(\Omega)$ where

$$u^+(x) = \max\{u(x), 0\} \tag{2.2.9}$$

$$u^-(x) = \max\{-u(x), 0\}.$$

Proof We apply the preceding theorem with $p = 2$ and $G(t) = |t|$. Thus $|u| \in H_0^1(\Omega)$ for $u \in H_0^1(\Omega)$. Now

$$u^+ = \frac{|u| + u}{2}, \; u^- = \frac{|u| - u}{2}$$

and so $u^+, u^- \in H_0^1(\Omega)$. ∎

Remark 2.2.2 The above corollary will be very useful in Chapter 3 when we consider properties of elliptic boundary value problems like the *maximum principle* and the *simplicity* of the *first eigenvalue* of *second order elliptic eigenvalue problems*. ∎

Theorem 2.2.6 Let $1 \leqslant p \leqslant \infty$ and $u \in W^{1,p}(\Omega) \cap C(\bar{\Omega})$. If $u = 0$ on $\partial\Omega$ then $u \in W_0^{1,p}(\Omega)$.

Proof: Step 1 Assume that supp(u) is a bounded set in $\bar{\Omega}$. Let $G \in C^1(\mathbb{R})$ with the following properties:

$$|G(t)| \leqslant |t| \text{ and } G(t) = \begin{cases} 0, & \text{if } |t| \leqslant 1. \\ t, & \text{if } |t| \geqslant 2 \end{cases} \tag{2.2.10}$$

Define the sequence

$$u_m = \frac{1}{m} G(mu) \tag{2.2.11}$$

Then by Theorem 2.2.3, $u_m \in W^{1,p}(\Omega)$. We claim that $u_m \to u$ in $W^{1,p}(\Omega)$. For since $u_m = u$ on the set $\{|u| \geqslant 2/m\}$,

$$\begin{aligned}
\left(\int_\Omega |u_m - u|^p\right)^{1/p} &= \left(\int_{|u|<2/m} |u_m - u|^p\right)^{1/p} \\
&\leqslant \left(\int_{|u|<2/m} |u_m|^p\right)^{1/p} + \left(\int_{|u|<2/m} |u|^p\right)^{1/p} \\
&\leqslant 2\left(\int_{0<|u|<2/m} |u|^p\right)^{1/p} \leqslant \frac{4}{m}\,(\text{meas}(\text{supp}(u)))^{1/p}
\end{aligned}$$

which tends to zero as $m \to \infty$. Thus $u_m \to u$ in $L^p(\Omega)$. Again for $1 \leqslant i \leqslant n$, using the chain rule (2.2.4)

$$\frac{\partial u_m}{\partial x_i} = G'(mu)\,\frac{\partial u}{\partial x_i}$$

We see that as soon as $m|u(x)| > 2$, $G'(mu) = 1$. Thus $\dfrac{\partial u_m}{\partial x_i} \to \dfrac{\partial u}{\partial x_i}$ pointwise and it follows by the Dominated Convergence Theorem that this convergence is also valid in $L^p(\Omega)$. Thus $u_m \to u$ in $W^{1,p}(\Omega)$.

Now, using (2.2.10) again, we have

$$\text{supp}(u_m) \subset \left\{x \in \Omega \,\middle|\, |u(x)| \geqslant \frac{1}{m}\right\}$$

and, as supp(u) is bounded and as u vanishes on $\partial\Omega$, it follows that supp(u_m) is a closed bounded set strictly contained in Ω, i.e. u_m has compact support in Ω. But then $u_m \in W_0^{1,p}(\Omega)$ by Theorem 2.2.4 and so as $W_0^{1,p}(\Omega)$ is a closed subspace, we have $u \in W_0^{1,p}(\Omega)$.

Step 2 If supp(u) is unbounded, let $\{\zeta_m\}$ be the cut-off sequence as in Step 3 of the proof of Theorem 2.1.3. Then $\zeta_m u \to u$ in $W^{1,p}(\Omega)$. Also $\zeta_m u$ has bounded support (in fact $\text{supp}(\zeta_m u) \subset B(0; m) \cap \overline{\Omega}$) and so by Step 1 above, $\zeta_m u \in W_0^{1,p}(\Omega)$. Hence $u \in W_0^{1,p}(\Omega)$. ∎

Remark 2.2.3 The above theorem says that continuous functions on $\overline{\Omega}$ which are in $W^{1,p}(\Omega)$ and vanish on the boundary are in fact in $W_0^{1,p}(\Omega)$. It is an important component of the *trace theorem* that $W_0^{1,p}(\Omega)$ is precisely the set of functions in $W^{1,p}(\Omega)$ for which the "boundary values" i.e. the traces, are zero. ∎

2.3 EXTENSION THEOREMS

As has been already mentioned it is often easy to prove properties of the Sobolev spaces $W^{m,p}(\Omega)$ starting with the case $\Omega = \mathbb{R}^n$. The corresponding results for a general Ω often follow easily if we have an extension

operator from Ω to $\mathbb{R}^n$. We also saw that if such operators exist, then we can prove results on the approximation by smooth functions.

First of all, extension operators are not unique. A domain Ω may have several such operators as we see from the following example.

Example 2.3.1 Let $\Omega = (0, 1) \subset \mathbb{R}$. Let $u \in W^{1,p}(\Omega)$. Then by Theorem 2.1.2, u is absolutely continuous. We also later saw that the value of u at any point was bounded uniformly by the $W^{1,p}$-norm of u (cf. (2.1.19)). Now consider the following function defined on $\mathbb{R}$:

$$\widetilde{u}_1(x) = \begin{cases} 0, \ x \leqslant -1, \quad x \geqslant 2 \\ (x+1)u(0), \ -1 \leqslant x \leqslant 0. \\ u(x), \qquad 0 \leqslant x \leqslant 1. \\ (2-x)u(1), \ 1 \leqslant x \leqslant 2. \end{cases} \tag{2.3.1}$$

Then $\widetilde{u}_1(x)$ is an absolutely continuous function (and hence its distribution derivative is its classical derivative); thus

$$\widetilde{u}_1'(x) = \begin{cases} 0, \qquad x < -1, x > 2 \\ u(0), \quad -1 < x < 0 \\ u'(x), \quad 0 < x < 1 \\ -u(1), \ 1 < x < 2. \end{cases} \tag{2.3.2}$$

and clearly $\widetilde{u}_1' \in L^p(\mathbb{R})$ and so $\widetilde{u}_1 \in W^{1,p}(\mathbb{R})$. Also by virtue of (2.1.19) it follows that $u \to \widetilde{u}_1$ is an extension operator from $W^{1,p}(\Omega)$ into $W^{1,p}(\mathbb{R})$. In the same way we can define several other such operators. For instance

$$\widetilde{u}_2(x) = \begin{cases} 0, \ x \leqslant -2, \quad x \geqslant 3 \\ \frac{1}{2}(x+2)u(0), \ -2 \leqslant x \leqslant 0. \\ u(x), \qquad 0 \leqslant x \leqslant 1 \\ \frac{1}{2}(3-x)u(1), \ 1 \leqslant x \leqslant 3. \end{cases} \tag{2.3.3}$$

defines yet another extension operator $u \to \widetilde{u}_2$ of $W^{1,p}(\Omega)$ into $W^{1,p}(\mathbb{R})$. ∎

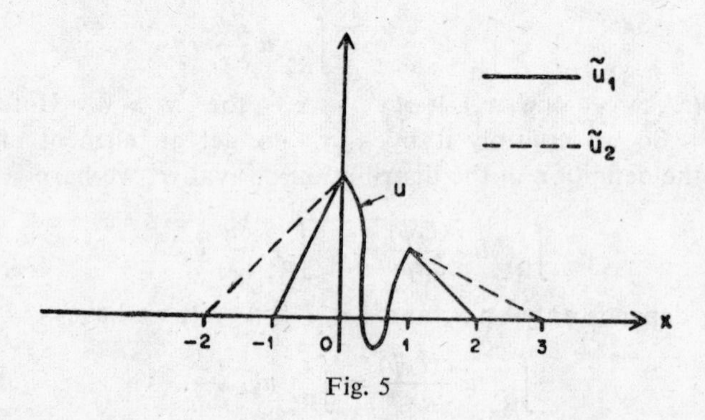

Fig. 5

One of the fundamental methods of providing extensions is **the method**

of reflection. We will now use it to show that the half-space has the extension property.

Notation: Let $x \in \mathbb{R}^n$, $x = (x_1, \ldots, x_n)$. We set $x' = (x_1, \ldots, x_{n-1})$ and write $x = (x', x_n)$. We then define

$$\mathbb{R}^n_+ = \{x \in \mathbb{R}^n \mid x_n > 0\}. \quad \blacksquare \tag{2.3.4}$$

Theorem 2.3.1 Let $u \in W^{1,p}(\mathbb{R}^n_+)$. Define u^* on $\mathbb{R}^n$ by

$$u^*(x) = \begin{cases} u(x', x_n), & x_n > 0. \\ u(x', -x_n), & x_n < 0. \end{cases}$$

Then $u^* \in W^{1,p}(\mathbb{R}^n)$ and further

$$\left.\begin{aligned} |u^*|_{0,p,\mathbb{R}^n} &\leqslant 2|u|_{0,p,\mathbb{R}^n_+} \\ |u^*|_{1,p,\mathbb{R}^n} &\leqslant 2|u|_{1,p,\mathbb{R}^n_+} \end{aligned}\right\} \tag{2.3.5}$$

In particular, $u \to u^*$ defines an extension operator of $W^{1,p}(\mathbb{R}^n_+)$ into $W^{1,p}(\mathbb{R}^n)$.

Proof: *Step 1* Let $\zeta \in \mathcal{E}(\mathbb{R})$ such that

$$\zeta(t) = \begin{cases} 0 & \text{if } t < 1 \\ 1 & \text{if } t > 2. \end{cases} \tag{2.3.6}$$

Define $\zeta_k(t) = \zeta(kt)$ so that

$$\zeta_k(t) = \begin{cases} 0 & \text{if } t < 1/k \\ 1 & \text{if } t > 2/k. \end{cases} \tag{2.3.7}$$

Step 2 Let $\phi \in \mathcal{D}(\mathbb{R}^n)$. Consider the integral

$$\int_{\mathbb{R}^n} u^* \frac{\partial \phi}{\partial x_i}$$

for $1 \leqslant i \leqslant n-1$. Then we can write this as

$$\int_{\mathbb{R}^n} u^* \frac{\partial \phi}{\partial x_i} = \int_{\mathbb{R}^n_+} u \frac{\partial \psi}{\partial x_i}$$

where $\psi(x', x_n) = \phi(x', x_n) + \phi(x', -x_n)$ for $x_n > 0$. Unfortunately $\psi \notin \mathcal{D}(\mathbb{R}^n_+)$. So we multiply it by $\zeta_k(x_n)$ to get an element of $\mathcal{D}(\mathbb{R}^n_+)$. Hence by the definition of the distribution derivative, we have

$$\int_{\mathbb{R}^n_+} u \frac{\partial(\zeta_k \psi)}{\partial x_i} = -\int_{\mathbb{R}^n_+} \frac{\partial u}{\partial x_i} \zeta_k \psi$$

Now as ζ_k depends only on x_n and $1 \leqslant i \leqslant n-1$, we have

$$\int_{\mathbb{R}^n_+} u \frac{\partial(\zeta_k \psi)}{\partial x_i} = \int_{\mathbb{R}^n_+} u \zeta_k \frac{\partial \psi}{\partial x_i}.$$

As $k \to \infty$, $\zeta_k(t) \to 1$ for all t and so by the Dominated Convergence

Theorem we get

$$\int_{\mathbb{R}^n_+} u\,\frac{\partial\psi}{\partial x_i} = -\int_{\mathbb{R}^n_+} \frac{\partial u}{\partial x_i}\,\psi, \text{ for } 1 \leqslant i \leqslant n-1.$$

Going back to the definition of ψ it follows that

$$\int_{\mathbb{R}^n} u^*\,\frac{\partial\phi}{\partial x_i} = -\int_{\mathbb{R}^n_+} \frac{\partial u}{\partial x_i}\,\phi(x', x_n) - \int_{\mathbb{R}^n_+} \frac{\partial u}{\partial x_i}\,\phi(x', -x_n). \qquad (2.3.8)$$

If we define

$$\left(\frac{\partial u}{\partial x_i}\right)^*(x', x_n) = \begin{cases} \dfrac{\partial u}{\partial x_i}(x', x_n), & x_n > 0 \\ \dfrac{\partial u}{\partial x_i}(x', -x_n), & x_n < 0 \end{cases}$$

then (2.3.8) yields

$$\int_{\mathbb{R}^n} u^*\,\frac{\partial\phi}{\partial x_i} = -\int_{\mathbb{R}^n} \left(\frac{\partial u}{\partial x_i}\right)^* \phi, \quad \text{for all } \phi \in \mathscr{D}(\mathbb{R}^n)$$

or

$$\frac{\partial u^*}{\partial x_i} = \left(\frac{\partial u}{\partial x_i}\right)^* \quad \text{for } 1 \leqslant i \leqslant n-1. \qquad (2.3.9)$$

Step 3 Let us now consider the case $i = n$. Then if $\phi \in \mathscr{D}(\mathbb{R}^n)$,

$$\int_{\mathbb{R}^n} u^*\,\frac{\partial\phi}{\partial x_n} = \int_{\mathbb{R}^n_+} u\,\frac{\partial\psi}{\partial x_n}$$

where now

$$\psi(x', x_n) = \phi(x', x_n) - \phi(x', -x_n), \text{ for } x_n > 0. \qquad (2.3.10)$$

Again we multiply by ζ_k and use the definition of the derivative:

$$\int_{\mathbb{R}^n_+} u\,\frac{\partial}{\partial x_n}(\zeta_k\psi) = -\int_{\mathbb{R}^n_+} \frac{\partial u}{\partial x_n}\cdot\zeta_k\psi$$

and

$$\frac{\partial}{\partial x_n}(\zeta_k\psi) = \zeta_k\,\frac{\partial\psi}{\partial x_n} + k\psi\zeta'(kx_n).$$

Let us now estimate the integral

$$k\int_{\mathbb{R}^n_+} u\psi\zeta'(kx_n).$$

Since $\psi(x', 0) = 0$, it follows that $|\psi(x', x_n)| \leqslant C|x_n|$, by the Mean Value Theorem. Also ζ' is a bounded function. Thus

$$\begin{aligned} \left|k\int_{\mathbb{R}^n_+} u\psi\zeta'(kx_n)\right| &= k\left|\int_{0<x_n<2/k} u\psi\zeta'(kx_n)\right| \\ &\leqslant kC\int_{0<x_n<2/k} |u|\,|x_n| \\ &< C\int_{0<x_n<2/k} |u| \end{aligned}$$

which tends to zero as $k \to \infty$. Hence we now have

$$\int_{\mathbb{R}^n_+} u \frac{\partial \psi}{\partial x_n} = -\int_{\mathbb{R}^n_+} \frac{\partial u}{\partial x_n} \psi$$

which will now give

$$\frac{\partial u^*}{\partial x_n} = \left(\frac{\partial u}{\partial x_n}\right)^\dagger \tag{2.3.11}$$

where for any function v on $\mathbb{R}^n_+$, $v^\dagger$ is defined by

$$v^\dagger(x', x_n) = \begin{cases} v(x', x_n), & x_n > 0 \\ -v(x', -x_n), & x_n < 0 \end{cases} \tag{2.3.12}$$

Clearly if $v \in L^p(\mathbb{R}^n_+)$ both v^* and $v^\dagger$ belong to $L^p(\mathbb{R}^n)$ and the result follows. ∎

Corollary Let $1 \leqslant p < \infty$. The restriction of elements of $\mathscr{D}(\mathbb{R}^n)$ to $\mathbb{R}^n_+$ are dense in $W^{1,p}(\mathbb{R}^n_+)$. In particular, $C^\infty(\overline{\mathbb{R}^n_+})$ is dense in $W^{1,p}(\mathbb{R}^n_+)$. ∎

Remark 2.3.1 The above proof is equally valid for sets like

$$Q_+ = \{x \in \mathbb{R}^n \mid |x'| < 1, 0 < x_n < 1\} \tag{2.3.13}$$

where $|x'|$ is the Euclidean norm of x' in $\mathbb{R}^{n-1}$. The method of reflection gives an extension operator from $W^{1,p}(Q_+)$ into $W^{1,p}(Q)$ where Q is the cylinder

$$Q = \{x \in \mathbb{R}^n \mid |x'| < 1, |x_n| < 1\} \quad ∎ \tag{2.3.14}$$

Example 2.3.2 (Brézis) The method of reflection can be used successively to provide an extension operator for domains like the square in $\mathbb{R}^2$ whose boundaries are not "smooth enough" (in a sense to be described presently). Let for example Ω be the unit square $(0, 1)\times(0, 1)$ in $\mathbb{R}^2$. By reflection along x-axis, we can extend a function in $W^{1,p}(\Omega)$ to $W^{1,p}(\Omega_1)$ where $\Omega_1 = (0, 1)\times(-1, 1)$. Now by reflection on the y-axis we can extend it to an element of $W^{1,p}(\Omega_2)$, where $\Omega_2 = (-1, 1)\times(-1, 1)$. Again by reflection on the line $y = 1$, we can extend this to an element of $W^{1,p}(\Omega_3)$, where $\Omega_3 = (-1, 1)\times(-1, 3)$. Finally reflect on the line $x = 1$ to extend it to

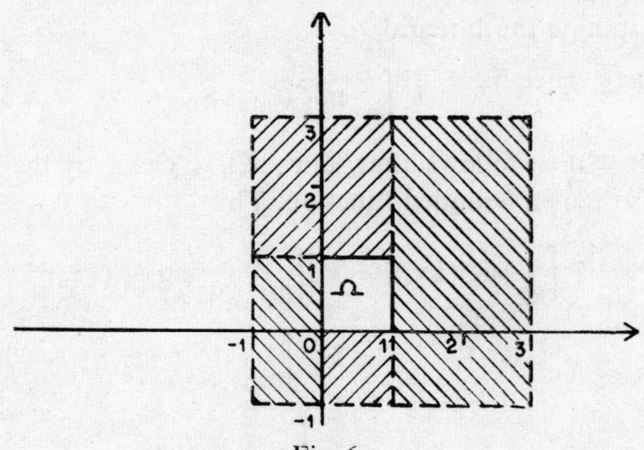

Fig. 6

the domain $\Omega_4 = (-1, 3)\times(-1, 3)$. Now let $\psi \in \mathscr{D}(\Omega_4)$ with $\psi \equiv 1$ on Ω. If u_4 is the extension of u to Ω_4 then $\tilde{u} = \psi u_4$ is an extension of u to $W^{1,p}(\mathbb{R}^n)$. (cf. Fig. 6). ∎

We now describe how we can use Theorem 2.3.1 combined with the notion of the partition of unity (cf. Chapter 1 and Appendix 1) to prove the existence of an extension operator for a "smooth" domain whose boundary is bounded.

Definition 2.3.1 We say that an open set Ω is of class C^k (k an integer $\geqslant 1$) if for every $x \in \partial\Omega$, there exists a neighbourhood U of x in $\mathbb{R}^n$ and a map $T : Q \to U$ such that

(i) T is a bijection

(ii) $T \in C^k(\bar{Q})$, $T^{-1} \in C^k(\bar{U})$

(iii) $T(Q_+) = U \cap \Omega$, $T(Q_0) = U \cap \partial\Omega$

where Q_+, Q are as in (2.3.13) and (2.3.14) respectively and

$$Q_0 = \{x \in Q \mid x_n = 0\}. \tag{2.3.15}$$

We say that Ω is of class C^∞ if it is of class C^k for every integer $k \geqslant 1$. ∎

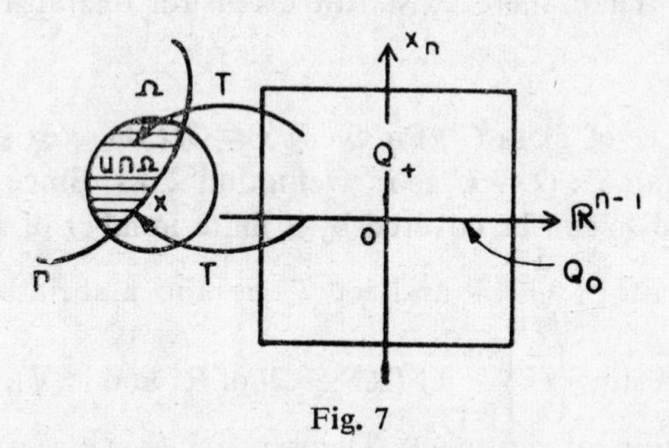

Fig. 7

Example 2.3.3 Open balls in $\mathbb{R}^n$ are examples of domains of class C^∞. So are complements of closed balls. A polygon in $\mathbb{R}^2$ is not even C^1 as the requirements of Definition 2.3.1 cannot be met at the vertices. ∎

Before we prove the next main result, we prove the following lemma which will be used in the proof of the Theorem 2.3.2.

Lemma 2.3.1 Let $\Omega \subset \mathbb{R}^n$ be an open set and let $u \in W^{1,p}(\Omega)$. If $K \subset \Omega$ is a closed set and u vanishes outside K, then the function

$$\tilde{u}(x) = \begin{cases} u(x), & x \in \Omega \\ 0, & x \in \mathbb{R}^n \setminus \Omega \end{cases} \tag{2.3.16}$$

is in $W^{1,p}(\mathbb{R}^n)$.

Proof Let $\phi \in \mathscr{D}(\mathbb{R}^n)$. Consider the integral

$$\int_{\mathbb{R}^n} \tilde{u} \frac{\partial \phi}{\partial x_i}.$$

Let $K_1 = K \cap \text{supp}(\phi)$. Then $K_1 \subset \Omega$ and is compact. Let $\psi \in \mathcal{D}(\Omega)$ such that $\psi \equiv 1$ on K_1. Now,

$$\int_{\mathbb{R}^n} \tilde{u} \frac{\partial \phi}{\partial x_i} = \int_\Omega u \frac{\partial \phi}{\partial x_i} = \int_{K_1} u \frac{\partial \phi}{\partial x_i} = \int_\Omega u\psi \frac{\partial \phi}{\partial x_i}$$

$$= \int_\Omega u \frac{\partial}{\partial x_i} (\psi\phi) - \int_\Omega u\phi \frac{\partial \psi}{\partial x_i}$$

But $\phi\psi \in \mathcal{D}(\Omega)$ and $\dfrac{\partial \psi}{\partial x_i} = 0$ on K_1. Thus

$$\int_{\mathbb{R}^n} \tilde{u} \frac{\partial \phi}{\partial x_i} = -\int_\Omega \frac{\partial u}{\partial x_i} \phi\psi = -\int_\Omega \frac{\partial u}{\partial x_i} \phi.$$

Thus it follows that

$$\frac{\partial \tilde{u}}{\partial x_i} = \widetilde{\frac{\partial u}{\partial x_i}} \in L^p(\mathbb{R}^n). \tag{2.3.17}$$

Hence the lemma. ∎

Theorem 2.3.2 Let $\Omega \subset \mathbb{R}^n$ be an open set of class C^1, whose boundary $\partial\Omega$ is bounded. Then there exists an extension operator $P : W^{1,p}(\Omega) \to W^{1,p}(\mathbb{R}^n)$.

Proof Since Ω is of class C^1, for every $x \in \partial\Omega$ there exists a neighbourhood U and a map $T : Q \to U$ as in Definition 2.3.1. Since $\partial\Omega$ is bounded, it is compact and so can be covered by a finite number of such neighbourhoods. Let then $\partial\Omega \subset \bigcup_{j=1}^{k} U_j$ and let T_j be the associated C^1-bijections. Consider the covering $\left\{U_j\right\}_{j=1}^{k} \cup \{\mathbb{R}^n \setminus \partial\Omega\}$ of $\mathbb{R}^n$ and let $\psi_1, \ldots \psi_k, \psi_0$ be an associated partition of unity (cf. Theorem 1.2.1), i.e. $\psi_j, j = 0, 1, \ldots, k$ are all C^∞ functions with $\text{supp}(\psi_j) \subset U_j$, $1 \leqslant j \leqslant k$, $\text{supp}(\psi_0) \subset \mathbb{R}^n \setminus \partial\Omega$ and $\sum_{j=0}^{k} \psi_j \equiv 1$, $0 \leqslant \psi_j \leqslant 1$, $0 \leqslant j \leqslant k$.

Thus we can write for $u \in W^{1,p}(\Omega)$

$$u = \sum_{j=0}^{k} \psi_j u.$$

Now $\psi_0 u$ is supported away from $\partial\Omega$, since $\text{supp}(\psi_0) \subset \mathbb{R}^n \setminus \partial\Omega$. Also $\psi_0 \in L^\infty(\mathbb{R}^n) \cap C^\infty(\mathbb{R}^n)$ (in fact $0 \leqslant \psi_0 \leqslant 1$) and as ψ_j are all compactly supported and for any i,

$$\frac{\partial \psi_0}{\partial x_i} = -\sum_{j=1}^{k} \frac{\partial \psi_j}{\partial x_i} \left(\text{since } \sum_{j=0}^{k} \psi_j \equiv 1\right).$$

We have that the derivatives of ψ_0 are also in $L^\infty(\mathbb{R}^n)$. Then (by the product rule for the distribution derivative, cf. (1.3.11)) it is easy to see that $\psi_0 u \in W^{1,p}(\Omega)$. Since it vanishes in a neighbourhood of $\partial\Omega$, by Lemma 2.3.1,

$\widetilde{\psi_0 u}$, the extension outside Ω by 0, belongs to $W^{1,p}(\mathbb{R}^n)$ and clearly

$$\|\widetilde{\psi_0 u}\|_{1,p,\mathbb{R}^n} \leqslant \|\psi_0 u\|_{1,p,\Omega} \leqslant C\|u\|_{1,p,\Omega}. \tag{2.3.18}$$

If $1 \leqslant j \leqslant k$, then consider $u|_{U_j \cap \Omega}$. Now define v_j on Q_+ by

$$v_j(y) = u(T_j(y)),\ y \in Q_+. \tag{2.3.19}$$

It is then easy to verify that (Exercise!) $v_j \in W^{1,p}(Q_+)$ and so can be extended by reflection to a function $v_j^* \in W^{1,p}(Q)$. Then again

$$w_j(x) = v_j^*(T_j^{-1}(x)),\ x \in U_j \tag{2.3.20}$$

is such that $w_j \in W^{1,p}(U_j)$ and $w_j = u_j$ on $U_j \cap \Omega$. Also

$$\|w_j\|_{1,p,U_j} \leqslant C_j\|u\|_{1,p,U_j \cap \Omega} \leqslant C_j\|u\|_{1,p,\Omega}.$$

Now $\psi_j w_j$ is compactly supported in U_j and so $\widetilde{\psi_j w_j}$, the extension by 0 outside U_j is in $W^{1,p}(\mathbb{R}^n)$ and again

$$\|\widetilde{\psi_j w_j}\|_{1,p,\mathbb{R}^n} \leqslant C_j\|u\|_{1,p,\Omega}. \tag{2.3.21}$$

Now the mapping

$$u \to Pu = \widetilde{\psi_0 u_0} + \sum_{j=1}^{k} \widetilde{\psi_j w_j} \tag{2.3.22}$$

defines an extension operator for $W^{1,p}(\Omega)$. ∎

Corollary: If Ω is of class C^1 and has $\partial\Omega$ bounded, then $C^\infty(\overline{\Omega})$ is dense in $W^{1,p}(\Omega)$, $1 \leqslant p < \infty$. ∎

Given an arbitrary element of $W^{1,p}(\Omega)$, the extension by zero does not belong to $W^{1,p}(\mathbb{R}^n)$. For example if $\Omega = (0, 1) \subset \mathbb{R}$ and $u(x) \equiv 1$ on Ω, then $\tilde{u} \in L^p(\mathbb{R}^n)$ but $\dfrac{d\tilde{u}}{dx} = \delta_0 - \delta_1$, which cannot be given by a locally integrable function. We have seen in Lemma 2.3.1 that if u is supported away from the boundary then the extension by zero provides an element of $W^{1,p}(\mathbb{R}^n)$. But such functions, if the support is compact, are in $W_0^{1,p}(\Omega)$. We will now show that the extension by zero is an extension operator on $W_0^{1,p}(\Omega)$ irrespective of the nature of Ω. Thus for functions in $W_0^{1,p}(\Omega)$ we always will have a canonical extension to $W^{1,p}(\mathbb{R}^n)$ which will help us to prove several important properties of these functions without supplementary hypotheses on the smoothness of Ω.

Theorem 2.3.3 Let $1 < p < \infty$ and $u \in W_0^{1,p}(\Omega)$, Ω an open set in $\mathbb{R}^n$. Then if $\tilde{u}$ denotes the extension of u by 0 outside Ω, $\tilde{u} \in W^{1,p}(\mathbb{R}^n)$. Further for any $1 \leqslant i \leqslant n$,

$$\frac{\partial \tilde{u}}{\partial x_i} = \widetilde{\frac{\partial u}{\partial x_i}}. \tag{2.3.23}$$

Proof: Let $u \in W_0^{1,p}(\Omega)$ and let $u_m \in \mathscr{D}(\Omega)$ such that $u_m \to u$ in $W^{1,p}(\Omega)$.

Let $\phi \in \mathcal{D}(\mathbb{R}^n)$. Now

$$\left|\int_\Omega u_m \frac{\partial \phi}{\partial x_i}\right| = \left|\int_\Omega \frac{\partial u_m}{\partial x_i}\phi\right| \leqslant \left|\frac{\partial u_m}{\partial x_i}\right|_{0,p,\Omega} |\phi|_{0,q,\Omega}$$
$$\leqslant |u_m|_{1,p,\Omega}|\phi|_{0,q,\mathbb{R}^n}$$

where q is the conjugate exponent of p (i.e. $p^{-1} + q^{-1} = 1$). When $m \to \infty$, we get

$$\left|\int_{\mathbb{R}^n} \tilde{u} \frac{\partial \phi}{\partial x_i}\right| = \left|\int_\Omega u \frac{\partial \phi}{\partial x_i}\right| \leqslant |u|_{1,p,\Omega}|\phi|_{0,q,\mathbb{R}^n}. \tag{2.3.24}$$

Now $\mathcal{D}(\mathbb{R}^n)$ is dense in $L^q(\mathbb{R}^n)$ (since we also have $1 < q < \infty$) and (2.3.24) tells us that the linear functional

$$\phi \to \int_{\mathbb{R}^n} \tilde{u} \frac{\partial \phi}{\partial x_i}$$

is continuous with respect to the $L^q(\mathbb{R}^n)$-norm. Hence by density, it extends uniquely to a continuous linear functional on $L^q(\mathbb{R}^n)$. By the Riesz Representation Theorem, since the dual of $L^q(\mathbb{R}^n)$ is $L^p(\mathbb{R}^n)$, there exists a function $v_i \in L^p(\mathbb{R}^n)$ such that

$$|v_i|_{0,p,\mathbb{R}^n} \leqslant |u|_{1,p,\Omega} \tag{2.3.25}$$

and

$$\int_{\mathbb{R}^n} v_i \phi = \int_{\mathbb{R}^n} \tilde{u} \frac{\partial \phi}{\partial x_i}. \tag{2.3.26}$$

In other words, $\dfrac{\partial \tilde{u}}{\partial x_i} = -v_i \in L^p(\mathbb{R}^n)$ for $1 \leqslant i \leqslant n$, i.e., $\tilde{u} \in W^{1,p}(\mathbb{R}^n)$. Now since $\left.\dfrac{\partial \tilde{u}}{\partial x_i}\right|_\Omega = \dfrac{\partial u}{\partial x_i}$ and $\left.\dfrac{\partial \tilde{u}}{\partial x_i}\right|_{\mathbb{R}^n \setminus \bar{\Omega}} = 0$, it follows that (2.3.23) holds since $\dfrac{\partial \tilde{u}}{\partial x_i}$ is an L^p-function which has been defined a.e. (except on $\partial\Omega$, which is of measure zero in $\mathbb{R}^n$) by these relations. ∎

We conclude this section with a very important property of the space $W_0^{1,p}(\Omega)$.

Theorem 2.3.4 **(Poincaré's Inequality).** Let Ω be a *bounded* open set in $\mathbb{R}^n$. Then there exists a positive constant $C = C(\Omega, p)$ such that

$$|u|_{0,p,\Omega} \leqslant C|u|_{1,p,\Omega} \quad \text{for every } u \in W_0^{1,p}(\Omega). \tag{2.3.27}$$

In particular, $u \to |u|_{1,p,\Omega}$ defines a *norm* on $W_0^{1,p}(\Omega)$, which is equivalent to the norm $\|\cdot\|_{1,p,\Omega}$. On $H_0^1(\Omega)$, the bilinear form

$$(u, v) \to \int_\Omega \sum_{i=1}^n \frac{\partial u}{\partial x_i}\frac{\partial v}{\partial x_i},$$

defines an inner-product giving rise to the norm $|\cdot|_{1,\Omega}$, equivalent to the norm $\|\cdot\|_{1,\Omega}$.

Proof: Let $\Omega = (-a, a)^n$, $a > 0$. Let $u \in \mathscr{D}(\Omega)$. Then

$$u(x) = \int_{-a}^{x_n} \frac{\partial u}{\partial x_n}(x', t)\, dt,\ x = (x', x_n), \text{since } u(x', -a) = 0. \quad (2.3.28)$$

Hence

$$|u(x)| \leqslant \left(\int_{-a}^{x_n} \left| \frac{\partial u}{\partial x_n}(x', t) \right|^p dt \right)^{1/p} |x_n + a|^{1/q},\ (p^{-1} + q^{-1} = 1)$$

or,

$$|u(x)|^p \leqslant |x_n + a|^{p/q} \int_{-a}^{a} \left| \frac{\partial u}{\partial x_n}(x', t) \right|^p dt$$

Integrating over x',

$$\int |u(x', x_n)|^p\, dx' \leqslant (2a)^{p/q} \int_\Omega \left| \frac{\partial u}{\partial x_n} \right|^p,$$

and so

$$\int_\Omega |u(x)|^p\, dx \leqslant (2a)^{p/q+1} \int_\Omega \left| \frac{\partial u}{\partial x_n} \right|^p,$$

which proves (2.3.27) for $u \in \mathscr{D}(\Omega)$. But $\mathscr{D}(\Omega)$ is dense in $W_0^{1,p}(\Omega)$ and the inequality follows for all $u \in W_0^{1,p}(\Omega)$. If Ω is not a 'box' let $\Omega \subset \tilde{\Omega}$, box of the form $(-a, a)^n$ and extend $u \in W_0^{1,p}(\Omega)$ by zero to get $\tilde{u} \in W_0^{1,p}(\tilde{\Omega})$ and apply the result which is available for $\tilde{\Omega}$ and the inequality follows. ∎

Remark 2.3.2 The Poincaré inequality will be fundamental in studying weak solutions of elliptic Dirichlet boundary value problems in the next chapter. ∎

Remark 2.3.3 Various generalizations are possible. On one hand it suffices that Ω is bounded in a particular direction, as can be seen from the proof above. On the other hand it is enough that functions vanish (in the trace sense) not on the whole boundary $\partial\Omega$ (which is the case of elements of $W_0^{1,p}(\Omega)$) but on a part of $\partial\Omega$ alone, that part having positive $(n-1)$-dimensional measure. The result is not valid if u does not vanish at all on the boundary; for instance if $u \equiv a$ non-zero constant then $u \in W^{1,p}(\Omega)$ and $|u|_{1,p,\Omega} = 0$ while $|u|_{0,p,\Omega} > 0$. ∎

Example 2.3.4 Poincaré's inequality is not true in truly unbounded domains. For instance, let $\Omega = \mathbb{R}^n$ and consider $\zeta \in \mathscr{D}(\mathbb{R}^n)$ such that $\zeta \equiv 1$ on $|x| \leqslant 1$ and 0 for $|x| \geqslant 2$ and $0 \leqslant \zeta \leqslant 1$. Let $\zeta_m(x) = \zeta(x/m)$. Then $|\zeta_m|_{1,p,\mathbb{R}^n} \to 0$ (if $n < p$) as $m \to \infty$ while $|\zeta_m|_{0,p,\mathbb{R}^n} \geqslant$ meas $(B(0;m))$ $\to \infty$ as $m \to \infty$. ∎

If $u \in W_0^{m,p}(\Omega)$ then $\dfrac{\partial u}{\partial x_i} \in W_0^{(m-1),p}(\Omega)$ as is immediate to see. Using this we can iterate Poincaré's inequality. For instance if $u \in W_0^{2,p}(\Omega)$,

then

$$\left|\frac{\partial u}{\partial x_i}\right|_{0,p,\Omega} \leqslant C\left|\frac{\partial u}{\partial x_i}\right|_{1,p,\Omega}; \quad 1 \leqslant i \leqslant n.$$

which yields

$$|u|_{1,p,\Omega} \leqslant C|u|_{2,p,\Omega}.$$

Also as $u \in W_0^{1,p}(\Omega)$ as well, we have

$$|u|_{0,p,\Omega} \leqslant C|u|_{1,p,\Omega} \leqslant C|u|_{2,p,\Omega}.$$

In general if $u \in W_0^{m,p}(\Omega)$, then $|\cdot|_{m,p,\Omega}$ is a *norm* equivalent to the usual norm $\|\cdot\|_{m,p,\Omega}$.

2.4 IMBEDDING THEOREMS

We saw in Section 2.1 that the space $W^{1,p}(I)$ can be imbedded in the space of absolutely continuous functions, when I is an open interval in $\mathbb{R}$. In this section we will investigate such imbedding properties of the spaces $W^{1,p}(\Omega)$, Ω an open set in $\mathbb{R}^n$.

We will first study the imbeddings of the spaces $W^{1,p}(\mathbb{R}^n)$. We have three different cases to analyse, viz. $p < n$, $p = n$ and $p > n$. To start with, let $1 \leqslant p < n$. Then we define the exponent p^* by

$$\frac{1}{p^*} = \frac{1}{p} - \frac{1}{n} \quad \text{or} \quad p^* = \frac{np}{n-p}. \tag{2.4.1}$$

Notice that $p^* > p$. Thus while we know that $W^{1,p}(\mathbb{R}^n) \to L^p(\mathbb{R}^n)$ it is indeed an additional piece of information if we know that functions in $W^{1,p}(\mathbb{R}^n)$ are smoother. In fact we will show that $W^{1,p}(\mathbb{R}^n)$ can be imbedded in $L^{p^*}(\mathbb{R}^n)$. Before we set about proving this, we need the following technical lemma.

Lemma 2.4.1 (*Gagliardo*). Let $n \geqslant 2$. Let $f_1, \ldots, f_n \in L^{n-1}(\mathbb{R}^{n-1})$. If $x \in \mathbb{R}^n$, define

$$\hat{x}_i = (x_1, \ldots, x_{i-1}, x_{i+1}, \ldots, x_n) \in \mathbb{R}^{n-1}, \ 1 \leqslant i \leqslant n. \tag{2.4.2}$$

Set

$$f(x) = f_1(\hat{x}_1) \ldots f_n(\hat{x}_n), \ x \in \mathbb{R}^n. \tag{2.4.3}$$

Then $f \in L^1(\mathbb{R}^n)$ and

$$|f|_{0,1,\mathbb{R}^n} \leqslant \prod_{i=1}^{n} |f_i|_{0,n-1,\mathbb{R}^{n-1}}. \tag{2.4.4}$$

Proof When $n = 2$ the result follows trivially by separating the variables in the double integral. Let $n = 3$. Now

$$\int_{\mathbb{R}} |f(x)|\, dx_3 = |f_3(x_1, x_2)| \int_{\mathbb{R}} |f_1(x_2, x_3)|\,|f_2(x_1, x_3)|\, dx_3$$

$$\leqslant |f_3(x_1, x_2)| \left(\int_{\mathbb{R}} |f_1(x_2, x_3)|^2\, dx_3\right)^{1/2} \left(\int_{\mathbb{R}} |f_2(x_1, x_3)|^2\, dx_3\right)^{1/2}$$

by the Cauchy-Schwarz inequality. Now if we integrate the above inequality with respect to x_1 and x_2 and apply the Cauchy-Schwarz inequality once again, we get

$$\int_{\mathbb{R}^3} |f(x)|\, dx \leqslant \left(\int |f_3(x_1, x_2)|^2\, dx_1\, dx_2\right)^{1/2}\left(\int |f_1(x_2, x_3)|^2\, dx_3\, dx_2\right)^{1/2} \times \left(\int |f_2(x_1, x_3)|^2\, dx_1\, dx_3\right)^{1/2}$$

which is precisely (2.4.4) when $n = 3$.

The general case follows by induction. Assume the result for n. Let $x \in \mathbb{R}^{n+1}$. Let us first fix x_{n+1}. Then by Holder's inequality,

$$\int_{\mathbb{R}^n} |f(x)|\, dx_1 \ldots dx_n \leqslant |f_{n+1}|_{0, n, \mathbb{R}^n} \left(\int |f_1 \ldots f_n|^{n'}\, dx_1 \ldots dx_n\right)^{1/n'}$$

where n' is the conjugate exponent of n; i.e., $n' = n/(n-1)$. Now treating x_{n+1} as a fixed parameter, the functions $|f_1|^{n'}, \ldots, |f_n|^{n'}$ are all in $L^{n-1}(\mathbb{R}^{n-1})$. Hence by the induction hypothesis,

$$\int |f_1|^{n'} \ldots |f_n|^{n'}\, dx_1 \ldots dx_n \leqslant \prod_{i=1}^{n} \||f_i|^{n'}|_{0, n-1, \mathbb{R}^{n-1}} = \prod_{i=1}^{n} |f_i|_{0, n, \mathbb{R}^{n-1}}^{n'}$$

since

$$\||f_i|^{n'}|_{0, n-1, \mathbb{R}^{n-1}} = \left(\int_{\mathbb{R}^{n-1}} |f_i|^{n'(n-1)}\right)^{1/(n-1)} = \left(\int_{\mathbb{R}^{n-1}} |f_i|^{n}\right)^{1/(n-1)} = |f_i|_{0, n, \mathbb{R}^{n-1}}^{n'}$$

Thus

$$\int |f(x)|\, dx_1 \ldots dx_n \leqslant |f_{n+1}|_{0, n, \mathbb{R}^n} \prod_{i=1}^{n} |f_i|_{0, n, \mathbb{R}^{n-1}}.$$

Now integrate both sides with respect to x_{n+1}. Observe that the map $x_{n+1} \to |f_i(\cdot, x_{n+1})|_{0, n, \mathbb{R}^{n-1}}$ is in $L^n(\mathbb{R})$ for each $1 \leqslant i \leqslant n$, since $f_i \in L^n(\mathbb{R}^n)$, to start with. Thus by the generalized Holder inequality we have that the product is in $L^1(\mathbb{R})$ and

$$\int_{\mathbb{R}^{n+1}} |f(x)|\, dx \leqslant |f_{n+1}|_{0, n, \mathbb{R}^n} \prod_{i=1}^{n} |f_i|_{0, n, \mathbb{R}^n}$$

which proves the lemma. ∎

Theorem 2.4.1 **(Sobolev's Inequality).** Let $1 \leqslant p < n$. Then there exists a constant $C = C(p, n) > 0$, such that

$$|u|_{0, p^*, \mathbb{R}^n} \leqslant C|u|_{1, p, \mathbb{R}^n} \tag{2.4.5}$$

for every $u \in W^{1,p}(\mathbb{R}^n)$. In particular, we have the continuous inclusion

$$W^{1,p}(\mathbb{R}^n) \to L^{p^*}(\mathbb{R}^n).$$

Proof: *Step 1.* Let $u \in \mathcal{D}(\mathbb{R}^n)$. Then if $1 \leqslant i \leqslant n$, we can write

$$|u(x)| \leqslant \int_{-\infty}^{\infty} \left|\frac{\partial u}{\partial x_i}(x_1, \ldots, x_{i-1}, t, x_{i+1}, \ldots, x_n)\right| dt = f_i(\hat{x}). \quad (2.4.6)$$

Hence

$$|u(x)|^n \leqslant \prod_{i=1}^{n} f_i(\hat{x}_i) \quad \text{or} \quad |u(x)|^{n/(n-1)} \leqslant \prod_{i=1}^{n} |f_i(\hat{x}_i)|^{1/(n-1)}.$$

Since u has compact support $\dfrac{\partial u}{\partial x_i}$ is integrable. It then follows that $|f_i|^{1/(n-1)} \in L^{n-1}(\mathbb{R}^{n-1})$ for each $1 \leqslant i \leqslant n$. Hence, by Lemma 2.4.1, we have

$$\int_{\mathbb{R}^n} |u|^{n/(n-1)} \leqslant \prod_{i=1}^{n} |f_i|_{0,1,\mathbb{R}^{n-1}}^{1/(n-1)} = \prod_{i=1}^{n} \left|\frac{\partial u}{\partial x_i}\right|_{0,1,\mathbb{R}^n}^{1/(n-1)}.$$

Notice that $n/(n-1) = 1^*$ (cf. (2.4.1)). Thus

$$|u|_{0,1^*,\mathbb{R}^n} \leqslant \prod_{i=1}^{n} \left|\frac{\partial u}{\partial x_i}\right|_{0,1,\mathbb{R}^n}^{1/n}. \quad (2.4.7)$$

Step 2 Let $1 \leqslant p < n$. Let $u \in \mathcal{D}(\mathbb{R}^n)$. Let $t \geqslant 1$ (to be chosen suitably) and consider the function $|u|^{t-1}u$. This function has compact support and is continuously differentiable. In fact

$$\frac{\partial}{\partial x_i}(|u|^{t-1}u) = t|u|^{t-1}\frac{\partial u}{\partial x_i}. \quad (2.4.8)$$

Also (2.4.6) and the subsequent analysis of Step 1 applies to this function as well (since we really only used the first derivative in that step). Thus applying (2.4.7) to $|u|^{t-1}u$ and simplifying, we get

$$\begin{aligned} |u|_{0,tn/(n-1),\mathbb{R}^n}^{t} &\leqslant t \prod_{i=1}^{n} \left||u|^{t-1}\frac{\partial u}{\partial x_i}\right|_{0,1,\mathbb{R}^n}^{1/n} \\ &\leqslant t|u|_{0,p'(t-1),\mathbb{R}^n}^{t-1} \prod_{i=1}^{n} \left|\frac{\partial u}{\partial x_i}\right|_{0,p,\mathbb{R}^n}^{1/n} \end{aligned} \quad (2.4.9)$$

by Holder's inequality, where p' is the conjugate exponent of p (i.e. $p^{-1} + (p')^{-1} = 1$). Now choose t such that

$$tn/(n-1) = p'(t-1).$$

Simplifying the above relation, we get

$$t = \frac{n-1}{(n/p)-1} = \left(\frac{n-1}{n}\right)p^*$$

which is greater than or equal to 1 since $p < n$. It now follows from (2.4.9) that

$$|u|_{0,p^*,\mathbb{R}^n} \leqslant \left(\frac{n-1}{n}\right)p^*|u|_{1,p,\mathbb{R}^n},\ u \in \mathcal{D}(\mathbb{R}^n) \quad (2.4.10)$$

which is (2.4.5) when $u \in \mathcal{D}(\mathbb{R}^n)$ and with $C(p, n) = \dfrac{(n-1)}{n}p^*$.

Step 3 Let $u \in W^{1,p}(\mathbb{R}^n)$. Then by Theorem 2.1.3, there exists a sequence $\{u_m\}$ in $\mathscr{D}(\mathbb{R}^n)$ such that $u_m \to u$ in $W^{1,p}(\mathbb{R}^n)$. Now by (2.4.10) we have

$$|u_m - u_l|_{0, p^*, \mathbb{R}^n} \leqslant C|u_m - u_l|_{1, p, \mathbb{R}^n}$$

and so $\{u_m\}$ is a Cauchy sequence in $L^{p^*}(\mathbb{R}^n)$. Hence as $\{u_m\}$ already converges to u in $L^p(\mathbb{R}^n)$, we must have that $u \in L^{p^*}(\mathbb{R}^n)$ and that $u_m \to u$ in $L^{p^*}(\mathbb{R}^n)$ as well. Now (2.4.5) follows from (2.4.10) by continuity and the theorem is proved. ∎

Corollary Let $1 \leqslant p < n$. Then we have the continuous inclusions

$$W^{1,p}(\mathbb{R}^n) \to L^q(\mathbb{R}^n) \quad \text{for all } q \in [p, p^*].$$

Proof Let $p \leqslant q \leqslant p^*$. Then we can choose $\alpha \in [0, 1]$ such that

$$\frac{1}{q} = \frac{\alpha}{p} + \frac{1-\alpha}{p^*}.$$

Now $|u|^{\alpha q} \in L^{p/\alpha q}(\mathbb{R}^n)$ and $|u|^{(1-\alpha)q} \in L^{p^*/(1-\alpha)q}(\mathbb{R}^n)$. Hence by the generalized Holder inequality, we get $u \in L^q(\mathbb{R}^n)$ and

$$\begin{aligned} |u|_{0, q, \mathbb{R}^n} &\leqslant |u|^{\alpha}_{0, p, \mathbb{R}^n} |u|^{1-\alpha}_{0, p^*, \mathbb{R}^n} \\ &\leqslant |u|_{0, p, \mathbb{R}^n} + |u|_{0, p^*, \mathbb{R}^n} \end{aligned}$$

by a standard generalization of the inequalities connecting arithmetic and geometric means. Now by Theorem 2.4.1, we get

$$|u|_{0, q, \mathbb{R}^n} \leqslant |u|_{0, p, \mathbb{R}^n} + C|u|_{1, p, \mathbb{R}^n} \leqslant C\|u\|_{1, p, \mathbb{R}^n}. \quad ∎$$

Corollary Let $\Omega \subset \mathbb{R}^n$ be an open set. Let $u \in W_0^{1,p}(\Omega)$. Then $u \in L^q(\Omega)$ for $q \in [p, p^*]$ and there exists a constant $C = C(p, n) > 0$ such that

$$\left.\begin{aligned} |u|_{0, p^*, \Omega} &\leqslant C|u|_{1, p, \Omega} \quad (2.4.11) \\ |u|_{0, q, \Omega} &\leqslant C\|u\|_{1, p, \Omega} \quad (2.4.12) \end{aligned}\right\} \text{ for every } u \in W_0^{1,p}(\Omega).$$

Proof If $u \in W_0^{1,p}(\Omega)$, then $\tilde{u}$ its extension by zero outside Ω, belongs to $W^{1,p}(\mathbb{R}^n)$ and so by the Theorem 2.4.1, and its corollary, is in $L^q(\mathbb{R}^n)$ for all $q \in [p, p^*]$. Thus $u \in L^q(\Omega)$ for $q \in [p, p^*]$. The inequalities (2.4.11) and (2.4.12) follow from the corresponding inequalities for $\tilde{u}$. ∎

Remark 2.4.1 The constant C which occurs in the Sobolev inequality (2.4.5) can be taken as $\left(\frac{n-1}{n}\right)p^*$. However the best constant is known, and is quite complicated to express (cf. Aubin [1], Lieb [1] and Talenti [1]). See also the comments at the end of this chapter. ∎

We now turn to the case $p = n$.

Theorem 2.4.2 Let $\Omega \subset \mathbb{R}^n$ be an open set. Then

$$W_0^{1,p}(\Omega) \subset L^q(\Omega) \quad \text{for all } q \in [n, \infty).$$

Proof Again, it suffices to prove the result for $\Omega = \mathbb{R}^n$. Let $u \in \mathcal{D}(\mathbb{R}^n)$. We can still apply the inequality (2.4.9) with $p = n$, $t \geqslant 1$. Thus we have

$$|u|^t_{0,\, tn/(n-1),\, \mathbb{R}^n} \leqslant t|u|^{t-1}_{0,\, n(t-1)/(n-1),\, \mathbb{R}^n}|u|_{1,\, n,\, \mathbb{R}^n}.$$

Using the inequality $(a+b)^t \geqslant ta^{t-1}b$ for $a, b \geqslant 0$, we get

$$|u|_{0,\, tn/(n-1),\, \mathbb{R}^n} \leqslant |u|_{0,\, n(t-1)/(n-1),\, \mathbb{R}^n} + |u|_{1,\, n,\, \mathbb{R}^n}. \tag{2.4.13}$$

Set $t = n$. Then

$$|u|_{0,\, n^2/(n-1),\, \mathbb{R}^n} \leqslant |u|_{0,\, n,\, \mathbb{R}^n} + |u|_{1,\, n,\, \mathbb{R}^n} = \|u\|_{1,\, n,\, \mathbb{R}^n}. \tag{2.4.14}$$

Now proceeding as in corollary we deduce that for all $q \in [n, n^2/(n-1)]$, $u \in L^q(\mathbb{R}^n)$ and that

$$|u|_{0,\, q,\, \mathbb{R}^n} \leqslant C\|u\|_{1,\, n,\, \mathbb{R}^n}. \tag{2.4.15}$$

Now we can repeat this argument with $t = n+1$ in (2.4.13), using (2.4.14) to get $u \in L^q(\mathbb{R}^n)$ for $q \in \left[\dfrac{n^2}{n-1}, \dfrac{n(n+1)}{n-1}\right]$ and that

$$|u|_{0,\, q,\, \mathbb{R}^n} \leqslant C\|u\|_{1,\, n,\, \mathbb{R}^n}. \tag{2.4.16}$$

We iterate this argument with $t = n+2, n+3, \ldots$ to show that $u \in L^q(\mathbb{R}^n)$ for all $q \in [n, \infty)$ and that

$$|u|_{0,\, q,\, \mathbb{R}^n} \leqslant C\|u\|_{1,\, n,\, \mathbb{R}^n},\ u \in \mathcal{D}(\mathbb{R}^n).$$

Once again the result follows for general $u \in W^{1,n}(\mathbb{R}^n)$ by density. ∎

Finally we come to the case $p > n$.

Theorem 2.4.3 Let $p > n$. Then we have the continuous inclusion

$$W^{1,\, p}(\mathbb{R}^n) \to L^\infty(\mathbb{R}^n).$$

Further there exists a constant $C = C(p, n) > 0$ such that

$$|u(x) - u(y)| \leqslant C|x-y|^\alpha |u|_{1,\, p,\, \mathbb{R}^n} \quad \text{a.e. in } \mathbb{R}^n \tag{2.4.17}$$

for every $u \in W^{1,\, p}(\mathbb{R}^n)$, where $\alpha = 1 - (n/p)$. Again, if Ω is an open set in $\mathbb{R}^n$, the same conclusions hold for $W^{1,\, p}_0(\Omega)$.

Proof: *Step 1* Let $u \in \mathcal{D}(\mathbb{R}^n)$. Let Q be a cube of side r containing the origin and with sides parallel to the coordinate axes. Let $x \in Q$. We have

$$u(x) - u(0) = \int_0^1 \frac{d}{dt}(u(tx))\, dt$$

and hence

$$|u(x) - u(0)| \leqslant \int_0^1 \sum_{i=1}^n |x_i| \left|\frac{\partial u}{\partial x_i}(tx)\right| dt. \tag{2.4.18}$$

Let $\bar{u}$ denote the average of u over Q, i.e.

$$\bar{u} = \frac{1}{\text{meas}\,(Q)} \int_Q u(y)\, dy.$$

It now follows from (2.4.18) that

$$|\bar{u} - u(0)| \leqslant \frac{r}{\text{meas }(Q)} \int_Q dx \sum_{i=1}^{n} \int_0^1 \left|\frac{\partial u}{\partial x_i}(tx)\right| dt$$

$$= \frac{1}{r^{n-1}} \int_0^1 dt \int_Q \sum_{i=1}^{n} \left|\frac{\partial u}{\partial x_i}(tx)\right| dx$$

$$= \frac{1}{r^{n-1}} \int_0^1 dt \int_{tQ} \sum_{i=1}^{n} \left|\frac{\partial u}{\partial x_i}(y)\right| t^{-n}\, dy.$$

But $tQ \subset Q$ for $0 \leqslant t \leqslant 1$ and further, by Holder's inequality,

$$\int_{tQ} \left|\frac{\partial u}{\partial x_i}(y)\right| dy \leqslant \left(\int_Q \left|\frac{\partial u}{\partial x_i}\right|^p\right)^{1/p} (\text{meas }(tQ))^{1/p'}$$

where p' is the conjugate exponent of p. Thus

$$|\bar{u} - u(0)| \leqslant \frac{1}{r^{n-1}} |u|_{1,p,Q} r^{n/p'} \int_0^1 t^{n/p'-n}\, dt$$

$$= \frac{r^{1-(n/p)}}{1-(n/p)} |u|_{1,p,Q}.$$

By translation, this inequality is valid for any cube Q in $\mathbb{R}^n$ with sides parallel to the coordinate axes, of length r, and for any $x \in Q$. Hence for any such Q and any $x \in Q$, we have

$$|\bar{u} - u(x)| \leqslant \frac{r^{1-(n/p)}}{1-(n/p)} |u|_{1,p,Q} \tag{2.4.19}$$

and so if $x, y \in Q$,

$$|u(x) - u(y)| \leqslant \frac{2r^{1-(n/p)}}{1-(n/p)} |u|_{1,p,Q}. \tag{2.4.20}$$

Given $x, y \in \mathbb{R}^n$ we can always find such a cube Q containing x, y and of side $r = 2|x-y|$. Substituting this in (2.4.20) we deduce (2.4.17) for $u \in \mathscr{D}(\mathbb{R}^n)$.

If $u \in W^{1,p}(\mathbb{R}^n)$ we construct a sequence $\{u_m\}$ in $\mathscr{D}(\mathbb{R}^n)$ such that $u_m \to u$ in $W^{1,p}(\mathbb{R}^n)$. Then (at least for a subsequence) $u_m \to u$ a.e. in $\mathbb{R}^n$. This then establishes (2.4.17) for all $u \in W^{1,p}(\mathbb{R}^n)$.

Step 2. If $u \in \mathscr{D}(\mathbb{R}^n)$, then again by (2.4.19), we have

$$|u(x)| \leqslant |\bar{u}| + C|u|_{1,p,Q}$$

$$\leqslant C\|u\|_{1,p,Q} \leqslant C\|u\|_{1,p,\mathbb{R}^n}$$

and the result extends to all of $W^{1,p}(\mathbb{R}^n)$ by density. This proves that the inclusion of $W^{1,p}(\mathbb{R}^n)$ in $L^\infty(\mathbb{R}^n)$ is continuous.

The results for $W_0^{1,p}(\Omega)$ follow as usual by using $\tilde{u}$ the extension to $\mathbb{R}^n$ by zero outside Ω of $u \in W_0^{1,p}(\Omega)$. ∎

As already mentioned previously (cf. Section 2.1) we will regard elements of $W^{1,p}(\mathbb{R}^n)$, when $p > n$, via their representatives which are continuous

functions satisfying (2.4.17). Thus we will say that elements of $W^{1,p}(\mathbb{R}^n)$ are *Holder continuous functions* of exponent $\alpha = 1 - (n/p)$.

Theorem 2.4.4 Let $\Omega = \mathbb{R}^n_+$ or an open set of class C^1 with bounded boundary $\partial\Omega$. Then we have the continuous inclusions

(i) if $1 \leqslant p < n$, $W^{1,p}(\Omega) \to L^{p^*}(\Omega)$,

(ii) if $p = n$, $W^{1,n}(\Omega) \to L^q(\Omega)$ for all $q \in [n, \infty)$,

(iii) if $p > n$, $W^{1,p}(\Omega) \to L^\infty(\Omega)$

and further, in the latter case, u is Holder continuous of exponent $\alpha = 1 - (n/p)$. In particular,

$$W^{1,p}(\Omega) \subset C(\bar{\Omega}),\ p > n.$$

Proof: If Ω is as stated, there exists an extension operator $P : W^{1,p}(\Omega) \to W^{1,p}(\mathbb{R}^n)$. Now the theorem is obvious. ∎

In the case $p = n$, we do not have, in general, $W^{1,p}(\Omega) \subset L^\infty(\Omega)$, as shown in the following example.

Example 2.4.1 Let $\Omega = B(0; \frac{1}{2}) \subset \mathbb{R}^2$. Let $r = |x| = \left(\sum_{i=1}^{n} |x_i|^2\right)^{1/2}$. Define

$$u(x) = \log(\log(2/r)),\ x \in \Omega. \tag{2.4.21}$$

Then $u \notin L^\infty(\Omega)$ because of the singularity at the origin. However we will show that $u \in H^1(\Omega)$ (so that $p = 2 = n$). First of all, $u \in L^2(\Omega)$, for

$$\int_\Omega |u|^2 = \int_0^{2\pi} d\theta \int_0^{1/2} r\,(\log\log(2/r))^2\, dr$$

and a simple application of L'Hospital's rule will show that the integrand is a bounded and continuous function on $(0, \frac{1}{2})$ and thus the integral is finite.

We now show that the distribution derivatives are just the classical derivatives which are defined on $\Omega \setminus \{0\}$. To see this, let $\Omega_\epsilon = \{x \mid \epsilon < r < \frac{1}{2}\}$ and if $\phi \in \mathscr{D}(\Omega)$, we have

$$\frac{\partial u}{\partial x_1}(\phi) = -\int_\Omega u\,\frac{\partial \phi}{\partial x_1} = -\lim_{\epsilon\to 0}\int_{\Omega_\epsilon} u\,\frac{\partial \phi}{\partial x_1}.$$

If we set u_{x_1} as the classical partial derivative on Ω_ϵ then by Green's theorem,

$$-\int_{\Omega_\epsilon} u\,\frac{\partial \phi}{\partial x_1} = \int_{\Omega_\epsilon} u_{x_1}\phi - \int_{r=\epsilon} u\phi\nu_1\, ds$$

where $\nu = (\nu_1, \nu_2)$ is the unit outward normal to Ω_ϵ on $\{r = \epsilon\}$. But

$$\left|\int_{r=\epsilon} u\phi\nu_1\right| \leqslant \int_0^{2\pi} |u|\,|\phi|\epsilon\, d\theta \leqslant C2\pi\epsilon \log\log(2/\epsilon)$$

which tends to zero as $\epsilon \to 0$. Thus

$$\frac{\partial u}{\partial x_1}(\phi) = \int_\Omega u_{x_1}\phi$$

which proves the claim. Now,

$$\frac{\partial u}{\partial x_1} = \frac{\cos\theta}{r\log(2/r)}, \text{ where } x_1 = r\cos\theta.$$

Again it is easy to check that this function is in $L^2(\Omega)$. The same analysis applies to $\frac{\partial u}{\partial x_2}$. Thus $u \in H^1(\Omega)$. ∎

For generalizations of the above example, see Adams [1]. Other examples can be found in Brezis [1].

Let us now assume that $p < \frac{n}{2}$ and that $u \in W^{2,p}(\mathbb{R}^n)$. Then both u, and $\frac{\partial u}{\partial x_i}$, $1 \leqslant i \leqslant n$ are all in $W^{1,p}(\mathbb{R}^n)$. Since $p < \frac{n}{2} < n$, we have that u, $\frac{\partial u}{\partial x_i} \in L^{p^*}(\mathbb{R}^n)$. Thus $u \in W^{1,p^*}(\mathbb{R}^n)$. Now $p^* = \frac{np}{n-p} < n$, since $p < \frac{n}{2}$. Thus again we have that $u \in L^{(p^*)^*}(\mathbb{R}^n)$. But it is easy to see that $(p^*)^*$ is given by

$$\frac{1}{(p^*)^*} = \frac{1}{p} - \frac{2}{n}.$$

Iterating this we can easily deduce the following result.

Theorem 2.4.5 Let $m \geqslant 1$ be an integer and $1 \leqslant p < \infty$. Then

(i) if $\frac{1}{p} - \frac{m}{n} > 0$, $W^{m,p}(\mathbb{R}^n) \subset L^q(\mathbb{R}^n)$, $\frac{1}{q} = \frac{1}{p} - \frac{m}{n}$,

(ii) if $\frac{1}{p} - \frac{m}{n} = 0$, $W^{m,p}(\mathbb{R}^n) \subset L^q(\mathbb{R}^n)$, for $q \in [p, \infty)$,

(iii) if $\frac{1}{p} - \frac{m}{n} < 0$, $W^{m,p}(\mathbb{R}^n) \subset L^\infty(\mathbb{R}^n)$,

and in the latter case, i.e. when $m > (n/p)$ if we set

$$k = \left[m - \frac{n}{p}\right], \ \theta = \left(m - \frac{n}{p}\right) - k,$$

[.], denoting the integral part of a real number, we have

$$|D^\alpha u|_{0,\infty,\mathbb{R}^n} \leqslant C\|u\|_{m,p,\mathbb{R}^n} \quad \text{for } |\alpha| \leqslant k \tag{2.4.22}$$

and

$$|D^\alpha u(x) - D^\alpha u(y)| \leqslant C|x-y|^\theta \|u\|_{m,p,\mathbb{R}^n} \text{ a.e. } (x,y), \tag{2.4.23}$$

for $|\alpha| = k$. (If $|\alpha| < k$, (2.4.23) holds with $\theta = 1$ by virtue of (2.4.22)). In particular we have the continuous inclusion

$$W^{m,p}(\mathbb{R}^n) \to C^k(\mathbb{R}^n), \ m > (n/p).$$

The same results are valid for any open set $\Omega \subset \mathbb{R}^n$ for the spaces $W_0^{m,p}(\Omega)$, and for $W^{m,p}(\Omega)$ if $\Omega = \mathbb{R}^n_+$ or is of class C^1 with bounded boundary $\partial\Omega$. In these cases we have the continuous inclusion

$$W^{m,p}(\Omega) \to C^k(\overline{\Omega}) \quad \text{for } m > (n/p). \ ∎$$

Theorem 2.4.6 Let $p > n$. Then the space $W^{1,p}(\mathbb{R}^n)$ is a commutative Banach algebra.

Proof: To prove that $W^{1,p}(\mathbb{R}^n)$ is a Banach algebra, we need to show that if $u, v \in W^{1,p}(\mathbb{R}^n)$, then the pointwise product $uv \in W^{1,p}(\mathbb{R}^n)$ and that

$$\|uv\|_{1,p,\mathbb{R}^n} \leqslant C\|u\|_{1,p,\mathbb{R}^n}\|v\|_{1,p,\mathbb{R}^n}.$$

Then the result would follow (with the equivalent norm $C\|u\|_{1,p,\mathbb{R}^n}$).
Let $u, v \in \mathscr{D}(\mathbb{R}^n)$. Then $uv \in \mathscr{D}(\mathbb{R}^n)$ and by the product rule

$$\frac{\partial}{\partial x_i}(uv) = u\frac{\partial v}{\partial x_i} + \frac{\partial u}{\partial x_i}v \in \mathscr{D}(\mathbb{R}^n).$$

Let $u, v \in W^{1,p}(\mathbb{R}^n)$. Then since $p > n$, $u, v \in L^\infty(\mathbb{R}^n)$ and so $uv \in L^p(\mathbb{R}^n)$ as well. Also

$$u\frac{\partial v}{\partial x_i} + \frac{\partial u}{\partial x_i}v \in L^p(\mathbb{R}^n).$$

We show that this is the distribution derivative $\dfrac{\partial}{\partial x_i}(uv)$.
If $\phi \in \mathscr{D}(\mathbb{R}^n)$, then

$$\begin{aligned}\int_{\mathbb{R}^n} uv\frac{\partial \phi}{\partial x_i} &= \lim_{m\to\infty}\int_{\mathbb{R}^n} u_m v_m \frac{\partial \phi}{\partial x_i}\\ &= -\lim_{m\to\infty}\int_{\mathbb{R}^n}\left(\frac{\partial u_m}{\partial x_i}v_m + u_m\frac{\partial v_m}{\partial x_i}\right)\phi\\ &= -\int_{\mathbb{R}}\left(\frac{\partial u}{\partial x_i}v + u\frac{\partial v}{\partial x_i}\right)\phi\end{aligned}$$

by the Dominated Convergence Theorem, where $u_m \to u$, $v_m \to v$ in $W^{1,p}(\mathbb{R}^n)$, $u_m, v_m \in \mathscr{D}(\mathbb{R}^n)$. This proves that

$$\frac{\partial(uv)}{\partial x_i} = \frac{\partial u}{\partial x_i}v + u\frac{\partial v}{\partial x_i}.$$

Further by the continuity of the inclusion of $W^{1,p}(\mathbb{R}^n)$ into $L^\infty(\mathbb{R}^n)$, it follows that

$$\begin{aligned}\left|\frac{\partial}{\partial x_i}(uv)\right|_{0,p,\mathbb{R}^n} &\leqslant \left|\frac{\partial u}{\partial x_i}\right|_{0,p,\mathbb{R}^n}|v|_{0,\infty,\mathbb{R}^n} + |u|_{0,\infty,\mathbb{R}^n}\left|\frac{\partial v}{\partial x_i}\right|_{0,p,\mathbb{R}^n}\\ &\leqslant C\|u\|_{1,p,\mathbb{R}^n}\|v\|_{1,p,\mathbb{R}^n}\end{aligned}$$

and the theorem is proved. ∎

In the same way it can be proved that $W^{m,p}(\mathbb{R}^n)$ is a Banach algebra when $m > (n/p)$. Also the same result is true for $W^{m,p}(\Omega)$ when Ω is either $\mathbb{R}^n_+$ or of class C^1 with bounded boundary. It is clear that the Banach algebra has an identity (viz. the function $u \equiv 1$) if and only if Ω is bounded.

2.5 COMPACTNESS THEOREMS

It was seen in Section 2.1 that the inclusion of $W^{1,p}(I)$ in $C(\bar{I})$ was compact when I was a bounded interval. In view of the inclusion theorems proved in the previous section we now ask the question that which of these are compact. It will be quickly realised that unbounded domains do not admit of such compact imbeddings as illustrated by the following example.

Example 2.5.1 Let I be the interval (0, 1) of $\mathbb{R}$ and let $I_j = (j, j+1)$. Let f be a C^1 function with support in I. We define f_j to be the same function defined on I_j by translation. We can suitably normalize f to be such that

$$\|f\|_{1,p,I} = 1.$$

The same is then true for each f_j and thus $\{f_j\}$ is a bounded sequence in $W^{1,p}(\mathbb{R})$. Since f is C^1 and of compact support, it is in $L^q(\mathbb{R})$ for every $1 \leqslant q \leqslant \infty$. Further if

$$|f|_{0,q,\mathbb{R}} = |f|_{0,q,I} = a > 0,$$

then for any $j \neq k$,

$$|f_j - f_k|_{0,q,\mathbb{R}} = 2^{1/q}a$$

and so $\{f_i\}$ cannot have a convergent subsequence in $L^q(\mathbb{R})$. Thus none of the imbeddings of $W^{1,p}(\mathbb{R})$ into the spaces $L^q(\mathbb{R})$ can be compact. This example can be easily generalized to $\mathbb{R}^n$ and to open sets like a half-space. ∎

In view of this example, we will restrict our attention to bounded domains. Henceforth, in this section, Ω will stand for a *bounded open set* in $\mathbb{R}^n$. We first examine the types of spaces we have imbeddings into and look for criteria for related compactness in these spaces. From the imbedding theorems of the previous section we have to deal with spaces of continuous functions $C(\bar{\Omega})$ or the various Lebesgue spaces $L^p(\Omega)$, $1 \leqslant p < \infty$. In the spaces $C(\overline{\Omega})$, a bounded set is relatively compact if it is equicontinuous by the Ascoli-Arzela Theorem. A proof of this can be found in any standard text on general topology (cf. for example, Simmons [1]). We now establish a similar criterion in the L^p spaces, due to Frechet and Kolmogorov.

Theorem 2.5.1 Let $\Omega \subset \mathbb{R}^n$ be a bounded open set and let $\Omega' \subset\subset \Omega$. Let $\mathcal{F}$ be a bounded set in $L^p(\Omega)$, $1 \leqslant p < \infty$. Assume that for every $\epsilon > 0$ there exists a $\delta > 0$ such that

(i) $\delta < \text{dist}(\Omega', \mathbb{R}^n \setminus \Omega)$

(ii) For every $h \in \mathbb{R}^n$ with $|h| < \delta$

$$|\tau_{-h}f - f|_{0,p,\Omega'} < \epsilon \quad \text{for every } f \in \mathcal{F}, \tag{2.5.1}$$

where $\tau_y(f)(x) = f(x - y)$. Then $\mathcal{F}|_{\Omega'}$, the set of restrictions of elements of $\mathcal{F}$ to Ω', is relatively compact in $L^p(\Omega')$.

Proof: Since Ω is bounded, $\mathcal{F}$ is also bounded in $L^1(\Omega)$. Hence the extension by zero outside Ω gives a bounded set $\widetilde{\mathcal{F}}$ both in $L^p(\mathbb{R}^n)$ and $L^1(\mathbb{R}^n)$.

Step 1. Let $\epsilon_m \downarrow 0$. Let $\{\rho_{\epsilon_m}\}$ be the corresponding sequence of mollifiers. Now, if $\tilde{f}$ is the extension by zero of $f \in \mathcal{F}$,

$$|(\rho_{\epsilon_m} * \tilde{f})(x) - \tilde{f}(x)| \leqslant \int_{\mathbb{R}^n} |\tilde{f}(x - y) - \tilde{f}(x)|\rho_{\epsilon_m}(y)\, dy.$$

If p' is the conjugate exponent of p, we write $\rho_{\epsilon_m} = \rho_{\epsilon_m}^{1/p}\rho_{\epsilon_m}^{1/p'}$ and apply Holders inequality to the functions $|\tilde{f}(x - y) - \tilde{f}(x)|\rho_{\epsilon_m}^{1/p}$ and $\rho_{\epsilon_m}^{1/p'}$ to get

$$|(\rho_{\epsilon_m} * \tilde{f})(x) - \tilde{f}(x)| \leqslant \left(\int_{\mathbb{R}^n} |\tilde{f}(x - y) - \tilde{f}(x)|^p \rho_{\epsilon_m}(y)\, dy\right)^{1/p}$$

since $\int_{\mathbb{R}^n} \rho_{\epsilon_m}(y) = 1$. Thus, using (2.5.1) when $\epsilon_m < \delta$, we get

$$|\rho_{\epsilon_m} * \tilde{f} - \tilde{f}|_{0,p,\Omega'}^p \leqslant \int_{B(0;\,\epsilon_m)} \rho_{\epsilon_m}(y) \int_{\Omega'} |\tilde{f}(x - y) - \tilde{f}(x)|^p\, dx\, dy$$
$$\leqslant \epsilon^p.$$

Hence

$$|\rho_{\epsilon_m} * \tilde{f} - \tilde{f}|_{0,p,\Omega'} < \epsilon \quad \text{for every } f \in \mathcal{F}.$$

Step 2. Define

$$\mathcal{H}_m = \{\rho_{\epsilon_m} * \tilde{f}\,|_{\Omega'} \mid f \in \mathcal{F}\}.$$

Now

$$|\rho_{\epsilon_m} * \tilde{f}|_{0,\infty,\mathbb{R}^n} \leqslant |\rho_{\epsilon_m}|_{0,\infty,\mathbb{R}^n} |\tilde{f}|_{0,1,\mathbb{R}^n} \leqslant C_m \tag{2.5.3}$$

for every $f \in \mathcal{F}$. Also if x_1 and $x_2 \in \mathbb{R}^n$, then

$$|(\rho_{\epsilon_m} * \tilde{f})(x_1) - (\rho_{\epsilon_m} * \tilde{f})(x_2)| \leqslant |\tilde{f}|_{0,1,\mathbb{R}^n} |\rho'_{\epsilon_m}|_{0,\infty,\mathbb{R}^n} |x_1 - x_2|$$
$$\leqslant C'_m |x_2 - x_2|. \tag{2.5.4}$$

It follows from (2.5.3) and (2.5.4) that for a fixed m, $\mathcal{H}_m$ is a bounded and equicontinuous family in $C(\bar{\Omega}')$ and hence is realtively compact. It follows *a fortiori* that $\mathcal{H}_m$ is relatively compact in $L^p(\Omega')$ as well.

Step 3 Now let $\epsilon > 0$ and choose m large enough such that $\epsilon_m < \delta$, where δ is as in the statement of the theorem. Since $\mathcal{H}_m$ is relatively compact in $L^p(\Omega')$, it can be covered by a finite number of balls of radius ϵ. Then by virtue of (2.5.2) it follows that $\mathcal{F}|_{\Omega'}$ can be covered by the same number of balls of radius 2ϵ. Since ϵ is arbitrary, this proves that $\mathcal{F}|_{\Omega'}$ is totally bounded and hence is relatively compact in $L^p(\Omega')$. ∎

We use the above result to formulate a criterion for compactness in $L^p(\Omega)$ rather than a relatively compact subset, Ω'.

Theorem 2.5.2 Let $\Omega \subset \mathbb{R}^n$ be a bounded open set and let $1 \leqslant p < \infty$. Let $\mathcal{F}$ be a bounded set in $L^p(\Omega)$. Assume that

(i) for every $\epsilon > 0$ and for every $\Omega' \subset\subset \Omega$, there exists a $\delta > 0$ such that $\delta < \text{dist}(\Omega', \mathbb{R}^n \setminus \Omega)$ and

$$|\tau_{-h}f - f|_{0,p,\Omega'} < \epsilon \tag{2.5.5}$$

for all $h \in \mathbb{R}^n$ with $|h| < \delta$ and for all $f \in \mathcal{F}$;

(ii) for every $\epsilon > 0$ there exists $\Omega' \subset\subset \Omega$ such that

$$|f|_{0,p,\Omega/\bar{\Omega}'} < \epsilon \quad \text{for every } f \in \mathcal{F}. \tag{2.5.6}$$

Then, $\mathcal{F}$ is relatively compact in $L^p(\Omega)$.

Proof Let $\epsilon > 0$ be given and let Ω' be chosen such that condition (ii) above is satisfied. By virtue of condition (i) and Theorem 2.5.1, $\mathcal{F}|_{\Omega'}$ is relatively compact in $L^p(\Omega')$. Hence $\mathcal{F}|_{\Omega'}$ can be covered (in $L^p(\Omega')$) by a finite number of balls B_i of radius ϵ with centres, say, $g_i \in L^p(\Omega')$, $1 \leqslant i \leqslant k$. Let $\tilde{g}_i$ be the extension of g_i by zero. Then by virtue of (2.5.6),

$$\mathcal{F} \subset \bigcup_{i=1}^{k} B(\tilde{g}_i, 2\epsilon)$$

the balls now being defined in $L^p(\Omega)$. Thus $\mathcal{F}$ is totally bounded and hence is relatively compact in $L^p(\Omega)$. ∎

Before proving the compactness theorems for the spaces $W^{1,p}(\Omega)$ using the above result, we will need the following result.

Lemma 2.5.1 Let $1 \leqslant p \leqslant \infty$. Then if $u \in W^{1,p}(\Omega)$, we have

$$|\tau_{-h}u - u|_{0,p,\Omega'} \leqslant |h|\,|u|_{1,p,\Omega} \tag{2.5.7}$$

for every $\Omega' \subset\subset \Omega$ and for every $h \in \mathbb{R}^n$ with $|h| < \text{dist}(\Omega', \mathbb{R}^n \setminus \Omega)$.

Proof Assume first that $u \in \mathcal{D}(\mathbb{R}^n)$. If $h \in \mathbb{R}^n$, then

$$u(x+h) - u(x) = \int_0^1 \nabla u(x+th)\cdot h \, dt \tag{2.5.8}$$

where

$$\nabla u = \left(\frac{\partial u}{\partial x_1}, \ldots, \frac{\partial u}{\partial x_n}\right).$$

It follows from (2.5.7) that

$$\int_{\Omega'} |\tau_{-h}u - u|^p \leqslant |h|^p \int_0^1 dt \int_{\Omega'} |\nabla u(x+th)|^p \, dx$$

$$= |h|^p \int_0^1 dt \int_{\Omega'+th} |\nabla u(y)|^p \, dy.$$

If $|h| < \text{dist}(\Omega', \mathbb{R}^n \setminus \Omega)$ then there exists $\Omega'' \subset\subset \Omega$ such that $\Omega' + th \subset \Omega''$ for $t \in [0, 1]$. Then

$$|\tau_{-h}u - u|_{0,p,\Omega'}^{p} \leqslant |h|^p \int_{\Omega''} |\nabla u|^p. \tag{2.5.9}$$

If $1 \leqslant p < \infty$, and $u \in W^{1,p}(\Omega)$, then there exists $u_m \in \mathcal{D}(\mathbb{R}^n)$ such that $u_m \to u$ in $L^p(\Omega)$ and $\nabla u_m \to \nabla u$ in $(L^p(\Omega''))^n$ (cf. Theorem 2.2.1). Now we can apply (2.5.9) to each u_m and pass to the limit as $m \to \infty$ to get (2.5.7).

If $p = \infty$, and $u \in W^{1,\infty}(\Omega)$, then as Ω' is relatively compact choose Ω'' such that $\Omega' \subset \Omega'' \subset\subset \Omega$. Then $u|_{\Omega''} \in W^{1,q}(\Omega'')$ for all $q \in [1, \infty]$. Thus by the preceding arguments,

$$|\tau_{-h}u - u|_{0,q,\Omega'} \leqslant |h| \, |u|_{1,q,\Omega''}, \quad 1 \leqslant q < \infty$$

and as $q \to \infty$, we get

$$|\tau_{-h}u - u|_{0,\infty,\Omega'} \leqslant |h| \, |u|_{1,\infty,\Omega''} \leqslant |h| \, |u|_{1,\infty,\Omega}. \quad ∎$$

Remark 2.5.1 If $1 < p \leqslant \infty$, the converse is also true, i.e. if there exists a constant $C > 0$ such that for every $\Omega' \subset\subset \Omega$ and, for every $h \in \mathbb{R}^n$ such that $|h| < \text{dist}(\Omega', \mathbb{R}^n \setminus \Omega)$

$$|\tau_{-h}u - u|_{0,p,\Omega'} \leqslant C|h| \tag{2.5.10}$$

for a given $u \in L^p(\Omega)$, then $u \in W^{1,p}(\Omega)$ and $|u|_{1,p,\Omega} \leqslant C$. It is not true when $p = 1$. Functions which satisfy (2.5.10) for $p = 1$ form a class larger than $W^{1,1}(\Omega)$. They are known as functions of bounded variation (i.e. L^1 functions whose distribution derivatives are bounded measures) and such spaces of functions are called **BV spaces.** For a description of such spaces see Giusti [1] or Maz'ja [1]. See also the comments at the end of this chapter. ∎

Theorem 2.5.3 **(Rellich-Kondrasov).** Let $\Omega \subset \mathbb{R}^n$ be a bounded open set of class C^1. Then the following inclusions are compact.

(i) if $p < n$, $W^{1,p}(\Omega) \to L^q(\Omega)$, $1 \leqslant q < p^*$,

(ii) if $p = n$, $W^{1,n}(\Omega) \to L^q(\Omega)$, $1 \leqslant q < \infty$,

(iii) if $p > n$, $W^{1,p}(\Omega) \to C(\bar{\Omega})$.

Proof When $p > n$, as observed earlier the functions of $W^{1,p}(\Omega)$ are Holder continuous. If B is the unit ball in $W^{1,p}(\Omega)$, it follows from Theorem 2.4.4 and the analogue of the inequality (2.4.17) that the functions in B are uniformly bounded and equicontinuous in $C(\bar{\Omega})$. Thus B is relatively compact in $C(\bar{\Omega})$ by the Ascoli-Arzela Theorem.

Assume for the moment that the result is true for $p < n$. Notice that as $p \to n$, $p^* \to \infty$. Hence, since Ω is bounded $W^{1,n}(\Omega) \subset W^{1,n-\epsilon}(\Omega)$ for every $\epsilon > 0$ and given any $q < \infty$ we can find $\epsilon > 0$ such that $1 \leqslant q < (n - \epsilon)^*$. Hence by using the case for $p = n - \epsilon < n$, we deduce that $W^{1,n}(\Omega)$ is compactly imbedded in $L^q(\Omega)$ for any $1 \leqslant q < \infty$.

Thus the theorem will be proved if we prove it for the case $p < n$. Let B be the unit ball in $W^{1,p}(\Omega)$. We now verify conditions (i) and (ii) of Theorem 2.5.2. Let $1 \leqslant q < p^*$. Then choose α such that $0 < \alpha \leqslant 1$ and

$$\frac{1}{q} = \frac{\alpha}{1} + \frac{1-\alpha}{p^*}.$$

Then (as in the corollary to Theorem 2.4.1) if $u \in B$, $\Omega' \subset\subset \Omega$ and $h \in \mathbb{R}^n$ such that $|h| < \text{dist}(\Omega', \mathbb{R}^n \setminus \Omega)$,

$$\begin{aligned} |\tau_{-h}u - u|_{0,p,\Omega'} &\leqslant |\tau_{-h}u - u|^{\alpha}_{0,1,\Omega'}|\tau_{-h}u - u|^{1-\alpha}_{0,p^*,\Omega'} \\ &\leqslant (|h|^{\alpha}|u|^{\alpha}_{1,1,\Omega})(2|u|_{0,p^*,\Omega})^{1-\alpha} \\ &\leqslant C|h|^{\alpha} \end{aligned}$$

using Lemma 2.5.1. We choose h small enough such that $C|h|^{\alpha} < \epsilon$. This will verify (2.5.5).

Now if $u \in B$ and $\Omega' \subset\subset \Omega$, it follows by Holder's inequality that

$$\begin{aligned} |u|_{0,q,\Omega\setminus\bar{\Omega}'} &\leqslant |u|_{0,p^*,\Omega\setminus\bar{\Omega}'} \quad \text{meas}\,(\Omega\setminus\bar{\Omega}')^{1-(q/p^*)} \\ &\leqslant C\,\text{meas}\,(\Omega\setminus\bar{\Omega}')^{1-(q/p^*)} \end{aligned}$$

which can be made to be less than any given $\epsilon > 0$ by choosing $\Omega' \subset\subset \Omega$ to be 'as closely filling Ω' as needed. This verifies (2.5.6). Thus B is relatively compact in $L^q(\Omega)$ for $1 \leqslant q < p^*$ and the theorem is proved. ∎

If Ω is any bounded domain, the above mentioned theorem is valid for $W_0^{1,p}(\Omega)$.

Remark 2.5.2 Clearly the proof fails in the extreme case $q = p^*$ in the verification of (2.5.5). In fact the inclusions

$$W^{1,p}(\Omega) \text{ (or } W_0^{1,p}(\Omega)) \to L^{p^*}(\Omega)$$

are never compact. See Exercise 2.19 for counter examples. In case of $H_0^1(\Omega)$ (i.e. $p = 2$) we can advance the following argument. Assume that $H_0^1(\Omega)$ is compactly imbedded in $L^{2^*}(\Omega)$, where $2^* = 2n/(n-2)$. Consider the following minimization problem:

$$\inf_{\substack{v \in H_0^1(\Omega) \\ |v|_{0,2^*,\Omega}=1}} \{|v|^2_{1,\Omega}\}$$

The infimum is a positive quantity which is related to the best possible constant in Sobolev's inequality (cf. (2.4.5)). Let $\{u_m\}$ be a minimizing sequence in $H_0^1(\Omega)$. Then, as $|u_m|^2_{1,\Omega}$ is bounded, $\|u_m\|_{1,\Omega}$ is also uniformly bounded (by Poincaré's inequality; cf. (2.3.22)) and we can thus extract a weakly convergent subsequence, say $\{u_{m_k}\}$. But a compact operator converts it into a strongly convergent sequence and so $\{u_{m_k}\}$ converges strongly $L^{2^*}(\Omega)$ and thus if u is the limit, $|u|_{0,2^*,\Omega} = 1$. Then by Fatou's

lemma

$$\inf_{\substack{v \in H_0^1(\Omega) \\ |v|_{0, 2^*, \Omega}=1}} |v|_{1,\Omega}^2 \leqslant |u|_{1,\Omega}^2 \leqslant \lim_{m\to\infty} \inf |u_m|_{1,\Omega}^2 = \inf_{\substack{v \in H_0^1(\Omega) \\ |v|_{0, 2^*, \Omega}=1}} |v|_{1,\Omega}^2.$$

Thus the infimum is attained at $u \in H_0^1(\Omega)$. But it can be proved that (cf. Aubin [1] and Talenti [1]) the said infimum is never attained over a proper subset of $\mathbb{R}^n$. (See also the comments as the end of this chapter). Thus the inclusion is not compact. ∎

We will now derive some useful consequences of the Rellich-Kondrasov Theorem.

Theorem 2.5.4 Let $\Omega \subset \mathbb{R}^n$ be a bounded and connected open set of class C^1. Let $P_m(\Omega)$ denote the space of all polynomials in $x_1, \ldots, x_n$ over Ω and of degree $\leqslant m$. Let $v \in W^{m+1,p}(\Omega)$, $1 \leqslant p \leqslant \infty$ and set $\dot{v}$ the equivalence class of v in the quotient space $W^{m+1,p}(\Omega)/P_m(\Omega)$. This space is equipped with the usual norm

$$\|\dot{v}\|_{m+1,p,\Omega} = \inf_{\tilde{p} \in P_m(\Omega)} \|v + \tilde{p}\|_{m,p,\Omega}.$$

This norm is equivalent to $|v|_{m+1,p,\Omega}$.

Proof. First of all let us verify that $\dot{v} \to |v|_{m+1,p,\Omega}$ defines a norm on the quotient space. Indeed, to start with, it is well-defined. For, if $\dot{v}_1 = \dot{v}_2$, then v_1 and v_2 differ by a polynomial in $P_m(\Omega)$. Then $|v_1|_{m+1,p,\Omega} = |v_2|_{m+1,p,\Omega}$. Thus the semi-norm is well-defined. Again if $|v|_{m+1,p,\Omega} = 0$, then $v \in P_m(\Omega)$ and so $\dot{v} = 0$ (in the quotient space). Thus the semi-norm defines a norm on the quotient space. Now on one hand,

$$|v|_{m+1,p,\Omega} \leqslant \|v + \tilde{p}\|_{m+1,p,\Omega}$$

for any $\tilde{p} \in P_m(\Omega)$. Hence taking the infimum on the right hand side we get

$$|v|_{m+1,p,\Omega} \leqslant \|\dot{v}\|_{m+1,p,\Omega}, \tag{2.5.11}$$

Let us assume that the inequality in the reverse sense does not hold with any constant $C > 0$. Then we can find a sequence $\{\tilde{v}_k\}$ in $W^{m+1,p}(\Omega)$ such that

$$|\tilde{v}_k|_{m+1,p,\Omega} \leqslant \frac{1}{k} \|\dot{\tilde{v}}_k\|_{m+1,p,\Omega}.$$

Normalizing, we can assume the existence of a sequence $\{v_k\}$ in $W^{m+1,p}(\Omega)$ such that

$$\|\dot{v}_k\|_{m+1,p,\Omega} = 1, \ |v_k|_{m+1,p,\Omega} \to 0 \quad \text{as } k \to \infty. \tag{2.5.12}$$

By adding a polynomial in $P_m(\Omega)$ if necessary, we can assume, without loss of generality, that

$$\|v_k\|_{m+1,p,\Omega} \leqslant 2, \quad \text{for all } k. \tag{2.5.13}$$

It now follows from the Rellich-Kondrasov Theorem that $\{v_k\}$ which is bounded in $W^{m+1,p}(\Omega)$ has to be relatively compact in $W^{m,p}(\Omega)$. Thus we

can extract a convergent subsequence (again indexed by k for convenience). Thus let $v_k \to v$ in $W^{m,p}(\Omega)$. Let α be a multi-index such that $|\alpha| = m+1$ and let $\phi \in \mathcal{D}(\Omega)$. Then

$$0 = \lim_{k\to\infty} \int_\Omega D^\alpha v_k \phi = (-1)^{m+1} \lim_{k\to\infty} \int_\Omega v_k D^\alpha \phi$$

$$= (-1)^{m+1} \int_\Omega v D^\alpha \phi$$

and so $D^\alpha v = 0$ for every α such that $|\alpha| = m+1$. Since Ω is connected, this implies that $v \in P_m(\Omega)$ (cf. Comments at the end of Chapter 1). Further

$$\|v_k - v\|^p_{m+1,p,\Omega} = \|v_k - v\|^p_{m,p,\Omega} + |v_k - v|^p_{m+1,p,\Omega}$$

$$= \|v_k - v\|^p_{m,p,\Omega} + |v_k|^p_{m+1,p,\Omega}$$

since $v \in P_m(\Omega)$. But both terms on the right tend to zero as $k \to \infty$ since $v_k \to v$ in $W^{m,p}(\Omega)$ and by (2.5.12), $|v_k|_{m+1,p,\Omega} \to 0$ as $k \to \infty$. Thus $v_k \to v$ in $W^{m+1,p}(\Omega)$ as well. But then $\|\dot{v}_k\|_{m+1,p,\Omega} \to 0$ since $v \in P_m(\Omega)$ which contradicts (2.5.12). Thus there exists a constant $C > 0$ such that for every $v \in W^{m+1,p}(\Omega)$,

$$\|\dot{v}\|_{m+1,p,\Omega} \leqslant C|v|_{m+1,p,\Omega} \tag{2.5.14}$$

This completes the proof. ∎

Corollary. Let $\Pi : W^{m+1,p}(\Omega) \to V$ be a continuous linear operator, where V is a Banach space containing $W^{m+1,p}(\Omega)$. Assume further that

$$\Pi(\tilde{p}) = \tilde{p} \tag{2.5.15}$$

for every $\tilde{p} \in P_m(\Omega)$. Then there exists a constant $C > 0$ such that for every $u \in W^{m+1,p}(\Omega)$,

$$\|u - \Pi(u)\|_V \leqslant C|u|_{m+1,p,\Omega}. \tag{2.5.16}$$

Proof Let $\tilde{p}$ be any polynomial in $P_m(\Omega)$. Then for any $u \in W^{m+1,p}(\Omega)$,

$$\|u - \Pi(u)\|_V = \|(u+\tilde{p}) - \Pi(u+\tilde{p})\|_V \quad \text{(using (2.5.15))}$$

$$= \|(I - \Pi)(u+\tilde{p})\|_V$$

$$\leqslant \|I - \Pi\| \, \|u+\tilde{p}\|_{m+1,p,\Omega}.$$

Since $\tilde{p}$ was arbitrary, we get, setting $C_1 = \|I - \Pi\|$,

$$\|u - \Pi(u)\|_V \leqslant C_1 \inf_{\tilde{p}\in P_m(\Omega)} \|u+\tilde{p}\|_{m+1,p,\Omega} = C_1\|\dot{u}\|_{m+1,p,\Omega}$$

$$\leqslant C|u|_{m+1,p,\Omega}$$

by the previous theorem. ∎

Remark 2.5.3 If m were large enough, then by the Sobolev imbedding theorems, we can take V to be $C(\bar{\Omega})$ or even $C^k(\bar{\Omega})$ for some integer

$k \geqslant 0$. We can think of Π as a Lagrange or Hermite interpolation operator so that it leaves some polynomials invariant. Thus the above theorem gives us an estimate for the *interpolation error* in V. Such estimates are very important in numerical analysis, especially in the error analysis of the *finite element method*. ∎

A consequence of the above corollary is the following inequality. (We assume that Ω is bounded, connected and of class C^1).

Theorem 2.5.5 **(Poincaré-Wirtinger Inequality).** There exists a constant $C > 0$ such that for every $u \in W^{1,p}(\Omega)$, $1 \leqslant p \leqslant \infty$,

$$|u - \bar{u}|_{0,p,\Omega} \leqslant C|u|_{1,p,\Omega} \tag{2.5.17}$$

where

$$\bar{u} = \frac{1}{\text{meas}\,(\Omega)} \int_\Omega u$$

is the average of u over Ω. Further, if $p < n$, then

$$|u - \bar{u}|_{0,p^*,\Omega} \leqslant C|u|_{1,p,\Omega}. \tag{2.5.18}$$

Proof If we set $\Pi u = \bar{u}$ with $V = L^p(\Omega)$ in the previous theorem, we easily deduce (2.5.17) from (2.5.16) ($m = 0$). If $p < n$, then

$$\begin{aligned} |u - \bar{u}|_{0,p^*,\Omega} &\leqslant C\|u - \bar{u}\|_{1,p,\Omega} \\ &\leqslant C|u - \bar{u}|_{0,p,\Omega} + C|u - \bar{u}|_{1,p,\Omega} \\ &\leqslant C|u|_{1,p,\Omega} \end{aligned}$$

using (2.5.17) and the fact that $\bar{u} \in P_0(\Omega)$, i.e. is a constant. ∎

2.6 DUAL SPACES, FRACTIONAL ORDER SPACES AND TRACE SPACES

In this section we will try to define the Sobolev spaces for all real parameters s instead of just for the natural numbers m. We start with the negative integers.

Definition 2.6.1 Let $1 \leqslant p < \infty$. Let p' be the conjugate exponent of p. The dual of the space $W_0^{m,p}(\Omega)$ where $m \geqslant 1$ is an integer, is denoted by $W^{-m,p'}(\Omega)$. If $p = 2$, $H^{-m}(\Omega)$ is the dual of the space $H_0^m(\Omega)$. ∎

Remark 2.6.1 $H_0^m(\Omega)$ is a Hilbert space and so, by the Riesz Representation Theorem, can be identified with its own dual. However except when $m = 0$, i.e. in the case of $L^2(\Omega)$, we do not identify $H_0^m(\Omega)$ with its dual. We have the following dense and continuous inclusions:

$$H_0^1(\Omega) \to L^2(\Omega) \to H^{-1}(\Omega). \quad ∎$$

One can easily understand the presence of the conjugate exponent p' in the notation of the dual of $W_0^{1,p}(\Omega)$. However the designation of the dual space by the negative order requires explanation. Also we must wonder why the dual of $W_0^{m,p}(\Omega)$ and not $W^{m,p}(\Omega)$ bears this notation. The reason is the following: if $u \in W^{m,p}(\Omega)$ then its first derivatives belong to $W^{m-1,p}(\Omega)$. We would like this feature to be preserved for all the integers. Thus if $u \in L^p(\Omega)$, its first derivatives must belong to $W^{-1,p}(\Omega)$. But this can happen only with the above definition, as shown by the following result.

Theorem 2.6.1 Let $F \in W^{-1,p'}(\Omega)$. Then there exist functions $f_0, f_1, \ldots, f_n \in L^{p'}(\Omega)$ such that

$$F(v) = \int_\Omega f_0 v + \sum_{i=1}^{n} \int_\Omega f_i \frac{\partial v}{\partial x_i},\ v \in W_0^{1,p}(\Omega), \tag{2.6.1}$$

and

$$\|F\| = \max_{0 \leqslant i \leqslant n} |f_i|_{0,p',\Omega} \tag{2.6.2}$$

Further, if Ω is bounded, we may assume $f_0 = 0$.

Proof Let $E = (L^p(\Omega))^{n+1}$. Let $T: W_0^{1,p}(\Omega) \to E$ be the isometry given in (2.1.13). Let $G = T(W_0^{1,p}(\Omega))$ be the image in E and $S: G \to W_0^{1,p}(\Omega)$ be the inverse of T. Now consider the continuous linear functional

$$h \in G \to F(S(h)).$$

By the Hahn-Banach Theorem, this extends to a continuous linear functional Φ on all of E and by the Riesz Representation Theorem, there exists $f_0, f_1, \ldots, f_n \in L^{p'}(\Omega)$ such that

$$\Phi(h) = \int_\Omega f_0 h_0 + \sum_{i=1}^{n} \int_\Omega f_i h_i,\ h = (h_i)_{i=0}^n \in E$$

and

$$\|\phi\|_{E'} = \|F\|.$$

Clearly then $\|F\| = \max_{0 \leqslant i \leqslant n} |f_i|_{0,p',\Omega}$ and taking $h \in G$, we get

$$h = \left(u, \frac{\partial u}{\partial x_1}, \ldots, \frac{\partial u}{\partial x_n}\right)$$

for some $u \in W_0^{1,p}(\Omega)$ and

$$F(u) = \int_\Omega f_0 u + \sum_{i=1}^{n} \int_\Omega f_i \frac{\partial u}{\partial x_i}.$$

If Ω is bounded, by Poincaré's inequality (Theorem 2.3.4) we can assume that the gradient map

$$u \in W_0^{1,p}(\Omega) \to \nabla u = \left(\frac{\partial u}{\partial x_1}, \ldots, \frac{\partial u}{\partial x_n}\right)$$

is an isometry of $W_0^{1,p}(\Omega)$ into $(L^p(\Omega))^n$ and proceed as before. Thus we may assume $f_0 = 0$. ∎

Since $\mathscr{D}(\Omega)$ is dense in $W_0^{1,p}(\Omega)$, the linear functional is uniquely fixed once it is fixed on $\mathscr{D}(\Omega)$. However from (2.6.1) we can write for $\phi \in \mathscr{D}(\Omega)$,

$$F(\phi) = \int_\Omega f_0\phi - \sum_{i=1}^{n} \int_\Omega \frac{\partial f_i}{\partial x_i} \phi$$

and so F can be identified with the distribution

$$f_0 - \sum_{i=1}^{n} \frac{\partial f_i}{\partial x_i}. \tag{2.6.3}$$

The previous theorem is also true for the dual of the space $W^{1,p}(\Omega)$ (except that we cannot assume $f_0 = 0$ even when Ω is bounded) but the above identification with the distribution is not possible. We may say that the dual of $W^{1,p}(\Omega)$ consists of extensions to $W^{1,p}(\Omega)$ (by the Hahn-Banach Theorem) of distributions as in (2.6.3). The extension to $W^{1,p}(\Omega)$ is not unique while the extension to $W_0^{1,p}(\Omega)$ is unique and so we identify it with the distribution itself. This now explains our notation. Since the dual of $W_0^{1,p}(\Omega)$ is made up derivatives of $L^{p'}$ functions, we use the notation $W^{-1,p'}$ for this space. Similarly for the spaces $W^{-m,p'}(\Omega)$.

Let $m > n/p$. Then $W_0^{m,p}(\Omega) \subset C(\bar{\Omega})$. Thus point values are well-defined. If $x_0 \in \Omega$ then when $\phi \in \mathscr{D}(\Omega)$

$$\delta_{x_0}(\phi) = \phi(x_0)$$

and further

$$|\delta_{x_0}(\phi)| = |\phi(x_0)| \leqslant |\phi|_{0,\infty,\Omega} \leqslant C\|\phi\|_{m,p,\Omega},$$

Thus δ_{x_0} is continuous for the $W^{m,p}(\Omega)$-norm on $\mathscr{D}(\Omega)$ and so uniquely extends to $W_0^{m,p}(\Omega)$. Hence $\delta_{x_0} \in W^{-m,p'}(\Omega)$ when $m > n/p$. Thus given any domain Ω, the Dirac distribution will belong to a Sobolev space of sufficiently large *negative* order.

Sobolev spaces $W^{s,p}(\Omega)$ for non-integral s could be defined in a variety of ways. One such method is the following. Let $1 \leqslant p < \infty$ and $0 < s < 1$. Then we define

$$W^{s,p}(\Omega) = \left\{ u \in L^p(\Omega) \,\Big|\, \frac{|u(x) - u(y)|}{|x - y|^{s+(n/p)}} \in L^p(\Omega \times \Omega) \right\} \tag{2.6.4}$$

with the obvious norm. We set now for $s = m + \sigma$, $m \geqslant 0$, m an integer, $0 < \sigma < 1$,

$$W^{s,p}(\Omega) = \{u \in W^{m,p}(\Omega) \mid D^\alpha u \in W^{\sigma,p}(\Omega) \quad \text{for all } |\alpha| = m\} \tag{2.6.5}$$

We denote by $W_0^{s,p}(\Omega)$, the closure of $\mathscr{D}(\Omega)$ in $W^{s,p}(\Omega)$ and $W^{-s,p'}(\Omega)$ is dual of $W_0^{s,p}(\Omega)$.

If $p = 2$ we have the spaces $H^s(\Omega)$. If $\Omega = \mathbb{R}^n$, we have already encountered a definition of the space $H^s(\mathbb{R}^n)$ (cf. (2.1.11)). It can be shown that these definitions produce the same space for each $s > 0$. Again we have $H^s(\mathbb{R}^n)$

as the dual of $H^{-s}(\mathbb{R}^n)$ when $s < 0$. We now show that the definition via the Fourier transform holds for negative indices as well.

Theorem 2.6.2 Let $s > 0$ be a real number. Then

$$H^{-s}(\mathbb{R}^n) = \{u \in \mathcal{S}'(\mathbb{R}^n) \mid (1 + |\xi|^2)^{-s/2}\hat{u}(\xi) \in L^2(\mathbb{R}^n)\} \tag{2.6.6}$$

Proof We will give the proof when $s = 1$. Now if $u \in H^{-1}(\mathbb{R}^n)$ then

$$u = f_0 + \sum_{i=1}^{n} \frac{\partial f_i}{\partial x_i}, \quad f_0, f_1, \ldots, f_n \in L^2(\mathbb{R}^n)$$

by (2.6.1). Hence u is a tempered distribution and further

$$\hat{u} = \hat{f_0} + \sum_{i=1}^{n} (2\pi i)\xi_i \hat{f_i} \quad \text{(cf. (1.11.5))}.$$

Then clearly $(1 + |\xi|^2)^{-1/2}\hat{u} \in L^2(\mathbb{R}^n)$. This proves one inclusion in (2.6.6). To prove the reverse inclusion, let $u \in \mathcal{S}'(\mathbb{R}^n)$ such that $(1 + |\xi|^2)^{-1/2}\hat{u}(\xi) \in L^2(\mathbb{R}^n)$. Let $\phi \in \mathcal{D}(\mathbb{R}^n)$. Then there exists $\psi \in \mathcal{S}(\mathbb{R}^n)$ such that $\phi = \hat{\psi}$. Thus if we set

$$\kappa(\xi) = (1 + |\xi|^2)^{1/2}, \quad \kappa_{-1}(\xi) = (1 + |\xi|^2)^{-1/2} \tag{2.6.7}$$

both κ and κ_{-1} are C^∞ functions on $\mathbb{R}^n$ and we can write

$$\begin{aligned} u(\phi) &= u(\hat{\psi}) = \hat{u}(\psi) = (\kappa\kappa_{-1})\hat{u}(\psi) \\ &= \kappa_{-1}\hat{u}(\kappa\psi) \end{aligned}$$

But $\kappa_{-1}\hat{u} \in L^2(\mathbb{R}^n)$ and so

$$u(\phi) = \int_{\mathbb{R}^n} (1 + |\xi|^2)^{-1/2}\hat{u}(\xi)(1 + |\xi|^2)^{1/2}\psi(\xi)\, d\xi$$

and so

$$|u(\phi)| \leqslant |(1 + |\xi|^2)^{-1/2}\hat{u}(\xi)|_{0,\mathbb{R}^n}|(1 + |\xi|^2)^{1/2}\psi(\xi)|_{0,\mathbb{R}^n}.$$

But

$$\begin{aligned} \left|(1 + |\xi|^2)^{1/2}\psi(\xi)\right|^2_{0,\mathbb{R}^n} &= \int_{\mathbb{R}^n} (1 + |\xi|^2)\psi^2(\xi)\, d\xi \\ &= \int_{\mathbb{R}^n} (1 + |\xi|^2)\psi^2(-\xi)\, d\xi \\ &= \int_{\mathbb{R}^n} (1 + |\xi|^2)\hat{\hat{\psi}}^2(\xi)\, d\xi \\ &= \int_{\mathbb{R}^n} (1 + |\xi|^2)(\hat{\phi}(\xi))^2\, d\xi = \left\|\phi\right\|^2_{H^1(\mathbb{R}^n)}. \end{aligned}$$

Thus u defines a continuous linear functional on $H^1(\mathbb{R}^n)$ and so $u \in H^{-1}(\mathbb{R}^n)$. Also,

$$\|u\|_{H^{-1}(\mathbb{R}^n)} = |(1 + |\xi|^2)^{-1/2}\hat{u}(\xi)|_{0,\mathbb{R}^n}. \quad \blacksquare$$

Again we recover, in a different way, a result we had proved previously.

If δ is the Dirac distribution, we know that $\hat{\delta} \equiv 1$ (cf. (1.11.16)) and so $\delta \in H^{-s}(\mathbb{R}^n)$ if and only if $(1 + |\xi|^2)^{-s/2} \in L^2(\mathbb{R}^n)$. This is true for $s > n/2$ since the integral (after converting to polar coordinates)

$$\int_0^\infty \frac{r^{n-1}\, dr}{(1 + r^2)^s}$$

is finite only when $s > n/2$.

We may also define the Sobolev spaces $H^s(\Omega)$ for real s as the restrictions to Ω of elements of $H^s(\mathbb{R}^n)$. This will give rise to another class of Sobolev spaces. However if Ω is a "smooth set" then this will give the same spaces as got by using (2.6.4)–(2.6.5) when $p = 2$.

If $s_1 < s_2$ then $H^{s_2}(\mathbb{R}^n) \subset H^{s_1}(\mathbb{R}^n)$. If $s_1 < s < s_2$ again we have

$$H^{s_2}(\mathbb{R}^n) \subset H^s(\mathbb{R}^n) \subset H^{s_1}(\mathbb{R}^n). \tag{2.6.8}$$

If we write $s = \theta s_1 + (1 - \theta)s_2$, we can show that the space $H^s(\mathbb{R}^n)$ is an 'interpolation' of the spaces $H^{s_1}(\mathbb{R}^n)$ and $H^{s_2}(\mathbb{R}^n)$ in the sense of the theorem below.

Theorem 2.6.3 Let $s_1 < s_2$ and $s = \theta s_1 + (1 - \theta)s_2$, $\theta \in (0, 1)$. If $u \in H^{s_2}(\mathbb{R}^n)$, then

$$\|u\|_{H^s(\mathbb{R}^n)} \leqslant \left\|u\right\|^{\theta}_{H^{s_1}(\mathbb{R}^n)} \left\|u\right\|^{1-\theta}_{H^{s_2}(\mathbb{R}^n)} \tag{2.6.9}$$

Proof Let $u \in H^{s_2}(\mathbb{R}^n)$. Then $u \in H^s(\mathbb{R}^n)$ and $u \in H^{s_1}(\mathbb{R}^n)$ as well, by (2.6.8). Now

$$\int_{\mathbb{R}^n} (1 + |\xi|^2)^s |\hat{u}(\xi)|^2\, d\xi = \int_{\mathbb{R}^n} 1 + (|\xi|^2)^{\theta s_1 + (1-\theta)s_2} |\hat{u}(\xi)|^{2\theta + 2(1-\theta)}\, d\xi.$$

But $(1+|\xi|^2)^{\theta s_1}|\hat{u}(\xi)|^{2\theta} \in L^{1/\theta}(\mathbb{R}^n)$ and $(1+|\xi|)^{s_2(1-\theta)}|\hat{u}(\xi)|^{2(1-\theta)} \in L^{1/(1-\theta)}(\mathbb{R}^n)$. Further $(1/\theta)$ and $1/(1 - \theta)$ are conjugate exponents. Hence by Holder's inequality,

$$\int_{\mathbb{R}^n} (1 + |\xi|^2)^s |\hat{u}(\xi)|^2\, d\xi \leqslant \left(\int_{\mathbb{R}^n} (1 + |\xi|^2)^{s_1} |\hat{u}(\xi)|^2\, d\xi\right)^{\theta} \times \left(\int_{\mathbb{R}^n} (1 + |\xi|^2)^{s_2} |\hat{u}(\xi)|^2\, d\xi\right)^{1-\theta}$$

which yields (2.6.9) on taking square roots and using (2.1.10) to define the norms in $H^s(\mathbb{R}^n)$. ∎

Remark 2.6.2 Let $s_1 < s_2$ and $t_1 < t_2$. Let $\theta \in (0, 1)$ and set $s = \theta s_1 + (1 - \theta)s_2$, $t = \theta t_1 + (1 - \theta)t_2$. Assume that T is a linear operator such that

$$T \in \mathcal{L}(H^{s_1}(\mathbb{R}^n), H^{t_1}(\mathbb{R}^n)) \cap \mathcal{L}(H^{s_2}(\mathbb{R}^n), H^{t_2}(\mathbb{R}^n)) \tag{2.6.10}$$

Then it is true that $T \in \mathcal{L}(H^s(\mathbb{R}^n), H^t(\mathbb{R}^n))$ and we also have

$$\left\|T\right\|_{\mathcal{L}(H^s(\mathbb{R}^n), H^t(\mathbb{R}^n))} \leqslant \left\|T\right\|^{\theta}_{\mathcal{L}(H^{s_1}(\mathbb{R}^n), H^{t_1}(\mathbb{R}^n))} \left\|T\right\|^{1-\theta}_{\mathcal{L}(H^{s_2}(\mathbb{R}^n), H^{t_2}(\mathbb{R}^n))} \tag{2.6.11}$$

The proof of this fact needs more functional analytic tools than what we have developed and so is omitted. The interested reader can refer to Adams [1] or Lions and Magenes [1] for a complete account of the theory of interpolation spaces. ∎

We can use the notion of interpolation spaces to define the spaces $H^s(\Omega)$, when $\Omega \subset \mathbb{R}^n$, by interpolating between the integral order spaces $H^m(\Omega)$, $m \geqslant 0$, an integer. Again, if Ω is smooth enough these spaces will coincide with the two previous constructions.

We will now state an interesting result regarding the space $H^{-1}(\Omega)$ from which will follow a useful inequality.

Theorem 2.6.4 Let $\Omega = \mathbb{R}^n$, $\mathbb{R}^n_+$ or a bounded open set of class C^1. If $u \in H^{-1}(\Omega)$ such that $\dfrac{\partial u}{\partial x_i} \in H^{-1}(\Omega)$ for all $1 \leqslant i \leqslant n$, then $u \in L^2(\Omega)$.

Proof (Sketch) The proof is simple for the case $\Omega = \mathbb{R}^n$. For, if $u, \dfrac{\partial u}{\partial x_i} \in H^{-1}(\mathbb{R}^n)$, then by Theorem 2.6.2,

$$(1 + |\xi|^2)^{-1/2}\hat{u}(\xi),\ (1 + |\xi|^2)^{-1/2}\xi_i\hat{u}(\xi) \in L^2(\mathbb{R}^n).$$

Thus

$$\int_{\mathbb{R}^n} (1 + |\xi|^2)^{-1}\left(1 + \sum_{i=1}^{n} \xi_i^2\right)|\hat{u}(\xi)|^2\, d\xi < \infty$$

i.e. $\hat{u}(\xi) \in L^2(\mathbb{R}^n)$, or, $u \in L^2(\mathbb{R}^n)$. If we prove it for $\Omega = \mathbb{R}^n_+$, then the result can be proved for Ω of class C^1 using the local chart defining the boundary and the corresponding partition of unity (cf. Section 2.3). The proof of the case $\Omega = \mathbb{R}^n_+$ is rather long and beyond the scope of this book. For full details, see Duvaut and Lions [1]. ∎

Using the preceding theorem we will now prove an important inequality in the theory of elasticity. First we need a few notations. Let $V = (H^1(\Omega))^3$, where $\Omega \subset \mathbb{R}^3$ is a bounded open set of class C^1. If $v \in V$, let $v = (v_1, v_2, v_3)$ be its components. For $1 \leqslant i, j \leqslant 3$ we define

$$\epsilon_{ij}(v) = \tfrac{1}{2}\left(\frac{\partial v_i}{\partial x_j} + \frac{\partial v_j}{\partial x_i}\right). \tag{2.6.12}$$

We denote by $\|\cdot\|_V$ the usual product norm on V. Now we can prove the following result.

Theorem 2.6.5 **(Korn's Inequality)** Let Ω be a bounded open subset of $\mathbb{R}^3$, of class C^1. Then there exists a constant $C > 0$, depending only on Ω, such that

$$\int_{\Omega} \epsilon_{ij}(v)\epsilon_{ij}(v) + \int_{\Omega} v_i v_i \geqslant C\|v\|_V^2, \tag{2.6.13}$$

for every $v \in V$.

Note We have used the convention of summation with respect to repeated indices in (2.6.13). In long hand it should read as

$$\int_\Omega \sum_{i,j=1}^{3} |\epsilon_{ij}(v)|^2 + \int_\Omega \sum_{i=1}^{3} |v_i|^2 \geqslant C\|v\|_V^2.$$

Proof Let us define

$$E = \{v \in (L^2(\Omega))^3 | \epsilon_{ij}(v) \in L^2(\Omega) \quad \text{for all } 1 \leqslant i, j \leqslant 3\}.$$

Then E is a Hilbert space for the norm

$$\left(\int_\Omega \epsilon_{ij}(v)\epsilon_{ij}(v) + \int_\Omega v_i v_i\right)^{1/2}.$$

Now it is a simple matter to check that for $1 \leqslant i, j, k \leqslant 3$,

$$\frac{\partial^2 v_i}{\partial x_j \partial x_k} = \frac{\partial}{\partial x_j}(\epsilon_{ik}(v)) + \frac{\partial}{\partial x_k}(\epsilon_{ij}(v)) - \frac{\partial}{\partial x_i}(\epsilon_{jk}(v)). \tag{2.6.14}$$

Hence if $v \in E$, then $\frac{\partial^2 v_i}{\partial x_j \partial x_k} \in H^{-1}(\Omega)$ for all $1 \leqslant i, j, k \leqslant 3$. But $\frac{\partial v_i}{\partial x_k} \in H^{-1}(\Omega)$ as well. Hence by Theorem 2.6.4, $\frac{\partial v_i}{\partial x_k} \in L^2(\Omega)$ for all $1 \leqslant i, k \leqslant 3$ which means that $v \in (H^1(\Omega))^3$. Conversely if $v \in (H^1(\Omega))^3$, trivially $v \in E$. Thus we have the set theoretic equality

$$E = (H^1(\Omega))^3. \tag{2.6.15}$$

It is obvious that the identity map $\text{I} : (H^1(\Omega))^3 \to E$ is continuous with the respective topologies as above and as it is one-one and onto as well (by (2.6.15)) it follows from the Open Mapping Theorem that the two norms are equivalent, which yields (2.6.13). ∎

Korn's inequality will play a crucial role in the weak formulation of the equations of linear elasticity, as we shall see in the next chapter.

We now conclude this section by defining the trace spaces $H^s(\Gamma)$ where $\Gamma = \partial\Omega$, the boundary of a bounded open subset of $\mathbf{R}^n$, $n \geqslant 2$. We assume that Ω is of class C^∞ (though it is not really necessary to assume so much smoothness) and that it is locally on the same side of its boundary, i.e. we avoid domains which lie on both sides of the boundary such as the one portrayed in Fig. 8.

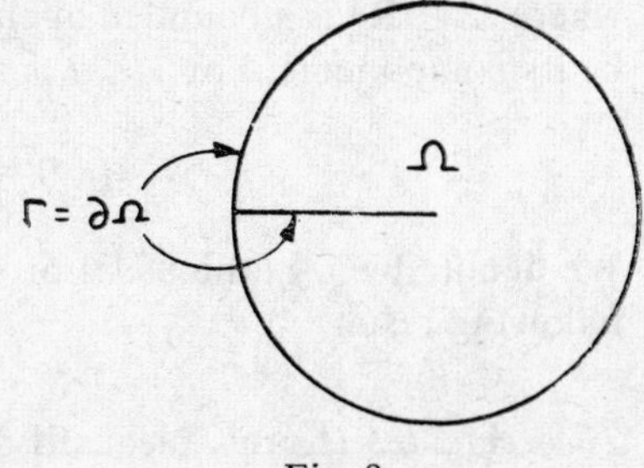

Fig. 8

Since $\Gamma = \partial\Omega$ is compact, it can be covered by a finite number of neighbourhoods $\{U_j\}_{j=1}^k$ and there exist associated C^∞ bijections $T_j : Q \to U_j$ such that $T_j(Q_+) = U_j \cap \Omega$ and $T_j(Q_0) = U_j \cap \Gamma$ where Q, Q_+ and Q_0 are as

defined in Section 2.3. We choose an associated C^∞ partition of unity $\{\psi_j\}$ on Γ subordinate to the cover $\{U_j\}$. Thus supp $(\psi_j) \subset U_j$ and $\sum_{j=1}^{k} \psi_j \equiv 1$ on Γ.

Now Γ is a $(n-1)$-dimensional manifold and can be provided with the $(n-1)$-dimensional surface measure induced on it from $\mathbb{R}^n$. Thus it is simple to define the space $L^2(\Gamma)$, viz. it is the space of square-integrable functions on Γ. Now given $u \in L^2(\Gamma)$ we write it as follows:

$$u = \sum_{j=1}^{k} \psi_j u.$$

Then the function $\psi_j u$ has support in U_j. Now consider the function

$$v_j(y', 0) = (\psi_j u)(T_j(y', 0)), \quad (y', 0) \in Q_0.$$

where $y \in \mathbb{R}^n$ is written as (y', y_n). Since v_j has compact support inside Q_0, we can extend it outside Q_0 by zero and thus consider v_j as defined on $\mathbb{R}^{n-1}$. It is easy to see that $u \to v_j$ is a continuous linear map of $L^2(\Gamma)$ into $L^2(\mathbb{R}^{n-1})$ and maps smooth functions to smooth functions. We can now define for $s \in \mathbb{R}$,

$$H^s(\Gamma) = \{u \mid v_j \in H^s(\mathbb{R}^{n-1}),\ 1 \leqslant j \leqslant k\}. \tag{2.6.16}$$

It can be checked that this definition is independent of the choice of the local chart $(U_j, T_j)_{j=1}^{k}$. It can also be proved that if $s < 0$ then

$$H^s(\Gamma) = (H^{-s}(\Gamma))' \tag{2.6.17}$$

and the space $H^s(\Gamma)$, $s = \theta s_1 + (1-\theta)s_2$, $\theta \in (0, 1)$ interpolates the spaces $H^{s_1}(\Gamma)$ and $H^{s_2}(\Gamma)$.

It is in these spaces that we will define the concept of boundary values of functions in $H^m(\Omega)$ in the next section and they are called the **trace spaces.**

2.7 TRACE THEORY

In this section we will assign a meaning to expressions like $u|_\Gamma$, $\dfrac{\partial u}{\partial \nu}$ (i.e. the normal derivative on Γ) when $u \in H^m(\Omega)$, Ω a bounded open set in $\mathbb{R}^n$ $(n \geqslant 2)$ with boundary Γ. As already remarked (cf. Section 2.1) it is not meaningful to talk of values of u on a set of measure zero when $u \in H^m(\Omega)$. Thus we are required to generalize the notion of boundary values to such functions.

We will start with the case $\Omega = \mathbb{R}^n_+$ (which, of course, is not bounded!) and completely study the situation here. The case of a smooth open set Ω will then follow by the use of local charts and a corresponding partition of unity.

The theory we will now develop can also be done for other values of p but we will restrict our attention to the case $p = 2$.

Theorem 2.7.1 Let $\Omega = \mathbb{R}^n_+$. Then there exists a continuous linear map $\gamma_0 : H^1(\mathbb{R}^n_+) \to L^2(\mathbb{R}^{n-1})$ which is such that if v is continuous on $\overline{\mathbb{R}^n_+}$ then

$$\gamma_0(v) = v|_{\mathbb{R}^{n-1}}. \tag{2.7.1}$$

Proof Let $v \in \mathcal{D}(\mathbb{R}^n)$. We denote $x \in \mathbb{R}^n$ by $x = (x', x_n)$ with $x' \in \mathbb{R}^{n-1}$. Now

$$\begin{aligned} |v(x', 0)|^2 &= -\int_0^\infty \frac{\partial}{\partial x_n} (|v(x', x_n)|^2)\, dx_n \\ &= -2 \int_0^\infty v(x', x_n) \frac{\partial v}{\partial x_n} (x', x_n)\, dx_n \\ &\leqslant \int_0^\infty \left[(v(x', x_n))^2 + \left(\frac{\partial v}{\partial x_n} (x', x_n) \right)^2 \right] dx_n. \end{aligned}$$

Integrating both sides of this relation with respect to x', we get

$$\left| v(x', 0) \right|^2_{0,\, \mathbb{R}^{n-1}} \leqslant \left\| v \right\|^2_{1,\, \mathbb{R}^n_+}.$$

Thus the map $v \to v|_{\mathbb{R}^{n-1}}$ is continuous on $\mathcal{D}(\mathbb{R}^n)$ with the $H^1(\mathbb{R}^n)$ topology. But the restriction of $\mathcal{D}(\mathbb{R}^n)$ functions to $\mathbb{R}^n_+$ are dense in $H^1(\mathbb{R}^n_+)$ by corollary to Theorem 2.3.1. Thus there exists a unique continuous extension to $H^1(\mathbb{R}^n_+)$ of this map. ∎

We now show that the range of the map γ_0, which is called the **trace map** (of order 0) is not the whole of $L^2(\mathbb{R}^{n-1})$. More precisely we have the following result.

Theorem 2.7.2 The range of the map γ_0 is the space $H^{1/2}(\mathbb{R}^{n-1})$.

Proof: *Step 1.* Let $w(x') = v(x', 0)$. We first relate the Fourier transform (in $\mathbb{R}^{n-1}$) of w to that of v(in $\mathbb{R}^n$). We denote by $\tilde{w}$ the Fourier transform of w in $\mathbb{R}^{n-1}$. By the Fourier inversion formula (1.10.2), if $v \in \mathcal{D}(\mathbb{R}^n)$,

$$\begin{aligned} v(x', 0) &= \int_{\mathbb{R}^n} \exp (2\pi i x' \cdot \xi') \hat{v}(\xi)\, d\xi,\ \xi = \xi', \xi_n \\ &= \int_{\mathbb{R}^{n-1}} \exp (2\pi i x' \cdot \xi') \left(\int_{-\infty}^\infty \hat{v}(\xi)\, d\xi_n \right) d\xi'. \end{aligned}$$

Now applying the same formula in $\mathbb{R}^{n-1}$, we get

$$v(x', 0) = w(x') = \int_{\mathbb{R}^{n-1}} \exp (2\pi i x' \cdot \xi') \tilde{w}(\xi')\, d\xi'.$$

By the uniqueness of this formula ($w, \tilde{w} \in \mathcal{S}(\mathbb{R}^{n-1})$ if $v \in \mathcal{D}(\mathbb{R}^n)$) we deduce that

$$\tilde{w}(\xi') = \int_{-\infty}^\infty \hat{v}(\xi)\, d\xi_n. \tag{2.7.2}$$

Step 2 To show that $v(x', 0) = w(x') \in H^{1/2}(\mathbb{R}^{n-1})$ we need to show that $(1 + |\xi'|^2)^{1/2}|\widetilde{w}(\xi')|^2$ is integrable. But

$$\begin{aligned}
&\int_{\mathbb{R}^{n-1}} (1 + |\xi'|^2)^{1/2}|\widetilde{w}(\xi')|^2 \, d\xi' \\
&= \int_{\mathbb{R}^{n-1}} (1 + |\xi'|^2)^{1/2}\left|\int_{-\infty}^{\infty} \hat{v}(\xi) \, d\xi_n\right|^2 d\xi' \\
&= \int_{\mathbb{R}^{n-1}} (1 + |\xi'|^2)^{1/2}\left|\int_{-\infty}^{\infty} \hat{v}(\xi)(1 + |\xi|^2)^{-1/2}(1 + |\xi|^2)^{1/2} \, d\xi_n\right|^2 d\xi' \\
&\leqslant \int_{\mathbb{R}^{n-1}} (1 + |\xi'|^2)^{1/2}\left[\int_{-\infty}^{\infty} (1 + |\xi|^2)|\hat{v}(\xi)|^2 \, d\xi_n \int_{-\infty}^{\infty} (1 + |\xi|^2)^{-1} \, d\xi_n\right] d\xi' \\
&= \pi \int_{\mathbb{R}^n} (1 + |\xi|^2)|\hat{v}(\xi)|^2 \, d\xi = \pi\|v\|^2_{H^1(\mathbb{R}^n)}
\end{aligned}$$

since

$$\int_{-\infty}^{\infty} (1 + |\xi|^2)^{-1} \, d\xi_n = \int_{-\infty}^{\infty} \frac{d\xi_n}{1 + |\xi'|^2 + \xi_n^2} = (1 + |\xi'|^2)^{-1/2}\pi.$$

Thus if $v \in \mathscr{D}(\mathbb{R}^n)$, $v(x', 0) \in H^{1/2}(\mathbb{R}^{n-1})$ and by density the result follows for $v \in H^1(\mathbb{R}^n_+)$.

Step 3 We now show that γ_0 is onto $H^{1/2}(\mathbb{R}^{n-1})$. Let $h(x') \in H^{1/2}(\mathbb{R}^{n-1})$. Let $\widetilde{h}(\xi')$ be its Fourier transform. We define $u(x', x_n)$ by

$$\widetilde{u}(\xi', x_n) = \exp\left[-(1 + |\xi'|)x_n\right]\widetilde{h}(\xi'). \tag{2.7.3}$$

We must first show that u then belongs to $H^1(\mathbb{R}^n_+)$. To see this extend u by zero outside $\overline{\mathbb{R}^n_+}$. Now

$$\begin{aligned}
\hat{u}(\xi) &= \int_{\mathbb{R}^n} \exp(-2\pi i x \cdot \xi)u(x) \, dx \\
&= \int_0^{\infty}\int_{\mathbb{R}^{n-1}} \exp(-2\pi i x' \cdot \xi')u(x', x_n) \exp(-2\pi i x_n\xi_n) \, dx' \, dx_n \\
&= \int_0^{\infty} \exp(-2\pi i x_n\xi_n)\widetilde{u}(\xi', x_n) \, dx_n \\
&= \widetilde{h}(\xi') \int_0^{\infty} \exp(-(1 + |\xi'| + 2\pi i\xi_n)x_n) \, dx_n \\
&= \frac{\widetilde{h}(\xi')}{1 + |\xi'| + 2\pi i\xi_n}.
\end{aligned}$$

Now,

$$\begin{aligned}
\int_{\mathbb{R}^n} (1 + |\xi'|^2)|\hat{u}(\xi)|^2 \, d\xi &= \int_{\mathbb{R}^n} \frac{(1 + |\xi'|)^2|\widetilde{h}(\xi')|^2}{(1 + |\xi'|)^2 + 4\pi^2\xi_n^2} \, d\xi \\
&\leqslant \int_{\mathbb{R}^n} \frac{(1 + |\xi'|^2)|\widetilde{h}(\xi')|^2}{1 + |\xi'|^2 + \xi_n^2} \, d\xi
\end{aligned}$$

$$= \pi \int_{\mathbb{R}^{n-1}} (1 + |\xi'|^2)^{1/2} |\widetilde{h}(\xi')|^2 \, d\xi' < +\infty$$

since $h \in H^{1/2}(\mathbb{R}^{n-1})$. This proves that u (extended by zero) and $\dfrac{\partial u}{\partial x_i}$, $1 \leqslant i \leqslant n-1$ are in $L^2(\mathbb{R}^n)$ and so u, $\dfrac{\partial u}{\partial x_i}$, $1 \leqslant i \leqslant n-1$ are all in $L^2(\mathbb{R}^n_+)$. For the case $i = n$, notice that by differentiating under the integral sign, we have

$$\left(\frac{\partial u}{\partial x_n}\right)^{\sim} (\xi', x_n) = \frac{\partial \widetilde{u}}{\partial x_n} (\xi', x_n) = -(1 + |\xi'|) \widetilde{u}(\xi', x_n).$$

Extend $\dfrac{\partial u}{\partial x_n}$ by zero outside $\mathbb{R}^n_+$. Then as before we get

$$\left(\frac{\partial u}{\partial x_n}\right)^{\wedge} (\xi) = \frac{-(1 + |\xi'|)\widetilde{h}(\xi')}{1 + |\xi'| + 2\pi_i \xi_n}.$$

Thus

$$\int_{\mathbb{R}^n} \left| \left(\frac{\partial u}{\partial x_n}\right)^{\wedge} (\xi) \right|^2 d\xi \leqslant 2 \int_{\mathbb{R}^n} \frac{(1 + |\xi'|^2)|\widetilde{h}(\xi')|^2}{1 + |\xi'|^2 + \xi_n^2} \, d\xi$$

$$= 2\pi \int_{\mathbb{R}^{n-1}} (1 + |\xi'|^2)^{1/2} |\widetilde{h}(\xi')|^2 \, d\xi' < \infty.$$

Thus $\dfrac{\partial u}{\partial x_n}$ (extended by zero) is in $L^2(\mathbb{R}^n)$ and so $\dfrac{\partial u}{\partial x_n} \in L^2(\mathbb{R}^n_+)$ and hence $u \in H^1(\mathbb{R}^n_+)$. Now (by the Fourier inversion formula)

$$\widetilde{u}(\xi', 0) = \widetilde{h}(\xi')$$

implies that $u(x', 0) = h(x')$ and so $\gamma_0(u) = h$. This completes the proof. ∎

Remark 2.7.1 With very minor modifications in the computations above we can easily prove that γ_0 maps $H^m(\mathbb{R}^n_+)$ onto $H^{m-1/2}(\mathbb{R}^{n-1})$. In the same way, if $u \in H^2(\mathbb{R}^n_+)$ we can imitate the proof of Theorem 2.7.1 to show that $\dfrac{\partial u}{\partial x_n}(x', 0)$ is in $L^2(\mathbb{R}^{n-1})$ and again that it in fact is in $H^{1/2}(\mathbb{R}^{n-1})$. We can then extend $-\dfrac{\partial u}{\partial x_n}(x', 0)$ to a continuous linear map $\gamma_1 : H^2(\mathbb{R}^n_+) \to L^2(\mathbb{R}^{n-1})$ whose range is $H^{1/2}(\mathbb{R}^{n-1})$. More generally we have a series of continuous linear maps $\{\gamma_j\}$ into $L^2(\mathbb{R}^{n-1})$ such that the map $\gamma = (\gamma_0, \gamma_1, \ldots, \gamma_{m-1})$ maps $H^m(\mathbb{R}^n_+)$ into $(L^2(\mathbb{R}^{n-1}))^m$ and the range in the space

$$\prod_{j=0}^{m-1} H^{m-j-1/2}(\mathbb{R}^{n-1}). \quad ∎$$

We now turn to the kernel of the map γ_0. We have already seen (cf. Theorem 2.2.5) that if u is continuous on $\bar{\Omega}$ and u vanishes on Γ, then $u \in H^1_0(\Omega)$. We now will show that $H^1_0(\mathbb{R}^n_+)$ is precisely the kernel of the map γ_0. Indeed as $\mathscr{D}(\mathbb{R}^n_+) \subset \ker(\gamma_0)$ which is a closed subspace of $H^1(\mathbb{R}^n_+)$

we already have $H_0^1(\mathbb{R}_+^n) \subset \ker(\gamma_0)$. The proof of the reverse inclusion is more delicate and involves a few steps.

Lemma 2.7.1 (Green's formula). Let $u, v \in H^1(\mathbb{R}_+^n)$. Then

$$\int_{\mathbb{R}_+^n} u\frac{\partial v}{\partial x_i} = -\int_{\mathbb{R}_+^n} \frac{\partial u}{\partial x_i} v \quad \text{if } 1 \leqslant i \leqslant (n-1) \tag{2.7.4}$$

$$\int_{\mathbb{R}_+^n} u\frac{\partial v}{\partial x_n} = -\int_{\mathbb{R}_+^n} \frac{\partial u}{\partial x_n} v - \int_{\mathbb{R}^{n-1}} \gamma_0(u)\gamma_0(v). \tag{2.7.5}$$

Proof If $u, v \in \mathscr{D}(\mathbb{R}^n)$, then the relations (2.7.4) and (2.7.5) follow by integration by parts. The general case follows by the density of the restrictions of functions of $\mathscr{D}(\mathbb{R}^n)$ in $H^1(\mathbb{R}_+^n)$ and the continuity of the map $\gamma_0 : H^1(\mathbb{R}_+^n) \to L^2(\mathbb{R}^{n-1})$. ∎

Corollary If $u, v \in H^1(\mathbb{R}_+^n)$ and at least one of them is in $\ker(\gamma_0)$ then (2.7.4) holds for all $1 \leqslant i \leqslant n$. ∎

Lemma 2.7.2 Let $v \in \ker(\gamma_0)$. Then its extension by zero outside $\mathbb{R}_+^n$, denoted $\tilde{v}$, is in $H^1(\mathbb{R}^n)$, and

$$\frac{\partial \tilde{v}}{\partial x_i} = \left(\frac{\partial v}{\partial x_i}\right)^{\sim}, \quad 1 \leqslant i \leqslant n. \tag{2.7.6}$$

Proof Let $\phi \in \mathscr{D}(\mathbb{R}^n)$. Then for $1 \leqslant i \leqslant n$,

$$\int_{\mathbb{R}^n} \tilde{v}\frac{\partial \phi}{\partial x_i} = \int_{\mathbb{R}_+^n} v\frac{\partial \phi}{\partial x_i} = -\int_{\mathbb{R}_+^n} \frac{\partial v}{\partial x_i}\phi = -\int_{\mathbb{R}^n}\left(\frac{\partial v}{\partial x_i}\right)^{\sim}\phi$$

by the preceding corollary. ∎

Let $h > 0$ and consider $\bar{h} = he_n \in \mathbb{R}^n$ where e_n is the unit vector

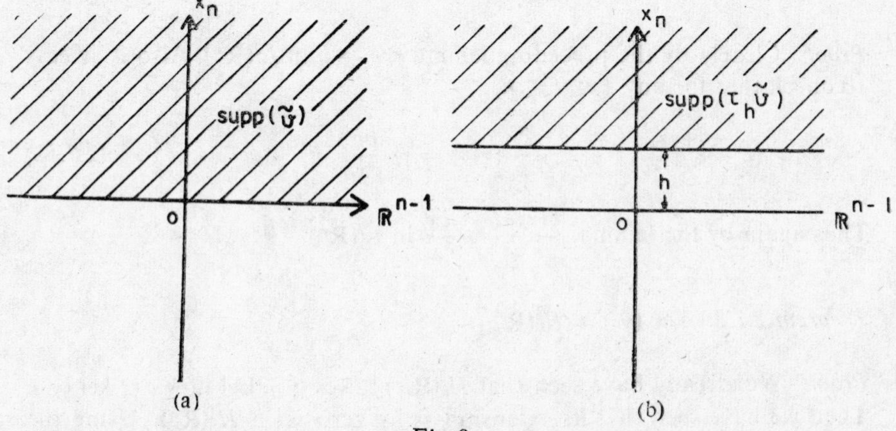

Fig. 9

$(0, 0, \ldots, 0, 1)$. Consider the function $\tau_{\bar{h}}\tilde{v}$, where $\tilde{v}$ is the extension by zero outside $\mathbb{R}^n_+$ of $v \in \ker(\gamma_0)$. Then $\tau_{\bar{h}}\tilde{v}$ vanishes for all $x \in \mathbb{R}^n$ such that $x_n < h$. (Recall that $\tau_{\bar{h}}\tilde{v}(x) = \tilde{v}(x - \bar{h})$.)

Lemma 2.7.3 Let $1 \leqslant p < \infty$ and $\bar{h} \in \mathbb{R}^n$. Then if $f \in L^p(\mathbb{R}^n)$

$$\lim_{\bar{h}\to 0} |\tau_{\bar{h}} f - f|_{0, p, \mathbb{R}^n} = 0. \tag{2.7.7}$$

Proof By the translation invariance of the Lebesgue measure, $\tau_{\bar{h}} f \in L^p(\mathbb{R}^n)$ as well. Let $\epsilon > 0$ be given and choose $\phi \in \mathcal{D}(\mathbb{R}^n)$ such that

$$|f - \phi|_{0, p, \mathbb{R}^n} < \epsilon/3. \tag{2.7.8}$$

Let $a > 0$ such that $\operatorname{supp}(\phi) \subset [-a, a]^n$. Since ϕ is uniformly continuous, there exists $\delta > 0$ such that $\delta < 1$ and if $|\bar{h}| < \delta$, we have

$$|\phi(x - \bar{h}) - \phi(x)| < \frac{\epsilon}{3}(2(a + 1))^{-(n/p)}.$$

Then

$$\int_{\mathbb{R}^n} |\phi(x - \bar{h}) - \phi(x)|^p dx = \int_{[-(a+1), (a+1)]^n} |\phi(x - \bar{h}) - \phi(x)|^p dx < \left(\frac{\epsilon}{3}\right)^p.$$

Thus for $|\bar{h}| < \delta$,

$$|\tau_{\bar{h}}\phi - \phi|_{0, p, \mathbb{R}^n} < \epsilon/3. \tag{2.7.9}$$

Finally, again by the translation invariance of the Lebesgue measure, we have

$$|\tau_{\bar{h}} f - \tau_{\bar{h}}\phi|_{0, p, \mathbb{R}^n} = |f - \phi|_{0, p, \mathbb{R}^n} < \epsilon/3. \tag{2.7.10}$$

The result now follows on combining (2.7.8), (2.7.9) and (2.7.10) by the triangle inequality. ∎

Corollary If $v \in H^1(\mathbb{R}^n)$ then

$$\lim_{h\to 0} \|\tau_{\bar{h}} v - v\|_{1, \mathbb{R}^n} = 0. \tag{2.7.11}$$

Proof Clearly by the preceding lemma $\tau_{\bar{h}} v \to v$ in $L^2(\mathbb{R}^n)$. Also it is easy to check that for any $1 \leqslant i \leqslant n$,

$$\frac{\partial}{\partial x_i}(\tau_{\bar{h}} v) = \tau_{\bar{h}} \frac{\partial v}{\partial x_i}. \tag{2.7.12}$$

Thus again by the lemma, $\dfrac{\partial(\tau_{\bar{h}} v)}{\partial x_i} \to \dfrac{\partial v}{\partial x_i}$ in $L^2(\mathbb{R}^n)$. ∎

Theorem 2.7.3 $\ker(\gamma_0) = H^1_0(\mathbb{R}^n_+)$.

Proof We already have seen that $H^1_0(\mathbb{R}^n_+) \subset \ker(\gamma_0)$. Let now $v \in \ker(\gamma_0)$. Then we have seen that its extension $\tilde{v}$ by zero is in $H^1(\mathbb{R}^n)$. Using the

cut-off functions $\{\zeta_k\}$ (cf. Theorem 2.1.3) we have that $\zeta_k\widetilde{v} \to \widetilde{v}$ as $k \to \infty$ in $H^1(\mathbb{R}^n)$. The functions $\zeta_k\widetilde{v}$ have compact support in $\mathbb{R}^n$ and vanish for $x_n < 0$. Now fix such a k so that

$$\|\widetilde{v} - \zeta_k\widetilde{v}\|_{1,\mathbb{R}^n} < \eta \tag{2.7.13}$$

where $\eta > 0$ is a given positive quantity. Again we can choose h small enough so that if $\bar{h} = he_n$, then

$$\|\tau_{\bar{h}}(\zeta_k\widetilde{v}) - \zeta_k\widetilde{v}\|_{1,\mathbb{R}^n} < \eta. \tag{2.7.14}$$

Now $\tau_{\bar{h}}(\zeta_k\widetilde{v})$ has compact support in $\mathbb{R}^n_+$ and vanishes for all $x \in \mathbb{R}^n$ with $x_n < h$. Let $\{\rho_\epsilon\}$ be the family of mollifers. If $\epsilon > 0$ is chosen small enough then $\rho_\epsilon * \tau_{\bar{h}}(\zeta_k\widetilde{v})$ will have support contained in the set

$$B(0;\epsilon) + K \cap \{x \mid x_n \geqslant h > 0\}$$

where $K = \operatorname{supp}(\tau_{\bar{h}}(\zeta_k\widetilde{v}))$ is compact. Thus

$$\rho_\epsilon * \tau_{\bar{h}}(\zeta_k\widetilde{v}) \in \mathscr{D}(\mathbb{R}^n_+)$$

and we know that as $\epsilon \downarrow 0$, $\rho_\epsilon * \tau_{\bar{h}}(\zeta_k\widetilde{v}) \to \tau_{\bar{h}}(\zeta_k\widetilde{v})$. Thus we can choose ϵ small enough such that

$$\|\rho_\epsilon * \tau_{\bar{h}}(\zeta_k\widetilde{v}) - \tau_{\bar{h}}(\zeta_k\widetilde{v})\|_{1,\mathbb{R}^n} < \eta. \tag{2.7.15}$$

Thus we have found a function $\phi_\eta \in \mathscr{D}(\mathbb{R}^n_+)$,

$$\phi_\eta = \rho_\epsilon * \tau_{\bar{h}}(\zeta_k\widetilde{v})$$

such that

$$\|\phi_\eta - v\|_{1,\mathbb{R}^n_+} \leqslant \|\phi_\eta - \widetilde{v}\|_{1,\mathbb{R}^n} < 3\eta.$$

Thus, as η is arbitrary, it follows that

$$\ker(\gamma_0) \subset \overline{\mathscr{D}(\mathbb{R}^n_+)} = H^1_0(\mathbb{R}^n_+).$$

This completes the proof. ∎

Remark 2.7.2 Similarly it can be proved that if $\gamma = (\gamma_0, \gamma_1, \ldots, \gamma_{m-1})$, then the kernel of γ in $H^m(\mathbb{R}^n_+)$ is precisely the set $H^m_0(\mathbb{R}^n_+)$. ∎

Let us now turn to the case of a bounded open set Ω of class C^1. Let $\left\{U_j, T_j\right\}_{j=1}^k$ be an associated local chart for the boundary Γ and let $\left\{\psi_j\right\}_{j=1}^k$ be a partition of unity subordinate to the cover $\{U_j\}$ of Γ. If $u \in H^1(\Omega)$, then $(\psi_j u|_{U_j \cap \Omega}) \circ T_j \in H^1(\mathbb{R}^n_+)$ and so we can define its trace as an element of $H^{1/2}(\mathbb{R}^{n-1})$. Coming back by T_j^{-1} we can define the trace on $U_j \cap \Gamma$. Piecing these together we get the trace $\gamma_0 u$ in $L^2(\Gamma)$ and the image (by definition of the spaces) will be precisely $H^{1/2}(\Gamma)$. Similarly if the boundary is smoother we can define the higher order traces γ_j. In particular we have the following result.

Theorem 2.7.4 **(Trace Theorem)** Let $\Omega \subset \mathbb{R}^n$ be a bounded open set of class

C^{m+1} with boundary Γ. Then there exists a trace map $\gamma = (\gamma_0, \gamma_1, \ldots, \gamma_{m-1})$ from $H^m(\Omega)$ into $(L^2(\Omega))^m$ such that

(i) If $v \in C^\infty(\bar{\Omega})$, then $\gamma_0(v) = v|_\Gamma$, $\gamma_1(v) = \frac{\partial v}{\partial \nu}|_\Gamma, \ldots,$ and $\gamma_{m-1}(v) = \frac{\partial^{m-1}}{\partial \nu^{m-1}}(v)|_\Gamma$, where ν is the unit exterior normal to the boundary Γ.

(ii) The range of γ is the space

$$\prod_{j=0}^{m-1} H^{m-j-1/2}(\Gamma).$$

(iii) The kernel of γ is $H_0^m(\Omega)$. ∎

The trace theorem above helps us to obtain Green's theorem for functions in $H^1(\Omega)$, Ω of class C^1. If $\nu(x)$ denotes the unit outernormal vector on the boundary Γ (which is defined uniquely a.e. on Γ), we denote its components along the coordinate axes by $\nu_i(x)$. Thus we write generically,

$$\nu = (\nu_1, \ldots, \nu_n).$$

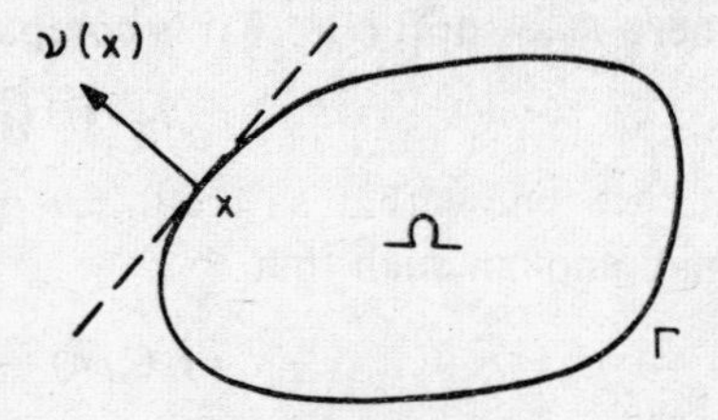

Fig. 10

For example if $\Omega = B(0; 1)$ then $\nu(x) = x$ for all $|x| = 1$. Thus $\nu_i(x) = x_i$ in this case. If $\Omega = B(0; R)$, then $\nu(x) = x/R$. If Ω has a part of its boundary, say, $x_n = 0$, then the unit outer normal is $\pm e_n$ depending on the side on which Ω lies.

Theorem 2.7.5 **(Green's Theorem,** or, **Green's Formula).** Let Ω be a bounded open set of $\mathbb{R}^n$ set of class C^1 lying on the same side of its boundary Γ. Let $u, v \in H^1(\Omega)$. Then for $1 \leqslant i \leqslant n$,

$$\int_\Omega u \frac{\partial v}{\partial x_i} = -\int_\Omega \frac{\partial u}{\partial x_i} v + \int_\Gamma (\gamma_0 u)(\gamma_0 v)\nu_i. \tag{2.7.16}$$

Proof By corollary to Theorem 2.3.2, $C^\infty(\bar{\Omega})$ is dense in $H^1(\Omega)$. If u_m, $v_m \in C^\infty(\bar{\Omega})$ then we have by the classical Green's Theorem

$$\int_\Omega u_m \frac{\partial v_m}{\partial x_i} = -\int_\Omega \frac{\partial u_m}{\partial x_i} v_m + \int_\Gamma u_m v_m \nu_i \tag{2.7.17}$$

and choosing $u_m \to u$, $v_m \to v$ in $H^1(\Omega)$ we deduce (2.7.16) by the continuity of the trace map γ_0. ∎

In future we will not write $\gamma_0 v$, but only v on Γ and understand it as the trace of v on Γ. Similarly if $\gamma_1 v$ is defined we will write it as $\frac{\partial v}{\partial \nu}$.

We list below a few simple consequences of Green's theorem which we will use often in the following chapters.

Setting $u \equiv 1$ and $v = v_i \in H^1(\Omega)$, we get from (2.7.16)

$$\int_\Omega \frac{\partial v_i}{\partial x_i} = \int_\Omega v_i \nu_i \tag{2.7.18}$$

and if $\bar{v} = (v_i) \in (H^1(\Omega))^n$, we get on summing with respect to i,

$$\int_\Omega \operatorname{div} \bar{v} = \int_\Gamma \bar{v} \cdot \nu \tag{2.7.19}$$

which is the Gauss Divergence Theorem. If we have $u \in H^2(\Omega)$ and use $\frac{\partial u}{\partial x_i}$ in place of u in (2.7.16), we get

$$\int_\Omega \frac{\partial u}{\partial x_i} \frac{\partial v}{\partial x_i} = -\int_\Omega \frac{\partial^2 u}{\partial x_i^2} v + \int_\Gamma \frac{\partial u}{\partial x_i} v \nu_i$$

If u were smooth then $\sum_{i=1}^{n} \frac{\partial u}{\partial x_i} \nu_i = \frac{\partial u}{\partial \nu}$. Thus by continuity of the trace γ_1, we get, for $u \in H^2(\Omega)$, $v \in H^1(\Omega)$,

$$\int_\Omega \nabla u \cdot \nabla v = -\int_\Omega (\Delta u) v + \int_\Gamma v \frac{\partial u}{\partial \nu}. \tag{2.7.20}$$

COMMENTS

1. *General Remarks* The spaces $W^{m,p}(\Omega)$ introduced in this chapter were studied by Sobolev [1] with several related spaces being studied by various authors, in particular Morrey [1] and Deny and Lions [1]. Various names like "Beppo-Levi Spaces" were given to these spaces. Also these spaces were introduced from different points of view like the completion of $C^\infty(\Omega)$ functions whose $\|\cdot\|_{m,p}$-norms were finite, or as restrictions of $W^{m,p}(\mathbb{R}^n)$ functions to Ω. It took the *Meyers-Serrin theorem* (cf. discussion following Theorem 2.2.2) to show that the former definition coincides with the one given in this chapter for any open set Ω. The latter, however, coincides with our definition only with additional smoothness hypotheses as shown in this chapter. In the same way fractional order spaces have been introduced from several points of view like interpolation theory (cf. Lions and Magenes [1] or Adams [1]) or by using first differences (as we have done in section 2.6) and which are sometimes also called the *Besov spaces*, or by using Second differences, leading to the *Nikolskii spaces*. When $p = 2$ the Besov spaces coincide with the interpolation spaces for all s and for other values of p only for s not an integer. However as we are interested mainly in the case $p = 2$ for the Hilbert theory of elliptic partial differential equations, this suffices for us.

Numerous generalizations of the Sobolev spaces have been studied in recent times. On one hand we have the *Orlicz-Sobolev* spaces based on the *Orlic z spaces* which are generalizations of the L^p-spaces (cf. Adams [1]). On the other hand, important from the point of view of non-linear partial differential equations and degenerate elliptic boundary value problems are the *weighted Sobolev spaces*. These are Sobolev type spaces

wherein the norm has a (non-negative) weight function in it. For instance if $\omega \geqslant 0$ is a weight function the corresponding $L^2(\Omega; \omega)$ space will be given by

$$L^2(\Omega; \omega) = \left\{ u \,\Big|\, \int_\Omega \omega |u|^2 < \infty \right\},$$

with the obvious associated norm. A vast literature exists on this subject. In particular see Baouendi [1], Baouendi and Goulaouic [1] or Geymonat and Grisvard [1].

2. *Types of Domains* Following Brezis [1], we have chosen to prove most of the results for domains in $\mathbb{R}^n$ which are of class C^1 (or more). However most of these results are available for a variety of analytic or geometric conditions like piecewise C^1, or *Lipschitz domains* (example: polygons in $\mathbb{R}^2$) or domains with *segment property*, or domains which satisfy the *cone condition*. For a description of these conditions and proofs of the corresponding results, see Adams [1].

3. *Important Inequalities* We have proved some of the important inequalities available in Sobolev spaces like those of Korn (2.6.13), Poincaré (2.3.22), Poincaré-Wirtinger (2.5.17–2.5.18) and Sobolev (2.4.5). There exist several other important inequalities like Hardy's Inequality and the family of interpolation inequalities known as the Gagliardo-Nirenberg Inequalities. The latter class of inequalities are especially useful in the study of nonlinear problems. For a description of some of these see Adams [**1**], Brezis [1], Friedman [1], Lions and Magenes [1] and Nirenberg [1].

4. *Sobolev's Inequality* If Ω is any domain in $\mathbb{R}^n$, the Sobolev's inequality (2.4.5) on Ω can be written as follows: if $1 \leqslant p < n$,

$$|u|_{1,p,\Omega} \geqslant C|u|_{0,p^*,\Omega},\ u \in W_0^{1,p}(\Omega),$$

where $C = C(n, p, \Omega)$. If we denote by $S(n, p, \Omega)$ the best possible constant (i.e. the largest possible) then the following are salient features of this constant:

(i) *$S(n, p, \Omega)$ does not depend on Ω but only on n and p.* This is a consequence of the fact that the ratio

$$|u|_{1,p,\Omega}/|u|_{0,p^*,\Omega}$$

is invariant under dilations: If $u \in W^{1,p}(\mathbb{R}^n)$ and $\lambda > 0$, define

$$u_\lambda(x) = u(\lambda x).$$

Then a simple calculation shows that

$$\frac{|u_\lambda|_{1,p,\mathbb{R}^n}}{|u_\lambda|_{0,p^*,\mathbb{R}^n}} = \lambda^{-\{1+(n/p^*)-(n/p)\}} \frac{|u|_{1,p,\mathbb{R}^n}}{|u|_{0,p^*,\mathbb{R}^n}}.$$

$$= |u|_{1,p,\mathbb{R}^n}/|u|_{0,p^*,\mathbb{R}^n}$$

since $1 + \dfrac{n}{p^*} - \dfrac{n}{p} = 0$.

Now assume, for instance, that $p = 2$ and $\Omega_m = B(0; m)$. Set

$$S_m^2 = \inf_{\substack{u \in H_0^1(\Omega_m) \\ |u|_{0,2^*,\Omega_m} = 1}} \{|u|_{1,\Omega_m}^2\}.$$

Then S_m is the best Sobolev constant for Ω_m. Since the extension by zero outside a domain preserves the H^1 nature of a H_0^1 function, it follows from the above characterization of S_m that if $m_1 > m_2$ (i.e. $\Omega_{m_2} \subset \Omega_{m_1}$) then $S_{m_1} < S_{m_2}$. However if $u \in H_0^1(\Omega_{m_1})$ then by choosing λ suitably we have $u_\lambda \in H_0^1(\Omega_{m_2})$ and vice-versa. Thus we have indeed that $S_{m_1} = S_{m_2}$ or S_m is independent of m! Thus the best constant is got just by considering

$$\inf_{\substack{u \in H^1(\mathbb{R}^n) \\ |u|_{0,2^*,\mathbb{R}^n} = 1}} \{|u|_{1,\mathbb{R}^n}^2\}.$$

Aubin [1] and Talenti [1] have shown for instance that the above infimum is achieved only for the functions $U(x)$, $U_\epsilon(x - x_0)$, $x_0 \in \mathbb{R}^n$, $\epsilon > 0$ where

$$U(x) = C(1 + |x|^2)^{-(n-2)/2}$$

$$U_\epsilon(x) = C_\epsilon(\epsilon + |x|^2)^{-(n-2)/2},$$

C and C_ϵ being (positive) normalization constants.

(ii) The infimum cannot be achieved except when $\Omega = \mathbb{R}^n$ for if $u \in H_0^1(\Omega)$, $\Omega \subsetneqq \mathbb{R}^n$ achieves the infimum on Ω then its extension by zero achieves the above infimum on $\mathbb{R}^n$. Then it must be of the form given above which never vanishes on $\mathbb{R}^n$. This fact was used by us to show the non-compactness of the imbedding $H_0^1(\Omega) \to L^{2^*}(\Omega)$ in Remark 2.5.2.

5. *Compact Imbeddings* We saw in Example 2.5.1 that for unbounded domains, the compactness of the Sobolev imbeddings are not valid. Indeed, a necessary condition for compactness is that Ω has to be *quasi-bounded*, i.e.

$$\lim_{\substack{|x| \to \infty \\ x \in \Omega}} \text{dist}\,(x, \partial\Omega) = 0.$$

If Ω were not quasibounded (in the sense above), there exists an infinity of disjoint balls in Ω of the same positive radius. Then we can imitate the construction of Example 2.5.1 to show that compactness cannot hold. In addition to this condition we need sufficient conditions of geometric or analytic nature to prove the compactness of the Sobolev imbeddings (cf. Adams [1]).

On the other hand, we can prove compactness of the imbeddings when we restrict our attention to some subspaces even in $\mathbb{R}^n$. For instance, it is known that if $H_r^1(\mathbb{R}^n)$ is the set of functions in $H^1(\mathbb{R}^n)$ which are *radial*, i.e. $u(x) = u(|x|)$ then it is compactly embedded in $L^q(\mathbb{R}^n)$, $2 < q < 2^*$ (cf. Exercise 2.20).

Similarly if Ω is an annulus, i.e. there exist $a, b > 0$ such that

$\Omega = \{x \in \mathbb{R}^n \mid a < |x| < b\}$, then the embedding of $H^1_{0,r}(\Omega)$ of radial functions is compact in $L^{2^*}(\Omega)$.

6. *BV Spaces* We saw in Lemma 2.5.1 that if $1 \leqslant p \leqslant \infty$ and $u \in W^{1,p}(\Omega)$ then $|\tau_{-h}u - u|_{0,p,\Omega'} \leqslant C|h|$ for $\Omega' \subset\subset \Omega$. In fact it can be shown that (cf. Exercise 2.8 of this chapter) that if $1 \leqslant p \leqslant \infty$ then for $u \in L^p(\Omega)$ the following are equivalent:

(i) $|\tau_{-h}u - u|_{0,p,\Omega'} \leqslant C|h|$, $\Omega' \subset\subset \Omega$, $|h| < \text{dist}\,(\Omega', \mathbb{R}^n \setminus \Omega)$;

(ii) there exists $C > 0$ such that for every $\phi \in \mathcal{D}(\Omega)$

$$\int_\Omega u \frac{\partial \phi}{\partial x_i} \leqslant C|\phi|_{0,p',\Omega}$$

where p' is the conjugate exponent.

Now if $1 < p < \infty$, the above property (ii) implies, by the Riesz Representation Theorem, that $u \in W^{1,p}(\Omega)$. This is no longer valid for $p = 1$. Thus the functions which satisfy property (i) (or, equivalently, property (ii)) above form a larger class than $W^{1,1}(\Omega)$ and are called $BV(\Omega)$-space functions (cf. Remark 2.5.1). These spaces play a more important role than the $W^{1,1}(\Omega)$ spaces. They occur in the theory of *minimal surfaces* (cf. Giusti [1]), in the study of *plasticity* (cf. Temam and Strang [1]) and in the study of *shock waves*. For a detailed description of these spaces, see Giusti [1] or Maz'ja [1].

7. *Trace Theory* The theory of traces and trace spaces can also be developed for the case $p \neq 2$. For instance if $u \in W^{1,p}(\Omega)$ then the trace is defined with values in $L^p(\Gamma)$. In fact the range of the trace operator is $W^{1-(1/p),p}(\Gamma)$ and the kernel is $W^{1,p}_0(\Omega)$. (See the references cited in Lions and Magenes [1]).

8. *Bibliography* This chapter is merely an introduction to the theory of Sobolev spaces on which there exists an abundant literature. As could have been observed upto now, the book by Adams [1] contains a very exhaustive treatment of the subject. We have followed the treatment of Brezis [1] in some parts of this chapter. A more recent treatise on the subject is the book by Maz'ja [1]. Other references are Agmon [1], Friedman [1] and Lions and Magenes [1].

EXERCISES

2.1 Let $\Omega = B(0; 1) \subset \mathbb{R}^2$. Show that the function $f : \Omega \to \mathbb{R}$ defined by

$$f(x) = (1 + |\log |x||)^k$$

belongs to $H^1(\Omega)$ when $0 < k < 1/2$.

2.2 Let $\Omega \subset \mathbb{R}^n$ be an open set. Let $\overline{T} = \left(T_i\right)_{i=1}^n$ be a family of distri-

butions on Ω; define

$$\operatorname{div} \overline{T} = \sum_{i=1}^{n} \frac{\partial T_i}{\partial x_i}.$$

Consider the space

$$H(\operatorname{div}; \Omega) = \{\bar{u} = (u_i) \in (L^2(\Omega))^n \mid \operatorname{div} \bar{u} \in L^2(\Omega)\},$$

provided with the norm

$$\|\bar{u}\| = \left(\sum_{i=1}^{n} |u_i|_{0,\Omega}^2 + |\operatorname{div} \bar{u}|_{0,\Omega}^2 \right)^{1/2}.$$

Show that $H(\operatorname{div}; \Omega)$ is a Hilbert space.

2.3 Let $\Omega = (a, b) \subset \mathbb{R}$. Let $a = x_0 < x_1 < \ldots < x_n = b$ be a partition of Ω. Let $I_k = (x_k, x_{k+1})$, $0 \leqslant k \leqslant n-1$. Let $f : \Omega \to \mathbb{R}$ be such that $f|_{I_k} \in H^1(I_k)$ for each $0 \leqslant k \leqslant n-1$. Show that $f \in H^1(\Omega)$, if, and only if, $f \in C(\overline{\Omega})$.

2.4 Let Ω be an open set in $\mathbb{R}^n$ and let $u \in H_0^1(\Omega)$. (Then we know that $|u| \in H_0^1(\Omega)$). With the definition

$$|v|_{1,\Omega}^2 = \int_\Omega \sum_{i=1}^{n} \left(\frac{\partial v}{\partial x_i} \right)^2 = \int_\Omega \nabla v \cdot \nabla v, \quad v \in H_0^1(\Omega),$$

show that

$$|\,|u|\,|_{1,\Omega} = |u|_{1,\Omega}.$$

2.5 (Product Rule). Let $\Omega \subset \mathbb{R}^n$ be an open set and let $1 \leqslant p \leqslant \infty$. Let $u, v \in W^{1,p}(\Omega) \cap L^\infty(\Omega)$. Then show that $uv \in W^{1,p}(\Omega) \cap L^\infty(\Omega)$ and that for $1 \leqslant i \leqslant n$,

$$\frac{\partial}{\partial x_i}(uv) = \frac{\partial u}{\partial x_i} v + u \frac{\partial v}{\partial x_i}.$$

2.6 Let Ω be a bounded open set and $\rho \in C^1(\overline{\Omega})$. Show that the mapping $v \to \rho v$ defines a continuous linear operator of $H^1(\Omega)$ into itself. If $\rho > 0$ on $\overline{\Omega}$ show that this operator is an isomorphism of $H^1(\Omega)$ onto itself.

2.7 Let $1 < p \leqslant \infty$ and $u \in L^p(\Omega)$, $\Omega \subset \mathbb{R}^n$ an open set. Show that if there exists a constant $C > 0$ such that

$$\left| \int_\Omega u \frac{\partial \phi}{\partial x_i} \right| \leqslant C |\phi|_{0,p',\Omega}, \text{ for every } \phi \in \mathcal{D}(\Omega) \text{ and } 1 \leqslant i \leqslant n,$$

p' the conjugate exponent of p, then $u \in W^{1,p}(\Omega)$.

2.8 Let $1 \leqslant p \leqslant \infty$ and $u \in L^p(\Omega)$. Assume that there exists a constant $C > 0$ such that for every $\Omega' \subset\subset \Omega$ and every $h \in \mathbb{R}^n$ with $|h| < \operatorname{dist}(\Omega', \mathbb{R}^n \setminus \Omega)$,

$$|\tau_{-h} u - u|_{0,p,\Omega'} \leqslant C|h|.$$

Then show that there exists a constant $C > 0$ such that

$$\left| \int_\Omega u \frac{\partial \phi}{\partial x_i} \right| \leqslant C|\phi|_{0, p', \Omega}, \quad 1 \leqslant i \leqslant n,$$

for every $\phi \in \mathscr{D}(\Omega)$. (In particular when $1 < p \leqslant \infty$, $u \in W^{1,p}(\Omega)$).

2.9 Let Ω_1 and Ω_2 be open sets in $\mathbb{R}^n$ such that $H : \Omega_2 \to \Omega_1$ is a bijection, $x = H(y)$ which satisfies

$$H \in C^1(\Omega_2),\ H^{-1} \in C^1(\Omega_1),\ J(H) \in (L^\infty(\Omega_2))^{n^2},\ J(H^{-1}) \in (L^\infty(\Omega_1))^{n^2},$$

where

$$J(H) = \left(\frac{\partial H_i}{\partial y_j} \right)_{1 \leqslant i, j \leqslant n}, \quad H = (H_1, \ldots, H_n).$$

Show that if $u \in W^{1,p}(\Omega_1)$, then $u \circ H \in W^{1,p}(\Omega_2)$ and that

$$\frac{\partial}{\partial y_j}(u \circ H)(y) = \sum_{i=1}^n \frac{\partial u}{\partial x_i}(H(y)) \frac{\partial H_i}{\partial y_j}(y), \quad 1 \leqslant j \leqslant n.$$

2.10 Using the above exercise and Example 2.3.2, show that every convex quadilateral K in $\mathbb{R}^2$ admits an extension operator $P : W^{1,p}(K) \to W^{1,p}(\mathbb{R}^2)$. Deduce the same property for any triangle K in $\mathbb{R}^2$.

2.11 Let Ω be an open set of class C^1 which is bounded. Show that

$$\bigcap_{m=0}^{\infty} W^{m,p}(\Omega) = C^\infty(\bar{\Omega})$$

2.12 (a) Let B_i, $i = 1, 2, 3$ be Hilbert spaces with the continuous inclusions

$$B_1 \to B_2 \to B_3.$$

Assume further that B_1 is compactly imbedded in B_2. Show that given any $\epsilon > 0$ there exists $C_\epsilon > 0$ such that for all $u \in B_1$,

$$\|u\|_{B_2} \leqslant \epsilon \|u\|_{B_1} + C_\epsilon \|u\|_{B_3}.$$

(b) Deduce that for a bounded open set Ω of class C^1, if $\epsilon > 0$ is given, there exists a $C_\epsilon > 0$ such that

$$\|u\|_{1,\Omega} \leqslant \epsilon \|u\|_{2,\Omega} + C_\epsilon |u|_{0,\Omega}$$

for all $u \in H^2(\Omega)$.

(c) Prove the same inequality for $\Omega = \mathbb{R}^n$ and show that in this C_ϵ can be chosen as $C\epsilon^{-1}$.

2.13 Let δ be the Dirac distribution on $\mathbb{R}$. Find functions $f_0, f_1 \in L^2(\mathbb{R})$ such that

$$\delta = f_0 + \frac{df_1}{dx}$$

(which is possible since $\delta \in H^{-1}(\mathbb{R})$.)

2.14 Let $u \in H_0^1(\mathbb{R}_+^n)$. Define $u^\dagger : \mathbb{R}^n \to \mathbb{R}$ by

$$u^\dagger(x', x_n) = \begin{cases} u(x', x_n), & \text{if } x_n > 0 \\ -u(x', -x_n), & \text{if } x_n < 0 \end{cases}$$

Show that $u^\dagger \in H^1(\mathbb{R}^n)$ and compute its first derivatives in terms of those of u.

2.15 Let Ω be a bounded open set of $\mathbb{R}^n$ such that $\bar{\Omega} = \bigcup_{i=1}^{k} \bar{\Omega}_i$, $\{\Omega_i\}_{1 \leqslant i \leqslant k}$ disjoint open subsets of Ω with piecewise smooth boundaries (so that Green's formula is valid in each of them). If $u \in C(\bar{\Omega})$ and $u|_{\Omega_i} \in H^1(\Omega_i)$ for each $1 \leqslant i \leqslant k$, show that $u \in H^1(\Omega)$. (This is a generalization of Exercise 2.3 above).

2.16 Let $\Omega \subset \mathbb{R}^n$ be a bounded open set. Show that if $u, v \in H_0^2(\Omega)$, then

$$\int_\Omega \Delta u \, \Delta u = \sum_{i,j=1}^{n} \int_\Omega \partial_{ij} u \partial_{ij} v \left(\partial_{ij} = \frac{\partial^2}{\partial x_i \, \partial x_j} \right).$$

Deduce that $u \to |\Delta u|_{0,\Omega}$ defines a norm on $H_0^2(\Omega)$ equivalent to the usual norm.

2.17 Show that if Ω is a bounded open set in $\mathbb{R}^n$, the map $u \to |\Delta u|_{0,\Omega}$ defines a norm on the space $H^2(\Omega) \cap H_0^1(\Omega)$.

2.18 Let $\Omega \subset \mathbb{R}^2$ be a bounded open set. For $u, v \in H^2(\Omega)$ define

$$[u, v] = \partial_{11} u \, \partial_{22} v + \partial_{22} u \, \partial_{11} v - 2\partial_{12} u \, \partial_{12} v.$$

For $u, v, w \in H_0^2(\Omega)$ define

$$B(u, v, w) = \int_\Omega [u, v] w.$$

Show that the map is well-defined and that there exists a constant $C > 0$ such that

$$|B(u, v, w)| \leqslant C \|u\|_{2,\Omega} \|v\|_{2,\Omega} \|w\|_{2,\Omega}$$

for all $u, v, w \in H_0^2(\Omega)$. Show also that if $u_n \to u$ and $v_n \to v$ weakly in $H_0^2(\Omega)$ then

$$B(u_n, v_n, w) \to B(u, v, w)$$

as $n \to \infty$, for all $w \in H_0^2(\Omega)$.

2.19 Let $\Omega \subset \mathbb{R}^n$ be a convex and bounded open set containing the origin and set

$$\lambda\Omega = \{\lambda x \mid x \in \Omega\} \subset \Omega \quad (\text{if } \lambda < 1).$$

Let $u \in W_0^{1,p}(\Omega)$ such that $|u|_{0,p^*,\Omega} = 1$. Define for $k \geqslant 1$,

$$u_k(x) = \begin{cases} k^{(n-p)/p} & u(kx), \ x \in (1/k)\Omega, \\ 0 & x \in \Omega \setminus (1/k)\Omega. \end{cases}$$

Show that $|u_k|_{1,p,\Omega} = |u|_{1,p,\Omega}$ and that $|u_k|_{0,p^*,\Omega} = |u|_{0,p^*,\Omega} = 1$ for all integers $k \geqslant 1$. Deduce that the imbedding $W^{1,p}(\Omega) \to L^{p^*}(\Omega)$ is not compact.

2.20. (a) Let $H_r^1(\mathbb{R}^n)$ be the space of functions in $H^1(\mathbb{R}^n)$ which are radial. Show that if $u \in H_r^1(\mathbb{R}^n)$ then u can be approximated in $H^1(\mathbb{R}^n)$ by radial functions in $\mathcal{D}$.

(b) Let $u \in \mathcal{D}$ and be radial. Using the relation

$$r^{n-1}|u(r)|^2 = -\int_r^\infty \frac{d}{ds}(s^{n-1}|u(s)|^2)\, ds,$$

prove that

$$|u(x)| \leqslant \left(\frac{2}{\omega_n}\right)^{1/2} |x|^{-(n-1)/2} |u|_{0,\mathbb{R}^n}^{1/2} |u|_{1,\mathbb{R}^n}^{1/2}$$

for $x \in \mathbb{R}^n$, where ω_n is the surface measure of the unit sphere in $\mathbb{R}^n$. Deduce the same inequality (a.e.) for every $u \in H_r^1(\mathbb{R}^n)$.

(c) Let $2 < q < 2^* = 2n/(n-2)$. Show that if $u_m \to 0$ in $H^1(\mathbb{R}^n)$ weakly and $u_m \in H_r^1(\mathbb{R}^n)$ for all m, then $u_m \to 0$ in $L^q(\mathbb{R}^n)$ strongly, i.e., the inclusion

$$H_r^1(\mathbb{R}^n) \to L^q(\mathbb{R}^n)$$

is compact.

(d) Let $u \in H_r^1(\mathbb{R}^n)$ with $|u|_{0,\mathbb{R}^n} = 1$. Define

$$u_k(x) = k^{-n/2} u(x/k).$$

Show that $|u_k|_{1,\mathbb{R}^n} \to 0$ and that $|u_k|_{0,\mathbb{R}^n} = 1$. Deduce that the inclusion

$$H_r^1(\mathbb{R}^n) \to L^2(\mathbb{R}^n)$$

is not compact.

2.21. If δ_h denotes the Dirac distribution concentrated at $h \in \mathbb{R}^n$, show that

$$\|\delta_x - \delta_y\|_{H^{-s}(\mathbb{R}^n)} \leqslant C_\alpha |x-y|^\alpha, \quad x, y \in \mathbb{R}^n,$$

where $s = \frac{n}{2} + \alpha$ with $0 < \alpha < 1$. Deduce that for $u \in H^s(\mathbb{R}^n)$,

$$|u(x) - u(y)| \leqslant C_\alpha |x-y|^\alpha \|u\|_{s,\mathbb{R}^n}$$

(cf. (2.4.17) and (2.4.23)).

THREE

Weak Solutions of Elliptic Boundary Value Problems

3.1 SOME ABSTRACT VARIATIONAL PROBLEMS

Most partial differential equations in Engineering and Physics arise out of a *variational principle*. We usually have a class of *admissible solutions* (say, displacements) and an energy functional associated to these admissible functions; we seek to minimize the energy to identify the solution of the problem. The corresponding *Euler equation* gives the partial differential equation we started with. We will see, in the next section, several examples of this general situation. In this section we will discuss some abstract variational problems which will form the basis of our study of elliptic boundary value problems.

One of the classical results in Functional Analysis is the minimization of the norm (or distance) in a Hilbert space.

Theorem 3.1.1 Let H be a (real) Hilbert space and $K \subset H$ a closed convex set. Let $x \in H$. Then there exists a unique $y \in K$ such that

$$\|x - y\| = \min_{z \in K} \|x - z\|. \tag{3.1.1}$$

Further y can be characterized by:

$$y \in K, (x - y, z - y) \leqslant 0 \quad \text{for all } z \in K. \tag{3.1.2}$$

Proof: Let $d = \inf_{z \in K} \|x - z\| \geqslant 0$. Let $\{y_m\}$ be a minimizing sequence in K, i.e. $y_m \in K$ and $\|x - y_m\| \to d$. Clearly then $\{y_m\}$ is a bounded sequence in H and so we can extract a weakly convergent subsequence. Let $y_m \to y$ weakly in H. Then since the norm is a weakly lower semi-continuous function on H, we have

$$\|x - y\| \leqslant \liminf_{m \to \infty} \|x - y_m\| = d$$

But K is closed and convex and so is weakly closed. Thus $y \in K$ and so $\|x - y\| \geqslant d$. Thus $\|x - y\| = d$. Thus we have proved the existence of $y \in K$ satisfying (3.1.1). Also if y' were another such element then by the

convexity of K we have $\frac{y+y'}{2} \in K$. By the parallellogram identity

$$\left\|\frac{y+y'}{2} - x\right\|^2 = \frac{\|y-x\|^2}{2} + \frac{\|y'-x\|^2}{2} - \left\|\frac{y-y'}{2}\right\|^2 < d^2$$

a contradiction to the definition of y, y' and d. Thus the element y is uniquely defined in K.

We will now prove the characterization (3.1.2). Let $x \in H$, $y \in K$ satisfy (3.1.1). Let $z \in K$. For $0 < t < 1$, $tz + (1-t)y \in K$ by convexity. Hence from (3.1.1),

$$\|x - y\| \leqslant \|x - (1-t)y - tz\| = \|(x-y) - t(z-y)\|$$

Thus

$$\|x-y\|^2 \leqslant \|x-y\|^2 - 2t(x-y, z-y) + t^2\|z-y\|^2$$

or

$$(x-y, z-y) \leqslant \frac{t}{2}\|z-y\|^2$$

which yields (3.1.2) on letting $t \to 0$. Conversely, if $x \in H$ and $y \in K$ satisfy (3.1.2), then for $z \in K$,

$$\|x-y\|^2 - \|x-z\|^2 = 2(x-y, z-y) - \|z-y\|^2 \leqslant 0.$$

Thus x and y satisfy (3.1.1). ∎

Remark 3.1.1 If $H = \mathbb{R}^2$ and K is a closed convex subset of $\mathbb{R}^2$ then geometrically the characterization (3.1.2) means that the line joining x with its 'projection' y and the line joining the projection y and any other point of K always make an obtuse angle. See Fig. 11. ∎

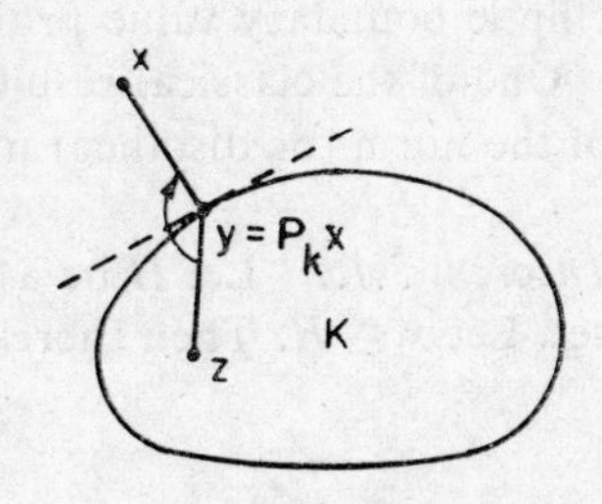

Fig. 11

Corollary: Let H be a Hilbert space, K a closed convex set in H. Let $P_K : H \to K$ be the map $x \to y$ defined by the preceding theorem. Then for $x_1, x_2 \in H$ we have

$$\|P_K x_1 - P_K x_2\| \leqslant \|x_1 - x_2\|. \tag{3.1.3}$$

Proof: We have

$$(x_1 - P_K x_1, P_K x_2 - P_K x_1) \leqslant 0$$

and

$$(x_2 - P_K x_2, P_K x_1 - P_K x_2) \leqslant 0.$$

Adding, we get

$$((x_1 - x_2) - (P_K x_1 - P_K x_2), (P_K x_2 - P_K x_1)) \leqslant 0$$

or

$$\|P_K x_2 - P_K x_1\|^2 \leqslant (x_2 - x_1, P_K x_2 - P_K x_1) \leqslant \|x_2 - x_1\| \, \|P_K x_2 - P_K x_1\|$$

by the Cauchy-Schwarz inequality, which leads to (3.1.3). ∎

Definition 3.1.1 Let $a: H \times H \to \mathbb{R}$ be a bilinear form. It is said to be continuous if there exists a constant $M > 0$ such that

$$\|a(u, v)\| \leqslant M\|u\| \, \|v\|, \quad \text{for all } u, v \in H. \tag{3.1.4}$$

It is said to be H-elliptic if there exists a constant $\alpha > 0$ such that

$$a(v, v) \geqslant \alpha\|v\|^2, \quad \text{for every } v \in H. \quad ∎ \tag{3.1.5}$$

Example 3.1.1 Let $H = \mathbb{R}^n$. Then every $n \times n$ matrix with real coefficients defines a continuous bilinear form on $\mathbb{R}^n$; if $A = (a_{ij})$, $1 \leqslant i, j \leqslant n$ is the matrix, then the corresponding bilinear form is given by

$$a(u, v) = v^T A u = \sum_{i,j=1}^{n} a_{ij} u_j v_i. \tag{3.1.6}$$

Clearly by the Cauchy-Schwarz inequality,

$$\begin{aligned} |a(u, v)| = |v^T A u| = |(v, Au)| &\leqslant \|v\| \, \|Au\| \\ &\leqslant \|A\| \, \|u\| \, \|v\|. \end{aligned}$$

If A is a symmetric and positive definite matrix then we know that

$$\sum_{i,j=1}^{n} a_{ij} v_i v_j \geqslant \alpha\|v\|^2 \tag{3.1.7}$$

which means that $a(. , .)$ is H-elliptic. ∎

We will have several more examples in the next section.

Theorem 3.1.2 Let $a(. , .)$ be a continuous, symmetric and H-elliptic bilinear form on a Hilbert space H. Let $K \subset H$ be a closed convex subset. Let $f \in H$. Then there exists a unique $u \in K$ such that

$$a(u, v - u) \geqslant (f, v - u), \quad \text{for every } v \in K. \tag{3.1.8}$$

Further u can be characterized by

$$\left.\begin{aligned} &u \in K \\ &J(u) = \min_{v \in K} J(v) \end{aligned}\right\} \tag{3.1.9}$$

where

$$J(v) = \tfrac{1}{2} a(v, v) - (f, v). \tag{3.1.10}$$

Proof Consider

$$\langle u, v \rangle = a(u, v), \quad \text{for } u, v \in H.$$

Then by the bilinearity and symmetry of $a(. , .)$ we have that $\langle u, v \rangle$ is an inner-product for H. Let

$$|||u|||^2 = \langle u, u \rangle = a(u, u). \tag{3.1.11}$$

Then by the continuity and H-ellipticity of $a(. , .)$, we get

$$\alpha\|u\|^2 \leqslant |||u|||^2 \leqslant M\|u\|^2$$

and so the new norm is equivalent to the original one and so H is also a Hilbert space with respect to the new inner-product. Now by the Riesz Representation Theorem there exists $\tilde{f} \in H$ such that for any $v \in H$

$$a(\tilde{f}, v) = \langle \tilde{f}, v \rangle = (f, v). \tag{3.1.12}$$

Now consider

$$\begin{aligned}\tfrac{1}{2}|||v - \tilde{f}|||^2 &= \tfrac{1}{2}a(v - \tilde{f}, v - \tilde{f})\\ &= \tfrac{1}{2}a(v, v) - a(v, \tilde{f}) + \tfrac{1}{2}a(\tilde{f}, \tilde{f})\\ &= \tfrac{1}{2}a(v, v) - (f, v) + \tfrac{1}{2}|||\tilde{f}|||^2, \text{ by (3.1.12)}\\ &= J(v) + \tfrac{1}{2}|||\tilde{f}|||^2.\end{aligned}$$

Since $|||\tilde{f}|||^2$ is a constant, minimizing $J(v)$ over K is equivalent to minimizing $|||v - \tilde{f}|||^2$ over K, which by the characterization (3.1.2) of Theorem 3.1.1, yields a unique $u \in K$ such that

$$\langle \tilde{f} - u, v - u \rangle \leqslant 0, \quad \text{for every } v \in K$$

or

$$a(u, v - u) \geqslant (f, v - u), \quad \text{for every } v \in K. \quad \blacksquare$$

We now prove that the symmetry condition on the bilinear form $a(. , ,)$ can be relaxed. Of course, we will no longer be able to identify the problem with one of minimization.

Theorem 3.1.3 **(Stampacchia).** Let H be a Hilbert space and let $a(. , .)$ be a continuous and H-elliptic bilinear form on H. Then given $f \in H$, there exists a unique $u \in K$ such that

$$a(u, v - u) \geqslant (f, v - u) \tag{3.1.13}$$

for every $v \in K$.

Proof Let $u \in H$ be fixed. Consider the map

$$v \to a(u, v).$$

By the continuity of $a(. , .)$, it follows that the above is a continuous linear functional. Thus there exists $Au \in H$ such that

$$(Au, v) = a(u, v), \; v \in H.$$

The map $u \to Au$ is clearly linear by the bilinearity of $a(. , .)$. Further by (3.1.4) and (3.1.5)

$$\|Au\| \leqslant M\|u\| \quad (\text{thus } A \text{ is continuous}) \tag{3.1.14}$$

$$(Au, u) \geqslant \alpha\|u\|^2 \tag{3.1.15}$$

With this new notation, we seek $u \in K$ such that

$$(Au, v - u) \geqslant (f, v - u), \quad \text{for every } v \in K. \tag{3.1.16}$$

Let $\rho > 0$ be a positive constant to be chosen presently. Then (3.1.16) is equivalent to finding $u \in K$ such that

$$(\rho f - \rho Au + u - u, v - u) \leqslant 0,$$

for every $v \in K$. In other words, by (3.1.2) we seek $u \in K$ such that

$$u = P_K(\rho f - \rho Au + u).$$

Hence we are now looking for a fixed point of the continuous map $F: H \to H$ (whose range lies in K) defined by

$$F(v) = P_K(\rho f - \rho Av + v).$$

Now if $v_1, v_2 \in H$, we have

$$\begin{aligned}\|F(v_1) - F(v_2)\| &= \|P_K(\rho f - \rho Av_1 + v_1) - P_K(\rho f - \rho Av_2 + v_2)\| \\ &\leqslant \|(v_1 - v_2) - \rho A(v_1 - v_2)\|\end{aligned}$$

using the corollary to Theorem 3.1.1. Hence

$$\begin{aligned}\|F(v_1) - F(v_2)\|^2 &\leqslant \|v_1 - v_2\|^2 - 2\rho(A(v_1 - v_2), (v_1 - v_2)) + \rho^2\|A(v_1 - v_2)\|^2 \\ &\leqslant (1 - 2\rho\alpha + \rho^2 M^2)\|v_1 - v_2\|^2.\end{aligned}$$

If we choose ρ such that $0 < \rho < \dfrac{2\alpha}{M^2}$ then

$$\beta^2 = 1 - 2\rho\alpha + \rho^2 M^2 < 1.$$

Thus

$$\|F(v_1) - F(v_2)\| \leqslant \beta\|v_1 - v_2\|$$

with $0 < \beta < 1$ and so F is a contraction and thus by the Contraction Mapping Theorem, F has a unique fixed point u which must lie in K. ∎

Let $K = V$, a closed subspace of H, so that it is automatically convex. Let $u \in V$ be as in (3.1.13) for given $f \in H$. If $v \in V$ set $w = v + u$. Then

$$a(u, v) \geqslant (f, v)$$

for any $v \in V$. Apply this to $-v \in V$. Hence

$$a(u, v) = (f, v), \quad v \in V.$$

In particular this is true for $V = H$. Thus we have proved the following.

Theorem 3.1.4 **(Lax-Milgram)** Let V be a Hilbert space and $a(. , .)$ a continuous V-elliptic bilinear form. Then given $f \in V$, there exists a unique $u \in V$ such that

$$a(u, v) = (f, v), \quad \text{for every } v \in V. \tag{3.1.17}$$

If $a(. , .)$ is also symmetric then the functional $J : V \to \mathbb{R}$ defined by

$$J(v) = \tfrac{1}{2}a(v, v) - (f, v) \tag{3.1.18}$$

attains its minimum at u. ∎

Remark 3.1.2 When $a(.,.)$ is symmetric and $J : V \to \mathbb{R}$ is given by (3.1.18) then the global minimum of J is attained at $u \in V$ which satisfies (3.1.17). These can be considered as the *Euler equations* of the unconstrained optimization problem. When we have a constrained optimization problem, i.e. we minimize J over a closed convex subset K, then we cannot expect the equations to result, but only the inequalities (3.1.8). They are called **variational inequalities.** ∎

We will conclude this section with yet another abstract variational problem.

Theorem 3.1.5 **(Babuska-Brezzi)** Let Σ, V be Hilbert spaces and $a : \Sigma \times \Sigma \to \mathbb{R}$, $b : \Sigma \times V \to \mathbb{R}$, bilinear forms which are continuous. Let

$$Z = \{\sigma \in \Sigma \mid b(\sigma, v) = 0, \quad \text{for every } v \in V\}. \tag{3.1.19}$$

Assume that $a(.,.)$ is Z-elliptic, i.e. there exists a constant $\alpha > 0$ such that

$$a(\sigma, \sigma) \geqslant \alpha \|\sigma\|_\Sigma^2, \quad \text{for every } \sigma \in Z. \tag{3.1.20}$$

Assume further that there exists a constant $\beta > 0$ such that

$$\sup_{\tau \in \Sigma} \frac{b(\tau, v)}{\|\tau\|_\Sigma} \geqslant \beta \|v\|_V, \quad \text{for every } v \in V. \tag{3.1.21}$$

Then if $\kappa \in \Sigma$ and $l \in V$, there exists a unique pair $(\sigma, u) \in \Sigma \times V$ such that

$$a(\sigma, \tau) + b(\tau, u) = (\kappa, \tau), \quad \text{for every } \tau \in \Sigma. \tag{3.1.22}$$

$$b(\sigma, v) = (l, v), \quad \text{for every } v \in V. \tag{3.1.23}$$

Proof Define $A : \Sigma \to \Sigma$, $B : \Sigma \to V$ by

$$(A\sigma, \tau) = a(\sigma, \tau),\ \tau \in \Sigma$$
$$(B\sigma, v) = b(\sigma, v),\ v \in V.$$

It is clear that A and B are well-defined and continuous linear operators. Then (3.1.22)–(3.1.23) is equivalent to the system

$$A\sigma + B^* u = \kappa \tag{3.1.24}$$

$$B\sigma = l \tag{3.1.25}$$

where $B^* : V \to \Sigma$ is the adjoint of B. Now, by virtue of (3.1.21) we deduce that

$$\|B^* v\| \geqslant \beta \|v\|, \quad \text{for every } v \in V. \tag{3.1.26}$$

Thus B^* has closed range and is one-one; as we are dealing with continuous linear operators, it follows that B is onto. Now choose $\sigma_1 \in \Sigma$ such that $B\sigma_1 = l$. Now by the Lax-Milgram Theorem (Theorem 3.1.4) we have that there exists a unique $\sigma_0 \in Z$ such that

$$a(\sigma_0, \tau) = (\kappa, \tau) - a(\sigma_1, \tau) \tag{3.1.27}$$

for every $\tau \in Z$, i.e. $A\sigma_0 = \kappa - A\sigma_1$ *when restricted to* Z. Now $Z = \text{Ker } B$ and so $B\sigma_0 = 0$. Hence if $\sigma = \sigma_0 + \sigma_1$ we have $B\sigma = l$. Further, by (3.1.27) $\kappa - A\sigma \in Z^\perp$, the orthogonal complement of Z in Σ. But we know that

$$\text{Range } B^* = (\text{Ker } B)^\perp = Z^\perp.$$

Thus there exists $u \in V$ such that $B^*u = \kappa - A\sigma$. Thus (u, σ) satisfy (3.1.24)–(3.1.25) or, equivalently, (3.1.22)–(3.1.23). This proves the existence of σ and u.

To prove the uniqueness, assume (σ_1, u_1) and (σ_2, u_2) are two solutions. Let $\tilde{\sigma} = \sigma_1 - \sigma_2$, $\tilde{u} = u_1 - u_2$. Then

$$a(\tilde{\sigma}, \tau) + b(\tau, \tilde{u}) = 0, \quad \text{for every } \tau \in \Sigma \tag{3.1.28}$$

$$b(\tilde{\sigma}, v) = 0, \quad \text{for every } v \in V. \tag{3.1.29}$$

Setting $\tau = \tilde{\sigma}$ in (3.1.28) and using (3.1.29) we get

$$a(\tilde{\sigma}, \tilde{\sigma}) = 0, \tilde{\sigma} \in Z$$

from which it follows (by (3.1.20)) that $\tilde{\sigma} = 0$. Then from (3.1.28) we get that $B^*\tilde{u} = 0$ and by (3.1.26) we deduce that $\tilde{u} = 0$. Hence $\sigma_1 = \sigma_2$ and $u_1 = u_2$. ∎

Remark 3.13 The condition (3.1.21) is known as the *Babuska-Brezzi condition* or sometimes as the *inf-sup condition*. ∎

Remark 3.1.4 If the bilinear form $a(\cdot\,, \cdot)$ is symmetric then we consider the constrained optimization problem: Find $\sigma \in \Sigma$ such that $B\tau = l$ and

$$J(\sigma) = \min_{B\tau = l} J(\tau),$$

where

$$J(\tau) = \tfrac{1}{2}a(\tau, \tau) - (\kappa, \tau).$$

Then we introduce the Lagrangian

$$\{\tfrac{1}{2}a(\tau, \tau) - (\kappa, \tau)\} + \{b(\tau, v) - (l, v)\}$$

and look for a saddle point. The corresponding set of equations will turn out to be exactly the system (3.1.22)–(3.1.23).

Another way to look at this system is from the point of view of elasticity. The equations (3.1.22) can be viewed as a *constitutive law* (for instance the relation between the stress σ and the displacement u via the strain; example: *Hooke's law*). The equations (3.1.23) correspond to the *equilibrium equations*. ∎

In the next section we will see numerous applications of the theorems of Lax-Milgram, Babuska-Brezzi and Stampacchia. While the Lax-Milgram formulation is the starting point for the so called direct finite element methods, the Babuska-Brezzi theorem will provide the basis for the mixed finite element methods. These considerations will be taken up in Section 3.7.

3.2 EXAMPLES OF ELLIPTIC BOUNDARY VALUE PROBLEMS

In this section we will present several examples of elliptic boundary value problems. In each case we will state what we mean by a **weak solution** of the problem and study its existence and uniqueness using the results of the preceding section.

3.2.1 The Dirichlet Problem for Second Order Elliptic Operators

Let $\Omega \subset \mathbb{R}^n$ be a bounded open set. Let us consider the problem

$$\left.\begin{aligned} -\Delta u &= f \quad \text{in } \Omega \\ u &= 0 \quad \text{on } \Gamma \end{aligned}\right\} \tag{3.2.1}$$

where $\Gamma = \partial\Omega$ and $f : \Omega \to \mathbb{R}$ is a given function. A **classical solution** u of (3.2.1) is a function $u \in C^2(\bar{\Omega})$ which satisfies (3.2.1) pointwise. Let us assume that u is a classical solution. If we multiply the equation (3.2.1) by $\phi \in \mathcal{D}(\Omega)$ and integrate we get

$$-\int_\Omega \Delta u \cdot \phi = \int_\Omega f\phi.$$

Applying Green's Theorem (cf. (2.7.20)) and using the fact that $\phi = 0$ on Γ we deduce that

$$\int_\Omega \nabla u \cdot \nabla \phi = \int_\Omega f\phi. \tag{3.2.2}$$

Now since $u \in C^2(\bar{\Omega})$ and $u = 0$ on Γ, it follows from Theorem 2.7.3 that $u \in H_0^1(\Omega)$. Further $\mathcal{D}(\Omega)$ is dense in that space and both sides of (3.2.2) are continuous in ϕ with respect to the $H_0^1(\Omega)$-topology. Thus by density, it follows that u satisfies: $u \in H_0^1(\Omega)$ such that

$$\int_\Omega \nabla u \cdot \nabla v = \int_\Omega fv, \quad \text{for every } v \in H_0^1(\Omega). \tag{3.2.3}$$

Notice that in (3.2.3) we do not need any information on the second derivatives of u. Hence we say that if $u \in H_0^1(\Omega)$ satisfies (3.2.3) then u is a **weak solution** of (3.2.1). We have also seen that every classical solution is automatically a weak solution. We now prove that a weak solution always exists uniquely for certain classes of functions f.

Theorem 3.2.1 Let Ω be a bounded open set and $f \in L^2(\Omega)$. Then there exists a unique weak solution $u \in H_0^1(\Omega)$ satisfying (3.2.3). Further u can be characterized by: $u \in H_0^1(\Omega)$ such that

$$J(u) = \min_{v \in H_0^1(\Omega)} J(v) \tag{3.2.4}$$

where

$$J(v) = \tfrac{1}{2}\int_\Omega \nabla v \cdot \nabla v - \int_\Omega fv. \tag{3.2.5}$$

Proof We set $V = H^1(\Omega)$ and

$$a(u, v) = \int_\Omega \nabla u \cdot \nabla v$$

Clearly

$$|a(u, v)| \leqslant |u|_{1,\Omega}|v|_{1,\Omega} \leqslant \|u\|_{1,\Omega}\|v\|_{1,\Omega}.$$

Further, by Poincaré's inequality (cf. (2.3.22)) we have

$$a(v, v) = \int_\Omega \nabla v \cdot \nabla v = |v|^2_{1,\Omega} \geqslant C\|v\|^2_{1,\Omega}.$$

Thus $a(\cdot, \cdot)$ is continuous, symmetric and V-elliptic and the result follows from the Lax-Milgram Theorem 3.1.4. ∎

Remark 3.2.1 Compare this result with the discussion in Example 1.1.2. ∎

Remark 3.2.2 The same result is valid when $f \in H^{-1}(\Omega)$. We only need to replace $\int_\Omega f \cdot u$ by $\langle f, u \rangle$, which denotes the duality pairing between $H^{-1}(\Omega)$ and $H^1_0(\Omega)$, i.e. the action of the linear functional f on u. ∎

If we know *a priori* that $u \in C^2(\bar{\Omega})$, u a weak solution, then u has to be a classical solution. For then $u = 0$ on Γ and for $\phi \in \mathcal{D}(\Omega)$ we get back by Green's theorem

$$\int_\Omega -\Delta u \cdot \phi = \int_\Omega f\phi$$

i.e. $-\Delta u = f$ in the sense of distributions; also $\mathcal{D}(\Omega)$ is dense in $L^2(\Omega)$ and we can replace in this equation ϕ by any $L^2(\Omega)$-function. Thus $-\Delta u = f$ as $L^2(\Omega)$ functions and hence the equality holds a.e. But as $u \in C^2(\bar{\Omega})$, $-\Delta u, f$ are in $C(\bar{\Omega})$ and so the equality holds everywhere.

The question therefore is: when is a weak solution smooth enough? This is answered by a **regularity theorem** and we defer such investigations to Section 3.3. At this juncture we merely state for record that if $f \in L^2(\Omega)$ and Ω is a "reasonable domain" then the weak solution $u \in H^2(\Omega) \cap H^1_0(\Omega)$. If $f \in H^m(\Omega)$ and Ω of class C^{m+2}, then $u \in H^{m+2}(\Omega)$. Thus combining such information with the Sobolev imbedding theorems of Section 2.4, we can study whether a weak solution is classical or not.

We now turn to the inhomogeneous Dirichlet problem. Let $\Omega \subset \mathbb{R}^n$ be a bounded open set. Let f on Ω and g on Γ be given functions. Consider

$$\left.\begin{aligned} -\Delta u &= f \quad \text{in } \Omega \\ u &= g \quad \text{on } \Gamma \end{aligned}\right\} \tag{3.2.6}$$

As before using Green's theorem we would like to give a weak formulation in a subspace of $H^1(\Omega)$. If $u \in H^1(\Omega)$ then its trace on Γ lies in $H^{1/2}(\Gamma)$ (cf. Theorem 2.7.4). Thus we assume that $g \in H^{1/2}(\Gamma)$ and that

$f \in L^2(\Omega)$. Now again by the trace theorem, there exists $\widetilde{u} \in H^1(\Omega)$ such that

$$\widetilde{u}|_\Gamma = \gamma_0(\widetilde{u}) = g$$

Now define

$$K = \{v \in H^1(\Omega) \mid v - \widetilde{u} \in H_0^1(\Omega)\}. \tag{3.2.7}$$

It is easy to see that K is a closed convex set in $H^1(\Omega)$. We define a weak solution of (3.2.6) to be $u \in K$ such that

$$\int_\Omega \nabla u \cdot \nabla v = \int_\Omega fv, \quad \text{for every } v \in H_0^1(\Omega). \tag{3.2.8}$$

We leave it to the reader to check that every classical solution is weak and that every weak solution in $C^2(\bar{\Omega})$ is classical. Also choosing $v = \phi \in \mathcal{D}(\Omega)$ we see that a weak solution satisfies $-\Delta u = f$ in $\mathcal{D}'(\Omega)$ and also has $u = g$ on Γ. As for the existence of u, we set $u = \widetilde{u} + w$ and get from (3.2.8) : $w \in H_0^1(\Omega)$ such that

$$\int_\Omega \nabla w \cdot \nabla v = \int_\Omega fv - \int_\Omega \nabla\widetilde{u} \cdot \nabla v, \quad \text{for every } v \in H_0^1(\Omega) \tag{3.2.9}$$

which is of the form (3.2.3) with "f" in $H^{-1}(\Omega)$. Thus w exists uniquely and so u exists. Now we show that u is unique (this is not obvious from the uniqueness of w, since $\widetilde{u}$ is not uniquely defined!). If u_1 and u_2 are two weak solutions then $u_1 - u_2 \in H_0^1(\Omega)$ and

$$\int_\Omega \nabla(u_1 - u_2) \cdot \nabla v = 0, \quad \text{for every } v \in H_0^1(\Omega).$$

Set $v = u_1 - u_2$ to get $|u_1 - u_2|_{1,\Omega} = 0$ which, by Poincaré's inequality, implies that $u_1 = u_2$.

We finally show that u depends continuously on the data f and g. Indeed since the trace map is onto, we can choose a continuous right inverse to it. Thus there exists a constant $C > 0$ independent of $g \in H^{1/2}(\Gamma)$ and $\widetilde{u} \in H^1(\Omega)$ such that

$$\|\widetilde{u}\|_{1,\Omega} \leqslant C|g|_{1/2,\Gamma}.$$

Also setting $v = w$ in (3.2.9) we get

$$\begin{aligned} |w|_{1,\Omega}^2 &\leqslant |f|_{0,\Omega}|w|_{0,\Omega} + |\widetilde{u}|_{1,\Omega}|w|_{1,\Omega} \\ &\leqslant C|f|_{0,\Omega}|w|_{1,\Omega} + \|\widetilde{u}\|_{1,\Omega}|w|_{1,\Omega} \end{aligned}$$

by Poincaré's inequality. It thus follows that

$$|w|_{1,\Omega} \leqslant C|f|_{0,\Omega} + \|\widetilde{u}\|_{1,\Omega} \leqslant C(|f|_{0,\Omega} + |g|_{1/2,\Gamma}).$$

Again, by Poincaré's inequality, we get

$$\|w\|_{1,\Omega} \leqslant C(|f|_{0,\Omega} + |g|_{1/2,\Gamma})$$

which proves in turn that for $u = w + \widetilde{u}$

$$\|u\|_{1,\Omega} \leqslant C(|f|_{0,\Omega} + |g|_{1/2,\Gamma}) \tag{3.2.10}$$

which gives the continuous-dependence of the solution u in terms of the data.

Remark 3.2.3 Existence, uniqueness and continuous dependence of the solution on the data are the criteria for a **well-posed problem** in the sense of Hadamard. ▮

We can generalize the preceding considerations to cover the case of a *second order elliptic operator*. Let Ω in $\mathbb{R}^n$ be a bounded open set and $a_{ij}(\cdot) \in C^1(\bar{\Omega})$, for $1 \leqslant i, j \leqslant n$ which verify the *ellipticity condition*:

$$\sum_{i,j=1}^{n} a_{ij}(x)\xi_i\xi_j \geqslant \alpha|\xi|^2 \quad \text{for all } \xi \in \mathbb{R}^n,\ x \in \Omega, \tag{3.2.11}$$

with $\alpha > 0$ independent of x and ξ. Let $a_0(\cdot) \in C(\bar{\Omega})$. Consider the problem:

$$\left.\begin{aligned} -\sum_{i,j} \frac{\partial}{\partial x_i}\left(a_{ij}(x)\frac{\partial u}{\partial x_j}\right) + a_0(x)u = f \text{ in } \Omega \\ u = 0 \text{ on } \Gamma. \end{aligned}\right\} \tag{3.2.12}$$

The differential operator in the above equation is said to be in **divergence form.** If $f \in L^2(\Omega)$ *a* **weak solution** of (3.2.12) is given by $u \in H_0^1(\Omega)$ satisfying

$$\int_\Omega \sum_{i,j=1}^{n} a_{ij} \frac{\partial u}{\partial x_j}\frac{\partial v}{\partial x_i} + \int_\Omega a_0 uv = \int_\Omega fv, \quad \text{for every } v \in H_0^1(\Omega). \tag{3.2.13}$$

Again it is trivial to check that every classical solution is a weak solution and that a weak and smooth solution is classical. If $a_0(x) \geqslant 0$ for all $x \in \bar{\Omega}$ then the bilinear form

$$a(u, v) = \int_\Omega \sum_{i,j=1}^{n} a_{ij} \frac{\partial u}{\partial x_j}\frac{\partial v}{\partial x_i} + \int_\Omega a_0 uv$$

is $H_0^1(\Omega)$-elliptic by virtue of the ellipticity condition (3.2.11) and Poincaré's inequality. Thus by the Lax-Milgram Theorem it then has a unique solution which, *when* $a(.\,,.)$ *is symmetric* (i.e. $a_{ij}(x) = a_{ji}(x)$ for all i, j and for all $x \in \Omega$), minimizes the functional

$$J(v) = \tfrac{1}{2}\int_\Omega a_{ij} \frac{\partial v}{\partial x_j}\frac{\partial v}{\partial x_i} + \tfrac{1}{2}\int_\Omega a_0 v^2 - \int_\Omega fv \tag{3.2.14}$$

over all of $H_0^1(\Omega)$.

Notice that to define a weak solution via (3.2.13) it suffices to assume that a_{ij}, $a_0 \in L^\infty(\Omega)$ and of course that $f \in H^{-1}(\Omega)$.

More generally we can consider the following second order elliptic boundary value problem:

$$\left.\begin{aligned} -\sum_{i,j=1}^{n} \frac{\partial}{\partial x_i}\left(a_{ij}\frac{\partial u}{\partial x_j}\right) + \sum_{i=1}^{n} a_i \frac{\partial u}{\partial x_i} + a_0 u = f \text{ in } \Omega \\ u = 0 \text{ on } \Gamma. \end{aligned}\right\} \tag{3.2.15}$$

where $\{a_{ij}\}$ satisfy the ellipticity condition (3.2.11) and $a_i \in C(\bar{\Omega})$ for $0 \leqslant i \leqslant n$. A **weak solution** is a $u \in H_0^1(\Omega)$ satisfying

$$a(u, v) = \int_\Omega fv, \quad \text{for every } v \in H_0^1(\Omega) \tag{3.2.16}$$

where for $u, v \in H_0^1(\Omega)$,

$$a(u, v) = \int_\Omega \sum_{i,j=1}^n a_{ij} \frac{\partial u}{\partial x_j} \frac{\partial v}{\partial x_i} + \int_\Omega \sum_{i=1}^n a_i \frac{\partial u}{\partial x_i} v + \int_\Omega a_0 uv. \tag{3.2.17}$$

This form is not always symmetric. It is sometimes $H_0^1(\Omega)$-elliptic in which case a unique solution exists by the Lax-Milgram Theorem. We however can prove the following result:

Theorem 3.2.2 The set of weak solutions of (3.2.15) is a finite dimensional subspace of dimension, say d, of $H_0^1(\Omega)$ when $f = 0$. Further there exists a d-dimensional subspace, say, F, such that (3.2.16) has a solution for a given $f \in L^2(\Omega)$, if, and only if, $f \in F^\perp$, the orthogonal complement of F in $L^2(\Omega)$.

Proof Let $\lambda > 0$ be chosen large enough such that

$$a_0(x) + \lambda \geqslant \gamma > 0, \quad \text{for all } x \in \Omega.$$

Then

$$\begin{aligned} a(u, u) + \lambda \int_\Omega u^2 &\geqslant \alpha|u|_{1,\Omega}^2 - \beta|u|_{1,\Omega}|u|_{0,\Omega} + \gamma|u|_{0,\Omega}^2 \\ &= \alpha|u|_{1,\Omega}^2 + \left(\gamma^{1/2}|u|_{0,\Omega} - \frac{\beta\gamma^{-1/2}}{2}|u|_{1,\Omega}\right)^2 - \frac{\beta^2}{4\gamma}|u|_{1,\Omega}^2 \\ &\geqslant \left(\alpha - \frac{\beta^2}{4\gamma}\right)|u|_{1,\Omega}^2. \end{aligned}$$

Again choose λ large enough so that $\alpha - \left(\frac{\beta^2}{4\gamma}\right) > 0$. Then $a(u, v) + \lambda \int_\Omega uv$ will be $H_0^1(\Omega)$-elliptic and for given $f \in L^2(\Omega)$ there will be a unique solution $u = Gf \in H_0^1(\Omega)$ such that

$$a(Gf, v) + \lambda \int_\Omega (Gf)v = \int_\Omega fv$$

for every $v \in H_0^1(\Omega)$. The operator G maps $L^2(\Omega)$ into $H^1(\Omega)$ which is compactly imbedded in $L^2(\Omega)$ (by the Rellich-Kondrasov Theorem). Thus $f \to Gf$ is a compact operator of $L^2(\Omega)$ into itself. Now if u is a weak solution of (3.2.15) then clearly

$$u = G(f + \lambda u)$$

If we set $v = f + \lambda u$ then v must satisfy the equation

$$v - \lambda Gv = f \tag{3.2.18}$$

Now $\lambda > 0$ and G is compact. Thus $I - \lambda G$ is invertible unless λ^{-1} is an eigenvalue of G. If λ^{-1} is not an eigenvalue a unique solution exists for all

f (and $d = 0$). If λ^{-1} is an eigenvalue of G, it has finite geometric multiplicity (G is compact) and so when $f = 0$, the set of solutions is the corresponding finite dimensional eigenspace. Now by the *Fredholm alternative* solutions exist for given f if, and only if, f satisfies the compatibility condition viz, $f \in \text{Ker}\,(I - \lambda G^*)^\perp$ and $\text{Ker}\,(1 - \lambda G^*)$ has the same dimension as $\text{Ker}\,(1 - \lambda G) = d$. ∎

Remark 3.2.4 If we can show that $d = 0$ then we saw that for all $f \in L^2(\Omega)$ there exists a unique solution. For instance using a *maximum principle* it can be proved that if $a_0 \geqslant 0$ then whatever be $a_i \in L^\infty(\Omega)$, $1 \leqslant i \leqslant n$, $d = 0$. Thus without further hypotheses on $\{a_i\}$, $1 \leqslant i \leqslant n$, we have the existence of a unique weak solution to (3.2.5) when $a_0 \geqslant 0$ (cf. Gilbarg and Trudinger [1]). ∎

3.2.2 The Neumann Problem

Let $\Omega \subset \mathbb{R}^n$ be a bounded open set and let ν be the exterior unit normal. We denote by $\frac{\partial u}{\partial \nu}$ the exterior normal derivative of u on the boundary. Now consider the boundary value problem:

$$\left.\begin{aligned} -\Delta u + u &= f \quad \text{in } \Omega \\ \frac{\partial u}{\partial \nu} &= 0 \quad \text{on } \Gamma \end{aligned}\right\}. \tag{3.2.19}$$

By Green's formula if u is a classical solution, then $u \in H^1(\Omega)$ and further u satisfies

$$\int_\Omega \nabla u \cdot \nabla v + \int_\Omega uv = \int_\Omega fv, \tag{3.2.20}$$

for every $v \in H^1(\Omega)$. If $f \in L^2(\Omega)$ then we define a weak solution of (3.2.19) as $u \in H^1(\Omega)$ satisfying (3.2.20) for every $v \in H^1(\Omega)$. Since if

$$a(u, v) = \int_\Omega \nabla u \cdot \nabla v + \int_\Omega uv,$$

then $a(u, u) = \|u\|^2_{1,\Omega}$, we trivially have $H^1(\Omega)$-ellipticity and the existence of a unique solution follows from the Lax-Milgram Theorem, and it minimizes the functional

$$J(v) = \tfrac{1}{2}\int_\Omega \nabla v \cdot \nabla v + \tfrac{1}{2}\int_\Omega v^2 - \int_\Omega fv$$

over all of $H^1(\Omega)$.

If u is a weak solution and $u \in H^2(\Omega)$ then by retracing the passage from (3.2.19) to (3.2.20) we get for $v \in H^1(\Omega)$,

$$\int_\Omega -\Delta u \cdot v + \int_\Omega uv + \int_\Gamma \frac{\partial u}{\partial \nu} \cdot v = \int_\Omega fv. \tag{3.2.21}$$

Choosing $v \in \mathscr{D}(\Omega)$, the integral on Γ vanishes, and so in the sense of distributions $-\Delta u + u = f$. This equality also holds in $L^2(\Omega)$ since $\mathscr{D}(\Omega)$ is dense in $L^2(\Omega)$ and all the Ω-integrals are continuous with respect to v

in the $L^2(\Omega)$-topology. Thus for any $v \in H^1(\Omega)$ we now have

$$\int_\Gamma \frac{\partial u}{\partial \nu} v = 0$$

But $v|_\Gamma \in H^{1/2}(\Gamma)$ which is dense in $L^2(\Gamma)$ and so

$$\frac{\partial u}{\partial \nu} = 0 \quad \text{in } L^2(\Gamma).$$

Thus if $u \in H^2(\Omega)$ and is a weak solution, then u satisfies $-\Delta u + u = f$ a.e. and further $\frac{\partial u}{\partial \nu} = 0$. If further $u \in C^2(\bar{\Omega})$, then u will be a classical solution.

Remark 3.2.5 We must emphasize here an important difference between the Dirichlet and Neumann problems. In the Dirichlet problems, the boundary condition $u = 0$ (or $u = g$) had to be imposed *a priori* in the function space in which we wish to prove existence. On the other hand, for the Neumann Problem, we imposed no *a priori* conditions on the function space and it turns out that a weak solution automatically satisfies the boundary condition as shown above. Thus Dirichlet type boundary conditions are called **essential boundary conditions** and have to be taken into account while formulating the problem while Neumann type conditions are called **natural boundary conditions** and take care of themselves. ∎

We can also study the Neumann problem for a general second order elliptic operator as in (3.2.12) or (3.2.15). Here the natural boundary condition will take the form

$$\sum_{i,j=1}^{n} a_{ij} \frac{\partial u}{\partial x_j} \nu_i = 0 \tag{3.2.22}$$

and the sum on the left is called the *conormal derivative* associated to the differential operator and is denoted by $\frac{\partial u}{\partial \nu_A}$ if A stands for the differential operator.

Again we can study the inhomogeneous Neumann problem, say,

$$\left.\begin{aligned} -\Delta u + u &= f \quad \text{in } \Omega \\ \frac{\partial u}{\partial \nu} &= g \quad \text{on } \Gamma \end{aligned}\right\}. \tag{3.2.23}$$

The weak formulation will be: Find $u \in H^1(\Omega)$ such that

$$\int_\Omega \nabla u \cdot \nabla v + \int_\Omega uv = \int_\Omega fv - \langle g, v \rangle_\Gamma \tag{3.2.24}$$

where now $g \in H^{-1/2}(\Gamma)$ and $\langle . , . \rangle_\Gamma$ stands for the duality pairing between $H^{-1/2}(\Gamma)$ and $H^{1/2}(\Gamma)$. If $g \in L^2(\Gamma)$ then of course we may write $\int_\Gamma gv$. Again a unique weak solution always exists.

If we consider the Neumann problem

$$\left.\begin{aligned} -\Delta u &= f \quad \text{in } \Omega \\ \frac{\partial u}{\partial \nu} &= g \quad \text{on } \Gamma \end{aligned}\right\} \tag{3.2.25}$$

then f and g must satisfy a compatibility condition, viz.

$$\int_\Omega f + \int_\Gamma g = 0. \tag{3.2.26}$$

This comes from applying Green's formula (2.7.20) with u as above and $v \equiv 1$. Conversely, using the Fredholm alternative it can be shown that if f and g satisfy (3.2.26) then there exist an infinity of weak solutions to (3.2.25) and among them a unique u such that, in addition

$$\int_\Omega u = 0. \tag{3.2.27}$$

In the same vein we can consider weak formulations of other types of boundary value problems for second order elliptic operators. For instance we can consider a *Robin condition* $\left(\text{i.e. of the type } \frac{\partial u}{\partial \nu} + \alpha u = 0, \alpha > 0, \text{ on } \Gamma\right)$ or an *oblique derivative problem* i.e. with $\alpha_1 \frac{\partial u}{\partial \nu} + \alpha_2 \frac{\partial u}{\partial \tau} = 0$ on Γ, $\frac{\partial}{\partial \tau}$ being a tangential derivative. Some of these will be outlined in the exercises at the end of this Chapter. Another problem is the *mixed problem* of the following form:

$$\left.\begin{aligned} -\Delta u &= f \quad \text{in } \Omega \\ u &= 0 \quad \text{on } \Gamma_1 \\ \frac{\partial u}{\partial \nu} &= 0 \quad \text{on } \Gamma_2 \end{aligned}\right\} \tag{3.2.28}$$

where $\Gamma = \Gamma_1 \cup \Gamma_2$, $\Gamma_1 \cap \Gamma_2 = \phi$. If the surface measure of Γ_1 is strictly positive then Poincaré's inequality is still available for the space

$$V = \{v \in H^1(\Omega) | v|_{\Gamma_1} = 0\}.$$

Thus we can apply the Lax-Milgram Theorem to the problem: Find $u \in V$ such that for every $v \in V$

$$\int_\Omega \nabla u \cdot \nabla v = \int_\Omega fv. \tag{3.2.29}$$

This can be taken as the weak formulation of (3.2.28). Notice again that the essential condition on Γ_1 was imposed on V while nothing was done about the natural condition on Γ_2.

3.2.3 The Biharmonic Equation

Let $\Omega \subset \mathbb{R}^n$ be a bounded open set and let Δ be the Laplace operator. Then the biharmonic operator is given by Δ^2 and is thus a differential operator of the fourth order. The Dirichlet problem for the biharmonic

operator is as follows:

$$\left.\begin{aligned} \Delta^2 u &= f \quad \text{in } \Omega \\ u = \frac{\partial u}{\partial \nu} &= 0 \quad \text{on } \Gamma. \end{aligned}\right\} \tag{3.2.30}$$

Let $f \in L^2(\Omega)$. If $\phi \in \mathcal{D}(\Omega)$ then multiplying the first equation by ϕ and repeatedly using Green's formula we get, when $u \in C^4(\bar{\Omega})$, i.e. u is a classical solution,

$$\int_\Omega \Delta u \, \Delta \phi = \int_\Omega f\phi. \tag{3.2.31}$$

Now for the equation (3.2.31) to make sense it suffices that $u \in H^2(\Omega)$. Also in this case the traces $u|_\Gamma = \gamma_0(u)$ and $\left.\frac{\partial u}{\partial \nu}\right|_\Gamma = \gamma_1(u)$ are well defined. Since we require that they vanish on Γ, we have, by the trace theorem, $u \in H_0^2(\Omega)$. Also $\mathcal{D}(\Omega)$ is dense in $H_0^2(\Omega)$ and both sides of (3.2.31) are continuous with respect to ϕ in this topology. Hence a **weak solution** of (3.2.30) is given by

$$u \in H_0^2(\Omega), \int_\Omega \Delta u \Delta v = \int_\Omega fv, \quad \text{for every } v \in H_0^2(\Omega). \tag{3.2.32}$$

Consider the bilinear form

$$a(u, v) = \int_\Omega \Delta u \, \Delta v$$

on $H_0^2(\Omega)$. It is clearly continuous, by the Cauchy-Schwarz inequality:

$$|a(u, v)| \leqslant |\Delta u|_{0,\Omega} |\Delta v|_{0,\Omega} \leqslant \|u\|_{2,\Omega} \|v\|_{2,\Omega}. \tag{3.2.33}$$

Further we have seen that (cf. Exercise 2.16) $|\Delta u|_{0,\Omega}$ defines a norm on $H_0^2(\Omega)$ equivalent to the usual norm. Thus

$$a(u, u) = |\Delta u|_{0,\Omega}^2 \geqslant \alpha \|u\|_{2,\Omega}^2 \tag{3.2.34}$$

and so $a(\cdot\,,\,\cdot)$ is $H_0^2(\Omega)$-elliptic. Thus by the Lax-Milgram Theorem there exists a unique weak solution $u \in H_0^2(\Omega)$ which minimizes the functional

$$J(v) = \tfrac{1}{2} \int_\Omega |\Delta v|^2 - \int_\Omega fv$$

over all of $H_0^2(\Omega)$. We again have continuous dependence on the data; for, setting $v = u$ in (3.2.32), using (3.2.33) and (3.2.34), we get

$$\alpha \|u\|_{2,\Omega}^2 \leqslant \int_\Omega \Delta u \, \Delta u = \int_\Omega fu \leqslant |f|_{0,\Omega} |u|_{0,\Omega} \leqslant |f|_{0,\Omega} \|u\|_{2,\Omega}.$$

Thus

$$\|u\|_{2,\Omega} \leqslant \frac{1}{\alpha} |f|_{0,\Omega}. \tag{3.2.35}$$

As usual if u is a smooth solution and $u \in H^4(\Omega)$, then $\Delta^2 u = f$ in the sense of distributions and also as $L^2(\Omega)$ functions. If in addition we know that $u \in C^4(\bar{\Omega})$, then u will be a classical solution.

Remark 3.2.6 The formulation (3.2.32) also makes sense when

$f \in H^{-2}(\Omega)$. We replace $\int_\Omega fv$ by $\langle f, v\rangle$, the duality pairing between $H^{-2}(\Omega)$ and $H_0^2(\Omega)$. ∎

Remark 3.2.7 It was explained in Example 1.1.2 that the Dirichlet problem for the Laplace operator could be used to model the situation of a membrane subjected to a force and fixed along its edge ($\Omega \subset \mathbb{R}^2$). In the same way, the biharmonic operator describes the bending of a thin *clamped plate*. The plate itself is a three-dimensional body and is approximated by its middle surface which will be an open subset of $\mathbb{R}^2$. If we wish to consider what is known as a *simply supported plate* fixed along its edge, then the boundary condition $u = 0$ will be retained but $\partial u/\partial \nu = 0$ will be replaced by another condition. ∎

Let us now describe another weak formulation for the biharmonic problem when $f \in L^2(\Omega)$. Let us set $\sigma = -\Delta u \in L^2(\Omega)$. Then we can write

$$\int_\Omega \sigma\tau + \int_\Omega \Delta u \cdot \tau = 0, \quad \text{for every } \tau \in L^2(\Omega) \tag{3.2.36}$$

and (3.2.32) becomes

$$-\int_\Omega \sigma \Delta v = \int_\Omega fv, \quad \text{for every } v \in H_0^2(\Omega). \tag{3.2.37}$$

If we set $\Sigma = L^2(\Omega)$, $V = H_0^2(\Omega)$ and define

$$\left.\begin{aligned} a(\sigma, \tau) &= \int_\Omega \sigma\tau, \qquad \sigma, \tau \in \Sigma \\ b(\tau, v) &= \int_\Omega \Delta v \cdot \tau, \ \tau \in \Sigma, \ v \in V \end{aligned}\right\} \tag{3.2.38}$$

then the system (3.2.36)–(3.2.38) is as in the Babuska-Brezzi Theorem. The bilinear form $a(\cdot\,,\cdot)$ is $L^2(\Omega)$-elliptic and so trivially Z-elliptic for any subspace Z of $\Sigma = L^2(\Omega)$. We now check the Babuska-Brezzi condition.

$$\sup_{\tau\in\Sigma} \frac{b(\tau, v)}{|\tau|_{0,\Omega}} \geqslant \frac{\int_\Omega \Delta v\, \Delta v}{|\Delta v|_{0,\Omega}} = |\Delta v|_{0,\Omega} \geqslant \alpha \|v\|_{2,\Omega}.$$

Thus the conditions of Theorem 3.1.5 are satisfied and there exists a unique (σ, u) satisfying (3.2.36) and (3.2.37). Clearly then $u \in H_0^2(\Omega)$ is the solution of (3.2.32) and $\sigma = -\Delta u$.

This formulation has no intrinsic value as we have only increased the unknowns. However, let us now define $\widetilde{\Sigma} = H^1(\Omega)$ and $\widetilde{V} = H_0^1(\Omega)$. Define

$$\left.\begin{aligned} a(\sigma, \tau) &= \int_\Omega \sigma\tau \\ \widetilde{b}(\tau, v) &= -\int_\Omega \nabla\tau \cdot \nabla v. \end{aligned}\right\} \tag{3.2.39}$$

Now consider the problem: Find $(\tilde{\sigma}, \tilde{u}) \in \tilde{\Sigma} \times \tilde{V}$ such that

$$a(\tilde{\sigma}, \tau) + \tilde{b}(\tau, \tilde{u}) = 0 \quad \text{for every } \tau \in \tilde{\Sigma}, \tag{3.2.40}$$

$$-\tilde{b}(\tilde{\sigma}, v) = \int_\Omega fv \quad \text{for every } v \in \tilde{V}. \tag{3.2.41}$$

Again $\tilde{b}(\cdot\,,\cdot)$ satisfies the Babuska-Brezzi condition:

$$\sup_{\tau \in \tilde{\Sigma}} \frac{\tilde{b}(\tau, v)}{\|\tau\|_{1,\Omega}} \geqslant \frac{\int \nabla v \cdot \nabla v}{\|v\|_{1,\Omega}} = \frac{|v|_{1,\Omega}^2}{\|v\|_{1,\Omega}} \geqslant \beta \|v\|_{1,\Omega}$$

by the Poincaré inequality (Theorem 2.3.4). Unfortunately $a(\cdot\,,\cdot)$ is not Z-elliptic. Thus we cannot apply the Theorem 3.1.5 to prove existence of a solution. If a solution exists, it has to be unique as is easy to check. We now have the following result.

Theorem 3.2.3 Assume that the solution u of (3.2.32) satisfies the additional condition that $u \in H^3(\Omega)$. Then $(-\Delta u, u) = (\tilde{\sigma}, \tilde{u}) \in \tilde{\Sigma} \times \tilde{V}$ is the unique solution of (3.2.40)–(3.2.41).

Proof Clearly $(\sigma, u) = (-\Delta u, u)$ satisfies (3.2.36)–(3.2.37). But if $\tau \in H^1(\Omega)$

$$b(\tau, u) = \int_\Omega \Delta u \cdot \tau = -\int_\Omega \nabla u \cdot \nabla \tau = \tilde{b}(\tau, u).$$

Thus $\tilde{\sigma} = \sigma$ and $\tilde{u} = u$ satisfy (3.2.40). Now we have by (3.2.37)

$$-\int_\Omega \sigma\, \Delta v = \int_\Omega fv \quad \text{for every } v \in H_0^2(\Omega).$$

Again, by Green's formula we get, since $\sigma \in H^1(\Omega)$,

$$\int_\Omega \nabla\sigma\, \nabla v = \int_\Omega fv \quad \text{for every } v \in H_0^2(\Omega),$$

in particular for all $v = \phi \in \mathcal{D}(\Omega)$. But both sides of the above equation are continuous with respect to v for the $H_0^1(\Omega)$-topology and so the above relation is true for all $v \in H_0^1(\Omega)$ by the density of $\mathcal{D}(\Omega)$ in $H_0^1(\Omega)$. Hence the result. ∎

Remark 3.2.8 This latter formulation is very useful from the computational view point, as we shall see later. It is due to Ciarlet and Raviart [1]. It depends on the regularity result $u \in H^3(\Omega)$ for (3.2.32). If Ω were of class C^∞, we in fact have $u \in H^4(\Omega)$ when $f \in L^2(\Omega)$. If Ω is a polygon or a Lipschitz domain we do have $u \in H^3(\Omega)$ by a result of Kondratev [1]. ∎

3.2.4 The Elasticity System

Let $\Omega \subset \mathbb{R}^3$ be a bounded open set representing the volume occupied by

an elastic body. Let Γ be partitioned into two parts Γ_0 and Γ_1, with the surface measure of Γ_0 being *strictly postive*. Assume that the body is fixed along Γ_0. Assume that a body force $\bar{f} = (f_i)$, $1 \leqslant i \leqslant 3$, acts on the body and that a surface force $\bar{g} = (g_i)$, $1 \leqslant i \leqslant 3$ acts on Γ_1.

Let $\bar{u} = (u_i)$, $1 \leqslant i \leqslant 3$ be the displacement vector. Then the *strain tensor* (ϵ_{ij}) is defined by

$$\epsilon_{ij}(\bar{u}) = \tfrac{1}{2}\left(\frac{\partial u_j}{\partial x_i} + \frac{\partial u_i}{\partial x_j}\right), \; 1 \leqslant i, j \leqslant 3 \tag{3.2.42}$$

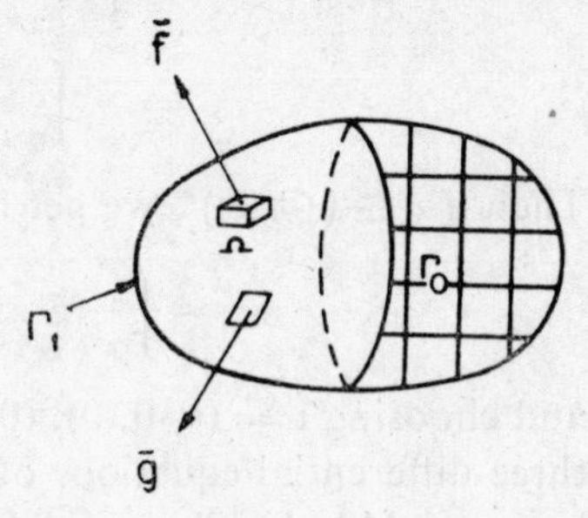

Fig. 12

If σ_{ij} is the *stress tensor* then we need a *constitutive law* to relate the two, which will describe the properties of the material of which the body is made. In this case we assume a linear law, viz. **Hooke's Law:**

$$\sigma_{ij}(\bar{u}) = \lambda\left(\sum_{k=1}^{3} \epsilon_{kk}(\bar{u})\right)\delta_{ij} + 2\mu\epsilon_{ij}(\bar{u}) \tag{3.2.43}$$

where $\lambda \geqslant 0$ and $\mu > 0$ are called *Lame's Coefficients*. Here δ_{ij} is the Kronecker symbol so that $\delta_{ij} = 0$ if $i \neq j$, and $\delta_{ij} = 1$ when $i = j$. The elasticity system consists of the following boundary value problem:

$$\left.\begin{aligned} -\sum_{j=1}^{3} \frac{\partial}{\partial x_j}(\sigma_{ij}(\bar{u})) &= f_i \text{ in } \Omega, \; 1 \leqslant i \leqslant 3 \\ \bar{u} &= 0 \text{ on } \Gamma_0 \\ \sum_{j=1}^{3} \sigma_{ij}(\bar{u})\nu_j &= g_i \text{ on } \Gamma_1, \; 1 \leqslant i \leqslant 3 \end{aligned}\right\} \tag{3.2.44}$$

Let us set $V = \{\bar{v} = (v_i)_{i=1}^{3} \in (H^1(\Omega))^3 \mid \bar{v} = 0 \text{ on } \Gamma_0\}$. Define

$$\begin{aligned} a(\bar{u}, \bar{v}) &= \int_\Omega \sum_{i,j=1}^{3} \sigma_{ij}(\bar{u})\epsilon_{ij}(\bar{v}) \\ &= \int_\Omega \left[\lambda \operatorname{div}(\bar{u}) \operatorname{div}(\bar{v}) + 2\mu \sum_{i,j=1}^{3} \epsilon_{ij}(\bar{u})\epsilon_{ij}(\bar{v})\right], \end{aligned}$$

the latter formula showing $a(\cdot,\cdot)$ to be symmetric. It is easily checked to be continuous. Let $f \in (L^2(\Omega))^3$ and $g \in (L^2(\Gamma))^3$. A weak solution of (3.2.44) is given by $\bar{u} \in V$ such that

$$a(\bar{u}, \bar{v}) = \int_\Omega \bar{f}\cdot\bar{v} + \int_{\Gamma_1} \bar{g}\cdot\bar{v}, \quad \text{for every } \bar{v} \in V. \tag{3.2.45}$$

If $\bar{u}$ is a weak solution which is smooth enough, we can recover (3.2.44). For if $\bar{v}$ were also smooth, then by Green's formula

$$a(\bar{u}, \bar{v}) = \int_\Omega \tfrac{1}{2} \sum_{i,j=1}^{3} \sigma_{ij}(\bar{u})\left(\frac{\partial v_i}{\partial x_j} + \frac{\partial v_j}{\partial x_i}\right)$$

$$= \int_\Omega \sum_{i,j=1}^{3} \sigma_{ij}(\bar{u}) \frac{\partial v_i}{\partial x_j} \quad \text{(by symmetry of } \sigma)$$

$$= -\int_\Omega \sum_{i,j=1}^{3} \frac{\partial}{\partial x_j} (\sigma_{ij}(\bar{u}))v_i + \int_{\Gamma_1} \sigma_{ij}(\bar{u})v_i\nu_j$$

Then if $v \in (\mathscr{D}(\Omega))^3$, we get that

$$-\int_\Omega \sum_{i,j=1}^{3} \frac{\partial}{\partial x_j} (\sigma_{ij}(\bar{u}))v_i = \int_\Omega \sum_{i=1}^{3} f_i v_i$$

and choosing $v = (v, 0, 0), (0, v, 0)$ or $(0, 0, v)$, $v \in \mathscr{D}(\Omega)$, we get that the three differential equations of (3.2.44) are satisfied in the sense of distributions. Also by choice of V, $\bar{u} = 0$ on Γ_0. Further the differential equations also hold in the L^2-sense and so we now get that

$$\sum_{i,j=1}^{3} \int_{\Gamma_1} \sigma_{ij} v_i \nu_j = \int_{\Gamma_1} g_i v_i$$

for all $\bar{v} \in V$, from which we deduce that the last three boundary conditions of (3.2.44) hold.

We now turn to the question of the existence of a weak solution. The existence and uniqueness will follow from the Lax-Milgram Theorem provided we prove the V-ellipticity of $a(\cdot, \cdot)$. This we now proceed to do.

Lemma 3.2.1 Let $\bar{v} \in V$ such that $\epsilon_{ij}(\bar{v}) = 0$ for all $1 \leqslant i, j \leqslant 3$. Then $\bar{v} = 0$.

Proof Step 1. If $\epsilon_{ij}(\bar{v}) = 0$ then we show that $\bar{v}$ is a **rigid displacement**, i.e., that there exist fixed vectors $\bar{a}$ and $\bar{b}$ in $\mathbb{R}^n$ such that

$$\bar{v}(x) = \bar{a} + \bar{b} \wedge \bar{x}$$

where '$\wedge$' denotes the vector cross product.

Indeed we now have the following relations:

$$\frac{\partial v_1}{\partial x_1} = \frac{\partial v_2}{\partial x_2} = \frac{\partial v_3}{\partial x_3} = 0; \frac{\partial v_1}{\partial x_2} + \frac{\partial v_2}{\partial x_1} = 0; \frac{\partial v_1}{\partial x_3} + \frac{\partial v_3}{\partial x_1} = 0; \frac{\partial v_2}{\partial x_3} + \frac{\partial v_3}{\partial x_2} = 0. \tag{3.2.46}$$

We may then assume that there exist functions $\Phi(x_1, x_2, x_3)$ and $\Psi(x_1, x_2, x_3)$ such that

$$v_1 = \frac{\partial \Phi}{\partial x_1}; \; v_2 = \frac{-\partial \Phi}{\partial x_2} = \frac{\partial \Psi}{\partial x_2}; \; v_3 = -\frac{\partial \Psi}{\partial x_3}.$$

But since $\partial_1 v_1 = 0$, $\partial_2 v_2 = 0$, it follows that Φ is at most linear in x_1 and x_2. Similarly Ψ is at most linear in x_2 and x_3. Thus we write

$$\Phi = (a + bx_1 + cx_2 + dx_1x_2 + ex_2x_3 + fx_3x_1)\phi(x_3)$$

$$\Psi = (a' + b'x_2 + c'x_3 + d'x_2x_3 + e'x_3x_1 + f'x_1x_2)\psi(x_1)$$

But $v_2 = -\dfrac{\partial \Phi}{\partial x_2} = \dfrac{\partial \Psi}{\partial x_2}$ which yields

$$(-c - dx_1 - ex_3)\phi(x_3) = (b' + d'x_3 + f'x_1)\psi(x_1).$$

The left hand side is linear in x_1 and the right hand side is linear in x_3. Hence ϕ and ψ must be constants. Without loss of generality assume that $\phi = \psi = 1$. Now we have

$$v_1 = b + dx_2 + fx_3$$
$$v_2 = -c - dx_1 - ex_3$$
$$v_3 = -c' + d'x_2 - e'x_1.$$

Now use $\epsilon_{13}(\bar{v}) = 0$ to see that $f = e'$ and $\epsilon_{23}(\bar{v}) = 0$ to see that $e = -d'$. Hence

$$v_3 = -c' + ex_2 - fx_1$$

and the result follows with $\bar{a} = (b, -c, -c')$ and $\bar{b} = (e, f, -d)$.

Step 2. Notice now that as $\bar{v} = 0$ on Γ_0, it has to vanish at least on 3 non-collinear points which shows that $\bar{b} = 0$ and $\bar{a} = 0$. Thus $\bar{v} \equiv 0$ on $\bar{\Omega}$. ∎

Theorem 3.2.4 The bilinear form $a(\cdot, \cdot)$ defined for the elasticity system is V-elliptic.

Proof We must show that there exists $\alpha > 0$ such that

$$a(\bar{v}, \bar{v}) \geqslant \alpha \|\bar{v}\|^2_{1, \Omega} \tag{3.2.47}$$

for all $\bar{v} \in V$, where

$$\|\bar{v}\|^2_{1,\Omega} = \sum_{i=1}^{3} \|v_i\|^2_{1,\Omega}.$$

If this were not true, we can find a sequence $\{\bar{v}_m\}$ in V such that

$$a(\bar{v}_m, \bar{v}_m) \leqslant \frac{1}{m} \|\bar{v}_m\|^2_{1.\Omega}$$

or after normalizing, that $\|\bar{v}_m\|_{1,\Omega} = 1$ for all m and $a(\bar{v}_m, \bar{v}_m) \to 0$. But this implies that $\sum_{i,j=1}^{n} \int_\Omega |\epsilon_{ij}(\bar{v}_m)|^2 \to 0$ as $m \to \infty$. Now $\{\bar{v}_m\}$ is bounded in V and by the Rellich-Kondrasov Theorem, has a subsequence strongly convergent in $(L^2(\Omega))^3$. Let $\bar{v} \in V$ such that $\bar{v}_m \to \bar{v}$ in $(L^2(\Omega))^3$. Then

$$0 \leqslant \sum_{i,j=1}^{3} \int_\Omega |\epsilon_{ij}(\bar{v})|^2 \leqslant \liminf_{m\to\infty} \sum_{i,j=1}^{3} \int_\Omega |\epsilon_{ij}(\bar{v}_m)|^2 = 0.$$

and so by Lemma 3.2.1, $\bar{v} = 0$.

On the other hand, by Korn's inequality (cf. (2.6.13)), there exists a constant $c > 0$ such that

$$\sum_{i,j=1}^{n} |\epsilon_{ij}(\bar{v}_m)|^2_{0,\Omega} + |\bar{v}_m|^2_{0,\Omega} \geqslant c\|\bar{v}_m\|^2_{1,\Omega} = c > 0.$$

Passing to the limit as $m \to \infty$ we get $0 \geqslant c > 0$, a contradiction. Thus (3.2.47) is proved. ∎

Thus the system of elasticity has a weak solution which achieves the

minimum over all V for the functional

$$J(\bar{v}) = \frac{1}{2}\int_{\Omega} \sum_{i,j=1}^{3} \sigma_{ij}(\bar{v})\epsilon_{ij}(\bar{v}) - \left(\int_{\Omega} \bar{f}\cdot\bar{v} + \int_{\Gamma_1} \bar{g}\cdot\bar{v}\right) \tag{3.2.48}$$

The functional $J(\bar{v})$ is the total energy of the elastic system and by minimizing it to get a weak solution, we have merely applied the well known *principle of virtual work*.

3.2.5 The Stokes' System

The Navier-Stokes equations in $\mathbb{R}^n$ describe the motion of a viscous incompressible fluid. The **incompressibility condition** (which comes from the equation of continuity) is given by

$$\text{div}\,(\bar{u}) = 0 \tag{3.2.49}$$

where $\bar{u} = (u_1, \ldots, u_n)$ is the velocity of the fluid at any given point. We now have the *constitutive law*

$$\sigma_{ij} = -P\delta_{ij} + 2\mu\epsilon_{ij}(\bar{u}), \tag{3.2.50}$$

where P is the pressure and μ the *viscosity* of the fluid. The viscosity is assumed to be constant. If ρ is the density of the fluid (also assumed constant), we get

$$p = P/\rho \quad \text{and} \quad \nu = \mu/\rho \tag{3.2.51}$$

and p and ν are called the kinematic pressure and kinematic viscosity respectively.

Assuming that we are looking at a steady state of the system and also neglecting non-linear effects, we can now use the equations of linear elasticity (3.2.44) to describe the motion. This leads us to the Stokes' equations or Stokes' system given by

$$\left.\begin{aligned} -2\nu \sum_{j=1}^{n} \frac{\partial}{\partial x_j}\,\epsilon_{ij}(\bar{u}) + \frac{\partial p}{\partial x_i} &= f_i, \quad 1 \leqslant i \leqslant n \\ \text{div}\,\bar{u} &= 0 \end{aligned}\right\}$$

But it is a simple computation to check that if div $(\bar{u}) = 0$, then

$$\sum_{j=1}^{n} \frac{\partial \epsilon_{ij}}{\partial x_j}(\bar{u}) = \frac{1}{2}\Delta u_i$$

Thus the Stokes' equations reduce to the form

$$\left.\begin{aligned} -\nu\Delta\bar{u} + \nabla p &= \bar{f} \quad \text{in} \quad \Omega \\ \text{div}\,(\bar{u}) &= 0. \end{aligned}\right\} \tag{3.2.52}$$

This is known as the velocity-pressure formulation and the existence theory we are now going to describe is based on a weak formulation studied by Girault and Raviart [1].

Let Ω be a bounded connected set with Lipschitz boundary Γ. Let $\bar{f} \in (H^{-1}(\Omega))^n$ and $g \in (H^{1/2}(\Gamma))^n$. Then we consider the following boundary

value problem:

$$\left.\begin{aligned}\left.\begin{aligned}-\nu\,\Delta\bar{u}+\nabla p&=\bar{f}\\ \operatorname{div}(\bar{u})&=0\end{aligned}\right\}&\quad\text{in}\quad\Omega\\ \bar{u}=\bar{g}&\quad\text{on}\quad\Gamma\end{aligned}\right\}\tag{3.2.53}$$

We will describe the weak formulation and prove the existence of a unique solution when $\bar{g}$ satisfies a certain compatibility condition. Then we will show that this solution satisfies (3.2.53) in the sense of distributions.

Notice first that if $(\bar{u}, p)$ is a solution of (3.2.53) so is $(\bar{u}, p + c)$ for any constant c. Thus p will be determined only upto a constant. We look for p in $L^2(\Omega)$. Thus to determine p uniquely we must work with $L^2(\Omega)/\mathbb{R}$ or, more conveniently, with

$$L_0^2(\Omega)=\left\{u\in L^2(\Omega)\;\middle|\;\int_\Omega u=0\right\}.\tag{3.2.54}$$

Before we can give the weak formulation, we need a few preliminary results.

Lemma 3.2.2 Let $\bar{g}\in(H^{1/2}(\Gamma))^n$ such that

$$\int_\Gamma \bar{g}\cdot\nu=0\tag{3.2.55}$$

Then there exists $\bar{u}\in(H^1(\Omega))^n$ such that

$$\operatorname{div}(\bar{u})=0,\ \bar{u}=\bar{g}\quad\text{on}\quad\Gamma\tag{3.2.56}$$

Proof We present a proof for the case $n = 2$.

Step 1 Let us assume that $\bar{g}\cdot\nu\equiv 0$ on Γ. If we can find a $\psi\in H^2(\Omega)$ such that

$$\frac{\partial\psi}{\partial\tau}=0\ \text{on}\ \Gamma,\quad \frac{\partial\psi}{\partial\nu}=-\bar{g}\cdot\tau\ \text{on}\ \Gamma$$

where τ is the unit tangent vector and $\dfrac{\partial}{\partial\tau}$ is the corresponding tangential derivative, then set $\bar{u}=\operatorname{curl}(\psi)$. Clearly $\operatorname{div}\bar{u}=0$. Further on Γ,

$$\bar{u}\cdot\nu=\frac{\partial\psi}{\partial\tau}=0$$

$$\bar{u}\cdot\tau=-\frac{\partial\psi}{\partial\nu}=\bar{g}\cdot\tau$$

and so $\bar{u}=\bar{g}$ on Γ. Thus $\bar{u}$ will be as required. Now $\bar{g}\cdot\tau\in H^{1/2}(\Gamma)$ and by the Trace Theorem (Theorem 2.7.4) there exists $\psi\in H^2(\Omega)$ such that $\psi=0$ on Γ and $\dfrac{\partial\psi}{\partial\nu}=-\bar{g}\cdot\tau$ on Γ. Hence $\dfrac{\partial\psi}{\partial\tau}=0$ and ψ is as required.

Step 2 In the general case consider the problem:

$$\left.\begin{aligned} -\Delta p &= 0 \quad \text{in } \Omega \\ \frac{\partial p}{\partial \nu} &= \bar{g}\cdot\nu \quad \text{on } \Gamma \end{aligned}\right\} \tag{3.2.57}$$

There does exist a $p \in H^1(\Omega)$ with $\int_\Omega p = 0$ since $\bar{g}\cdot\nu$ satisfies (3.2.55) which is the compatibility condition necessary and sufficient for the Neumann problem to have a solution (cf. (3.2.26)). Further if Γ is smooth enough we have by a regularity theorem (see Section 3.3) that $p \in H^2(\Omega)$. Now by Step 1, there exists $\bar{u}_1 \in (H^1(\Omega))^2$ such that div $(\bar{u}_1) = 0$ and

$$\bar{u}_1 = \bar{g} - \gamma_0(\nabla p) \text{ on } \Gamma$$

(so that $\bar{u}_1\cdot\nu \equiv 0$ on Γ). Now $\bar{u} = \bar{u}_1 + \nabla p$ is the required function. ∎

Lemma 3.2.3 Let

$$W = \{\bar{v} \in (H_0^1(\Omega))^n \mid \text{div}\,(\bar{v}) = 0\}.$$

Let $W^\perp$ be its orthogonal complement in $(H_0^1(\Omega))^n$, for the scalar product

$$\int_\Omega \nabla\bar{u}\cdot\nabla\bar{v} = \int_\Omega \sum_{i,j=1}^n \frac{\partial u_i}{\partial x_j}\frac{\partial v_i}{\partial x_j}. \tag{3.2.58}$$

Then the map $\bar{v} \to \text{div}\,(\bar{v})$ is an isomorphism of $W^\perp$ onto $L_0^2(\Omega)$.

Proof Let $\bar{v} \in (H_0^1(\Omega))^n$. Then by Green's formula

$$\int_\Omega \text{div}\,(\bar{v}) = \int_\Gamma \bar{v}\cdot\nu = 0.$$

Thus div $(\bar{v}) \in L_0^2(\Omega)$. The kernel of the given map is precisely W. Thus the divergence map is one-one on $W^\perp$. Let $f \in L_0^2(\Omega)$. By a regularity theorem for the Dirichlet problem for the Laplace operator (cf. Section 3.2.1) we know that there exists $\phi \in H^2(\Omega) \cap H_0^1(\Omega)$ such that

$$\Delta\phi = f \text{ in } \Omega.$$

Thus set $\bar{v}_1 = \nabla\phi \in (H^1(\Omega))^n$. Then

$$\text{div}\,(\bar{v}_1) = \sum_{i=1}^n \frac{\partial}{\partial x_i}\left(\frac{\partial\phi}{\partial x_i}\right) = \Delta\phi = f.$$

Further,

$$\int_\Gamma \bar{v}_1\cdot\nu = \int_\Omega \text{div}\,(\bar{v}_1) = \int_\Omega f = 0$$

since $f \in L_0^2(\Omega)$. Hence by Lemma 3.2.2, there exists $\bar{w}_1 \in (H^1(\Omega))^n$ such that div $(\bar{w}_1) = 0$ and $\bar{w}_1 = \bar{v}_1$ on Γ. Consider now

$$\bar{v} = \bar{v}_1 - \bar{w}_1.$$

Then div $(\bar{v}) = f$ and $\bar{v} = 0$ on Γ, i.e. $\bar{v} \in (H_0^1(\Omega))^n$. Thus the divergence map is onto. Hence

$$\text{div} : W^\perp \to L_0^2(\Omega)$$

is one-one, onto and continuous and hence an isomorphism by the Open Mapping Theorem. ∎

Corollary There exists a constant $C > 0$ such that

$$\sup_{\bar{v} \in (H_0^1(\Omega))^n} \int_\Omega \frac{\phi \operatorname{div}(\bar{v})}{|\bar{v}|_{1,\Omega}} \geqslant C|\phi|_{0,\Omega} \tag{3.2.59}$$

for all $\phi \in L_0^2(\Omega)$.

Proof By Lemma 3.2.3, choose $\bar{w} \in W^\perp$ such that $\operatorname{div}(\bar{w}) = \phi$ and $|\bar{w}|_{1,\Omega} \leqslant c_1|\phi|_{0,\Omega}$, which exists by virtue of the fact that div is an isomorphism of $W^\perp$ onto $L_0^2(\Omega)$. Now

$$\sup_{\bar{v} \in H_0^1(\Omega)} \frac{\int_\Omega \phi \operatorname{div}(\bar{v})}{|\bar{v}|_{1,\Omega}} \geqslant \frac{\int_\Omega \phi \operatorname{div}(\bar{w})}{|\bar{w}|_{1,\Omega}} = \frac{|\phi|_{0,\Omega}^2}{|\bar{w}|_{1,\Omega}} \geqslant \frac{1}{c_1}|\phi|_{0,\Omega}.$$

Hence the result. ∎

Let us now set $\Sigma = (H_0^1(\Omega))^n$ with the product norm $|\cdot|_{1,\Omega}$ and $V = L_0^2(\Omega)$ with the $L^2(\Omega)$-norm. Define

$$\left.\begin{aligned} a(\bar{u}, \bar{v}) &= 2\nu \int_\Omega \epsilon_{ij}(\bar{u})\epsilon_{ij}(\bar{v}) \\ b(\bar{v}, \phi) &= -\int_\Omega \phi \operatorname{div}(\bar{v}) \\ \chi(\bar{u}) &= \langle \bar{f}, \bar{u}\rangle - a(\bar{u}_0, \bar{v}) \\ l(\phi) &= 0, \text{ for all } \phi \in L_0^2(\Omega) \end{aligned}\right\} \tag{3.2.60}$$

where $\bar{u}_0 \in (H^1(\Omega))^n$ such that $\operatorname{div}(\bar{u}_0) = 0$ and $\bar{u}_0 = \bar{g}$ on Γ, which exists by Lemma 3.2.2. Here $\bar{f} \in (H^{-1}(\Omega))^n$. Now consider the system.

$$\left.\begin{aligned} a(\bar{w}, \bar{v}) + \bar{b}(\bar{v}, p) &= \chi(\bar{v}) \quad \text{for all } \bar{v} \in \Sigma \\ \bar{b}(\bar{w}, \phi) &= 0 \qquad \text{for all } \phi \in V \end{aligned}\right\} \tag{3.2.61}$$

Clearly $Z = W$. If $\bar{v} \in W$, then $\operatorname{div}(\bar{v}) = 0$ and so

$$\begin{aligned} a(\bar{v}, \bar{v}) &= 2\nu \int_\Omega \sum_{i,j=1}^n \epsilon_{ij}(\bar{v}) \frac{\partial v_i}{\partial x_j} = \nu \int_\Omega \sum_{i,j=1}^n \frac{\partial v_i}{\partial x_j}\frac{\partial v_i}{\partial x_j} \\ &= \nu|\bar{v}|_{1,\Omega}^2 \end{aligned}$$

Thus $a(\cdot, \cdot)$ is Z-elliptic. Next by (3.2.59) we see that $b(\cdot, \cdot)$ satisfies the Babuska-Brezzi condition. Hence by the Babuska-Brezzi Theorem there exists a unique pair $(\bar{w}, p) \in \Sigma \times V$ which is a solution of (3.2.61).

Now set $\bar{u} = \bar{w} + \bar{u}_0$. We show that the pair $(\bar{u}, p)$ verifies the Stokes' equations (3.2.52) in the sense of distributions. First of all $\bar{u} \in \bar{u}_0 = \bar{g}$ on Γ. Thus the boundary condition is satisfied. Also $\bar{w} \in (H_0^1(\Omega))^n$ and so $\operatorname{div}(\bar{w}) \in V = L_0^2(\Omega)$. But

$$\int_\Omega \operatorname{div}(\bar{w})\phi = 0$$

for all $\phi \in L_0^2(\Omega)$. Hence div $(\bar{w}) = 0$. Since div $(\bar{u}_0) = 0$, we have

$$\text{div } (\bar{u}) = 0.$$

Finally, let $\bar{\phi} \in \mathcal{D}(\Omega))^n$. We now have from (3.2.60) and (3.2.61),

$$a(\bar{u}, \bar{\phi}) + b(\bar{\phi}, p) = \langle \bar{f}, \phi \rangle.$$

Now,

$$a(\bar{u}, \bar{\phi}) = 2\nu \int_\Omega \sum_{i,j=1}^n \epsilon_{ij}(\bar{u}) \frac{\partial \phi_i}{\partial x_j}, \text{ by symmetry of } \epsilon_{ij}(\bar{\phi}),$$

$$= \nu \int_\Omega \sum_{i,j=1}^n \frac{\partial u_i}{\partial x_j} \frac{\partial \phi_i}{\partial x_j} = -\nu \sum_{i=1}^n \Delta u_i(\phi_i)$$

Also

$$b(\bar{\phi}, p) = -\int_\Omega \text{div } (\bar{\phi}) \cdot p = \sum_{i=1}^n \frac{\partial p}{\partial x_i} (\phi_i)$$

Thus we get that

$$(-\nu \Delta \bar{u} + \nabla p) \cdot \bar{\phi} = \langle \bar{f}, \bar{\phi} \rangle$$

or

$$-\nu \Delta \bar{u} + \nabla p = \bar{f} \text{ in } \mathcal{D}'(\Omega).$$

Hence the result.

3.2.6 Problems in Unbounded Domains

It is possible to study several of the problems described above in unbounded domains as well. In such cases not only do we need to specify the boundary conditions on $\partial\Omega$ but also the growth at infinity. For instance the condition that u vanishes at infinity will become $u \in H^1(\Omega)$ for the weak solution. Thus the problem

$$\left.\begin{aligned} -\Delta u + u &= f \quad \text{in } \mathbb{R}^n \\ u &\to 0 \quad \text{as } |x| \to +\infty \end{aligned}\right\} \tag{3.2.62}$$

has a weak solution $u \in H^1(\mathbb{R}^n)$ such that

$$\int_{\mathbb{R}^n} \nabla u \cdot \nabla v + \int_{\mathbb{R}^n} uv = \int_{\mathbb{R}^n} fv \tag{3.2.63}$$

for all $v \in H^1(\mathbb{R}^n)$ when $f \in L^2(\mathbb{R}^n)$.

Similarly if $f \in L^2(\mathbb{R}^n_+)$ and we consider

$$\left.\begin{aligned} -\Delta u + u &= f \quad && \text{in } \mathbb{R}^n \\ u(x', 0) &= 0 && x' \in \mathbb{R}^{n-1} \\ u &\to 0 && \text{as } |x| \to +\infty \end{aligned}\right\} \tag{3.2.64}$$

we can have a unique $u \in H_0^1(\mathbb{R}^n_+)$ such that

$$\int_{\mathbb{R}^n_+} \nabla u \cdot \nabla v + \int_{\mathbb{R}^n_+} uv = \int_{\mathbb{R}^n_+} fv, \quad \text{for every } v \in H_0^1(\mathbb{R}^n_+). \tag{3.2.65}$$

The existence and uniqueness of the weak solutions follow immediately from the Lax-Milgram Theorem.

3.3 REGULARITY OF WEAK SOLUTIONS

In the previous section we saw that a weak solution of an elliptic boundary value problem was posed in a space which always contained information only on derivatives of order lower than that of the differential operator. Thus second order problems were formulated in subspaces of $H^1(\Omega)$ which only carry information on the first order derivative; fourth order problems were formulated in $H^2(\Omega)$ or sometimes even in subspaces of $H^1(\Omega)$ and so on. When we ask ourselves the question whether a weak solution is also classical then we need to know information on the existence and continuity of higher order derivatives. The first step in this is to examine whether a weak solution will belong to a higher order Sobolev space. Then using the Sobolev inclusion theorems we can decide on the continuity of the function and its derivatives. A result of this nature is called a **regularity theorem.** The smoothness of the solution must obviously depend on at least two factors, viz., the smoothness of the data and the smoothness of the domain.

To keep the exposition simple, we will just outline the procedure for proving regularity results for second order elliptic problems. We will give a proof in two simple cases viz. when $\Omega = \mathbb{R}^n$ and $\Omega = \mathbb{R}^n_+$. For general domains, we will state the corresponding result and indicate the main steps. For details, see Agmon [1], Brezis [1] or Friedman [1].

Theorem 3.3.1 Let $u \in H^1(\mathbb{R}^n)$ such that

$$-\Delta u + u = f \in L^2(\mathbb{R}^n) \tag{3.3.1}$$

in the sense of distributions. Then $u \in H^2(\mathbb{R}^n)$ and

$$\|u\|_{2, \mathbb{R}^n} \leqslant C|f|_{0, \mathbb{R}^n} \tag{3.3.2}$$

where $C > 0$ is a constant independent of f. If $f \in H^m(\mathbb{R}^n)$ where $m \geqslant 0$ is an integer, then $u \in H^{m+2}(\mathbb{R}^n)$ and

$$\|u\|_{m+2, \mathbb{R}^n} \leqslant C\|f\|_{m, \mathbb{R}^n}. \tag{3.3.3}$$

Proof Since u, Δu are in $L^2(\mathbb{R}^n)$, we may apply the Fourier transform. Thus $\hat{u} \in L^2(\mathbb{R}^n)$ and

$$(\Delta u)^\wedge = \hat{u} - \hat{f} \in L^2(\mathbb{R}^n)$$

and so by (1.11.5), $|\xi|^2\hat{u}(\xi) \in L^2(\mathbb{R}^n)$. Thus $(1 + |\xi|^2)\hat{u}(\xi)$ is in $L^2(\mathbb{R}^n)$ and hence by definition $\hat{u} \in H^2(\mathbb{R}^n)$. Also

$$|(1 + |\xi|^2)\hat{u}(\xi)|_{0, \mathbb{R}^n} \leqslant C|f|_{0, \mathbb{R}^n}$$

from (3.3.1) and so (3.3.2) follows.

If $f \in H^1(\mathbb{R}^n)$ then for $1 \leqslant i \leqslant n$,

$$-\Delta\left(\frac{\partial u}{\partial x_i}\right) + \frac{\partial u}{\partial x_i} = \frac{\partial f}{\partial x_i} \in L^2(\mathbb{R}^n)$$

and so $\frac{\partial u}{\partial x_i} \in H^2(\mathbb{R}^n)$ and so $u \in H^3(\mathbb{R}^n)$. The case of a general m and the inequality (3.3.3) follow by recurrence. ∎

Theorem 3.3.2 Let $u \in H_0^1(\mathbb{R}_+^n)$ such that

$$\int_{\mathbb{R}_+^n} \nabla u \cdot \nabla v + \int_{\mathbb{R}_+^n} u \cdot v = \int_{\mathbb{R}_+^n} fv \tag{3.3.4}$$

for every $v \in H_0^1(\mathbb{R}_+^n)$, where $f \in L^2(\mathbb{R}_+^n)$. Then $u \in H^2(\mathbb{R}_+^n)$ and

$$\|u\|_{2, \mathbb{R}_+^n} \leqslant C|f|_{0, \mathbb{R}_+^n}. \tag{3.3.5}$$

Further if $f \in H^m(\mathbb{R}_+^n)$, then $u \in H^{m+2}(\mathbb{R}_+^n)$ and

$$\|u\|_{m+2, \mathbb{R}_+^n} \leqslant C\|f\|_{m, \mathbb{R}_+^n}. \tag{3.3.6}$$

Proof We define $u^\dagger$ on $\mathbb{R}^n$ by

$$u^\dagger(x', x_n) = \begin{cases} u(x', x_n), & x_n > 0 \\ -u(x', -x_n), & x_n < 0. \end{cases} \tag{3.3.7}$$

Then we can show that $u^\dagger \in H^1(\mathbb{R}^n)$ (cf. Exercise 2.14). Now we compute $-\Delta u^\dagger + u^\dagger$ in the sense of distributions. Let $\phi \in \mathcal{D}(\mathbb{R}^n)$.

$$\begin{aligned} -\int_{\mathbb{R}^n} u^\dagger \Delta\phi + \int_{\mathbb{R}^n} u^\dagger \phi &= -\int_{\mathbb{R}_+^n} u(x', x_n)\, \Delta\phi(x', x_n) + \int_{\mathbb{R}_-^n} u(x', -x_n)\, \Delta\phi(x', x_n) \\ &\quad + \int_{\mathbb{R}_+^n} u\phi - \int_{\mathbb{R}_-^n} u(x', -x_n)\phi(x', x_n) \\ &= -\int_{\mathbb{R}_+^n} u\, \Delta\chi + \int_{\mathbb{R}_+^n} u\chi \end{aligned}$$

where $\chi(x', x_n) = \phi(x', x_n) - \phi(x', -x_n)$, $x_n > 0$. But then $\chi(x', 0) = 0$ and so $\chi \in H_0^1(\mathbb{R}_+^n)$. Thus

$$-\int_{\mathbb{R}_+^n} u\, \Delta\chi + \int_{\mathbb{R}_+^n} u\chi = \int_{\mathbb{R}_+^n} \nabla u \cdot \nabla\chi + \int_{\mathbb{R}_+^n} u\chi = \int_{\mathbb{R}_+^n} f\chi = \int_{\mathbb{R}^n} f^\dagger \phi$$

where $f^\dagger = f$ for $x_n > 0$ and $f^\dagger(x', x_n) = -f(x', -x_n)$ for $x_n < 0$. Thus $-\Delta u^\dagger + u^\dagger = f^\dagger \in L^2(\mathbb{R}^n)$ and so $u^\dagger \in H^2(\mathbb{R}^n)$ with

$$\|u^\dagger\|_{2, \mathbb{R}^n} \leqslant C|f^\dagger|_{0, \mathbb{R}^n} \leqslant C|f|_{0, \mathbb{R}_+}.$$

Thus

$$\|u\|_{2, \mathbb{R}_+^n} \leqslant \|u^\dagger\|_{2, \mathbb{R}^n} \leqslant C|f|_{0, \mathbb{R}_+^n}.$$

Again if $f \in H^1(\mathbb{R}_+^n)$, then for $1 \leqslant i \leqslant n-1$,

$$-\Delta\left(\frac{\partial u}{\partial x_i}\right) + \left(\frac{\partial u}{\partial x_i}\right) = \frac{\partial f}{\partial x_i} \in L^2(\mathbb{R}_+^n)$$

and $\dfrac{\partial u}{\partial x_i} \in H_0^1(\mathbb{R}_+^n)$. (This is not true for $i = n$.) Thus

$$\frac{\partial u}{\partial x_i} \in H^2(\mathbb{R}_+^n), \quad 1 \leqslant i \leqslant n-1.$$

In particular

$$\frac{\partial^2 u}{\partial x_i\, \partial x_k} \in H^1(\mathbb{R}_+^n), \quad 1 \leqslant i \leqslant n-1, 1 \leqslant k \leqslant n.$$

Now

$$\frac{\partial^2 u}{\partial x_n^2} = u - f - \sum_{i=1}^{n-1} \frac{\partial^2 u}{u x_i^2} \in H^1(\mathbb{R}_+^n).$$

Thus all second derivatives are in $H^1(\mathbb{R}_+^n)$ and so u is in $H^3(\mathbb{R}_+^n)$. The general case now follows by recurrence of this argument.

Remark 3.3.1 If $u \in H^1(\mathbb{R}_+^n)$ satisfying (3.3.4) for all $v \in H^1(\mathbb{R}_+^n)$ then u is the weak solution of the Neumann Problem $-\Delta u + u = f$ in $\mathbb{R}_+^n$, $\dfrac{\partial u}{\partial \nu} = 0$ on $\mathbb{R}^{n-1}$. Again the conclusions of the preceding theorem are valid. We only need to use the extension by reflection u^* (cf. Theorem 2.3.1) in place of $u^\dagger$. ∎

For a general bounded open set, we have the following theorem:

Theorem 3.3.3 Let Ω be a bounded open set of class C^2. Let $f \in L^2(\Omega)$ and $u \in H_0^1(\Omega)$ such that

$$\int_\Omega \nabla u \cdot \nabla v + \int_\Omega uv = \int_\Omega fv, \quad \text{for every } v \in H_0^1(\Omega). \tag{3.3.8}$$

Then $u \in H^2(\Omega)$ and there is a constant $C > 0$ (depending only on Ω) such that

$$\|u\|_{2,\Omega} \leqslant C|f|_{0,\Omega}. \tag{3.3.9}$$

Further if Ω is of class C^{m+2} and $f \in H^m(\Omega)$, then $u \in H^{m+2}(\Omega)$, and

$$\|u\|_{m+2,\Omega} \leqslant C\|f\|_{m,\Omega}. \tag{3.3.10}$$

In particular if $m > n/2$, then $u \in C^2(\bar{\Omega})$. ∎

The proof of the above theorem (cf. Brezis [1]) proceeds in two stages. Let $\left\{U_j, T_j\right\}_{j=1}^k$ be a local chart associated to the boundary Γ. Consider the covering $\left\{U_j\right\}_{j=1}^k \cup \{\mathbb{R}^n \setminus \Gamma\}$ of $\mathbb{R}^n$ and the corresponding partition of unity, $\left\{\psi_j\right\}_{j=1}^k$ and ψ_0 (cf. Theorem 2.3.2). We write

$$u = \sum_{j=0}^{k} \psi_j u = \sum_{j=0}^{k} u_j. \tag{3.3.11}$$

Step 1 (*Regularity in the interior*) To show that $u_0 \in H^2(\Omega)$.

This is easily done. Since $\psi_0|_\Omega \in \mathscr{D}(\Omega)$, $\psi_0 u = u_0$, when extended by zero

outside Ω, will belong to $H^1(\mathbb{R}^n)$. Further we also have

$$-\Delta u_0 + u_0 = \psi_0 f - 2\nabla\psi_0 \cdot \nabla u - (\Delta\psi_0)u = g_0 \in L^2(\mathbb{R}^n). \quad (3.3.12)$$

Then we have $u_0 \in H^2(\mathbb{R}^n)$ and so u_0 in Ω is in $H^2(\Omega)$. To see (3.3.12), let $\phi \in \mathcal{D}(\mathbb{R}^n)$. Then $\psi_0\phi|_\Omega \in \mathcal{D}(\Omega)$. Hence

$$\begin{aligned}\int_{\mathbb{R}^n} g_0\phi &= \int_{\mathbb{R}^n} f\psi_0\phi - 2\int_{\mathbb{R}^n} (\nabla\psi_0 \cdot \nabla u)\phi - \int_{\mathbb{R}^n} \Delta\psi_0 u\phi \\ &= \int_\Omega f(\psi_0\phi) - 2\int_\Omega (\nabla\psi_0 \cdot \nabla u)\phi + \int_\Omega (\nabla\psi_0 \cdot \nabla u)\phi + \int_\Omega (\nabla\psi_0 \cdot \nabla\phi)u \\ &= \int_\Omega \nabla u \cdot \nabla(\psi_0\phi) + \int_\Omega u\psi_0\phi - \int_\Omega (\nabla\psi_0 \cdot \nabla u)\phi + \int_\Omega (\nabla\psi_0 \cdot \nabla\phi)u \\ &= \int_\Omega u\psi_0\phi + \int_\Omega (\nabla u \cdot \nabla\phi)\psi_0 + \int_\Omega (\nabla\psi_0 \cdot \nabla\phi)u \\ &= \int_{\mathbb{R}^n} u\psi_0\phi + \int_{\mathbb{R}^n} \nabla(u\psi_0) \cdot \nabla\phi\end{aligned}$$

which gives (3.3.12). We also have

$$\|u_0\|_{2,\Omega} \leqslant \|u_0\|_{2,\mathbb{R}^n} \leqslant |g_0|_{0,\mathbb{R}^n} \leqslant C(\|u\|_{1,\Omega} + |f|_{0,\Omega})$$

But from (3.3.8) we get, setting $v = u$, that

$$\|u\|_{1,\Omega} \leqslant |f|_{0,\Omega}.$$

Thus

$$\|u_0\|_{2,\Omega} \leqslant C|f|_{0,\Omega}. \quad (3.3.13)$$

Step 2 (*Regularity at the boundary*) To show that $u_i \in H^2(\Omega)$ $1 \leqslant i \leqslant k$.

In this case $u_i \in H_0^1(\Omega \cap U_i)$ and satisfies

$$-\Delta u_i = \psi_i f - \psi_i u - 2\nabla\psi_i \cdot \nabla u - (\Delta\psi_i)u = g_i \quad (3.3.14)$$

and $g \in L^2(\Omega \cap U_i)$, $|g|_{0,\Omega \cap U_i} \leqslant C|f|_{0,\Omega}$. Using the map T_i^{-1} we now transport this into an equation in Q_+. Unfortunately the operator $(-\Delta)$ will now be replaced (because of the change of coordinate, cf. Exercise 2.9) into a second order elliptic operator with variable coefficients on Q_+. We have to prove for this the analogue of Theorem 3.3.2 and then come back to $\Omega \cap U_l$ via T_i. The principle is always the same. We prove that all derivatives except those involving $\dfrac{\partial}{\partial x_n}$ are smooth enough and then prove the same for these using the equation. The calculations are somewhat tedious and the reader is referred to Brezis [1] or one of the other references cited above. ∎

The method used in Step 1 above to prove the interior regularity can be used to prove the hypoellipticity of the Laplace operator. (Another proof was indicated in the comments at the end of Chapter 1). Let Ω be an open set $f \in L^2(\Omega)$ and $u \in H_0^1(\Omega)$ such that $-\Delta u + u = f$ in Ω and let $x_0 \notin \text{sing.supp}\,(f)$. Then f is C^∞ in a neighbourhood of x_0 contained in

Ω, We can choose this neighbourhood to be a small ball around x_0 and thus if U is this neighbourhood, $f \in H^m(U)$ for all $m \geqslant 0$. Now choose $\theta_0 \in \mathcal{D}(U)$ such that $\theta_0 \equiv 1$ in a smaller neighbourhood of x_0. The function $\theta_0 u \in H_0^1(U)$ and its extension by zero is in $H^1(\mathbb{R}^n)$ and satisfies an equation similar to (3.3.12). Thus $\theta_0 u \in H^2(U)$ and by recurrence $\theta_0 u \in H^m(U)$ for all $m \geqslant 0$. Hence by Theorem 2.4.5, $\theta_0 u \in C^\infty(U)$. Thus u is C^∞ at x_0, and Δ is hypoelliptic. Thus the smoothness of u at a point in Ω depends only on the smoothness of Δu at the same point.

The regularity theorems of the type cited in Theorem 3.3.3 are also valid for the Neumann problem and also for more general second order elliptic operators. For details see the references given in this section, in particular Agmon [1] and Friedman [1].

3.4 AN EXAMPLE OF THE GALERKIN METHOD

The **Galerkin Method** is a method of approximation of solutions to abstract functional equations. It can also be used to prove the existence of solutions to an equation by proving the convergence of the approximate solutions in a suitable topology. We shall now illustrate this method via an example.

Let Ω be a bounded open set in $\mathbb{R}^n$. Consider the space $V = (H_0^1(\Omega))^2$ with its product norm. Define for $u = (u_1, u_2)$, $v = (v_1, v_2)$ in V, the bilinear form

$$a(u, v) = \int_\Omega u_1 v_1 + \int_\Omega u_2 v_2 + \int_\Omega \nabla u_2 \cdot \nabla v_1 - \int_\Omega \nabla u_1 \cdot \nabla v_2. \quad (3.4.1)$$

Clearly $a(\cdot\,, \cdot)$ is a continuous bilinear form on V by virtue of the Cauchy-Schwarz inequality. Unfortunately it is not V-elliptic; for

$$a(u, u) = \int_\Omega (u_1^2 + u_2^2) \quad (3.4.2)$$

which cannot be bounded from below by the V-norm. Thus if $f = (f_1, f_2) \in (L^2(\Omega))^2$, we cannot use the Lax-Milgram Theorem directly to prove the existence of a solution to the problem : $u \in V$ such that

$$a(u, v) = (f, v), \text{ for every } v \in V \quad (3.4.3)$$

where

$$(f, v) = \int_\Omega (f_1 v_1 + f_2 v_2). \quad (3.4.4)$$

We use therefore the Galerkin method to prove the existence of a solution to this problem. Since $H_0^1(\Omega)$ is a separable Hilbert space (cf. Theorem 2.1.1), so is V. Hence there exists a countable orthonormal basis for V, say $\{e_1, e_2, \ldots, e_m, \ldots\}$. Let us define

$$V_m = \text{span}\,\{e_1, \ldots, e_m\} \quad (3.4.5)$$

for each integer $m > 0$. The Galerkin method now proceeds in the following stages:

Step 1 Prove the existence of a solution $u_m \in V_m$ such that

$$a(u_m, v) = (f, v), \quad \text{for every } v \in V_m, \tag{3.4.6}$$

for each m.

Step 2 Show that the family $\{u_m\}$ of 'approximate solutions' is bounded in V independent of m. This is done via *a priori* estimates.

Step 3 From Step 2 it follows that a weakly convergent subsequence exists for $\{u_m\}$. Show that the weak limit is a solution to the problem.

Step 4 Prove uniqueness of the solution to the original problem and hence deduce that every convergent subsequence of $\{u_m\}$ has the same limit and so the entire sequence $\{u_m\}$ converges to the weak limit.

We now proceed to execute this programme.

Step 1. Since V_m is *finite dimensional* all norm topologies are equivalent on it. Endow V_m with the $(L^2(\Omega))^2$ topology. Then we see from (3.4.2) that a $(\cdot, \cdot)$ is V_m-elliptic. Hence by the Lax-Milgram Theorem, there exists a unique $u_m \in V_m$ satisfying (3.4.6). (In fact we need not use the Lax-Milgram Theorem at all; by virtue of (3.4.2), the matrix $(a(e_i, e_j))$, $1 \leqslant i, j \leqslant m$ is invertible, in fact positive definite. Hence the finite dimensional linear system (3.4.6) must have a unique solution.)

Step 2. Set $v = (u_m^{(2)}, -u_m^{(1)})$, where $u_m = (u_m^{(1)}, u_m^{(2)})$, in (3.4.6). Then we get

$$\begin{aligned}|u_m^{(2)}|_{1,\Omega}^2 + |u_m^{(1)}|_{1,\Omega}^2 &= \int_\Omega (f_1 u_m^{(2)} - f_2 u_m^{(1)}) \\ &\leqslant |f_1|_{0,\Omega} |u_m^{(2)}|_{0,\Omega} + |f_2|_{0,\Omega} |u_m^{(1)}|_{0,\Omega}. \\ &\leqslant C(|f_1|_{0,\Omega} |u_m^{(2)}|_{1,\Omega} + |f_2|_{0,\Omega} |u_m^{(1)}|_{1,\Omega})\end{aligned}$$

by Poincaré's inequality. Hence

$$|u_m^{(2)}|_{1,\Omega}^2 + |u_m^{(1)}|_{1,\Omega}^2 \leqslant C(|f_1|_{0,\Omega} + |f_2|_{0,\Omega})(|u_m^{(2)}|_{1,\Omega} + |u_m^{(1)}|_{1,\Omega})$$

Now using the inequality $(a+b)^2 \leqslant 2(a^2+b^2)$ we get

$$\tfrac{1}{2}(|u_m^{(2)}|_{1,\Omega} + |u_m^{(1)}|_{1,\Omega})^2 \leqslant C(|f_1|_{0,\Omega} + |f_2|_{0,\Omega})(|u_m^{(2)}|_{1,\Omega} + |u_m^{(1)}|_{1,\Omega})$$

and so

$$|u_m^{(2)}|_{1,\Omega} + |u_m^{(1)}|_{1,\Omega} \leqslant C(|f_1|_{0,\Omega} + |f_2|_{0,\Omega}) \tag{3.4.7}$$

Thus $\{u_m\}$ is bounded independent of m in V.

Step 3. Since V is a Hilbert space, there exists a subsequence $\{u_{m_k}\}$ of $\{u_m\}$ such that $u_{m_k} \to u \in V$ weakly. Now let $v \in V$. Then there exists $v_m \in V_m$ such that $v_m \to v$ in V strongly. In fact since

$$v = \sum_{k=1}^{\infty} (v, e_k)_V e_k \tag{3.4.8}$$

where $(\cdot, \cdot)_V$ is the inner-product in V, it suffices to choose

$$v_m = \sum_{k=1}^{m} (v, e_k)_V e_k. \tag{3.4.9}$$

Now, from (3.4.6)

$$a(u_m, v_m) = (f, v_m), \tag{3.4.10}$$

and as $u_m \to u$ weakly, and $v_m \to v$ strongly we can pass to the limit in (3.4.10) to get (3.4.3). Thus the problem has a solution u, viz. the weak limit of any convergent subsequence of $\{u_m\}$.

Step 4. The solution is unique. If $u^{(1)}$ and $u^{(2)}$ are two solutions, set $u = u^{(1)} - u^{(2)}$. Then

$$a(u, v) = 0, \text{ for every } v \in V.$$

Set $v = u$ to get from (3.4.2) that $u = 0$.

Thus the solution is uniquely defined and hence as every convergent subsequence has the same weak limit, the entire sequence $\{u_m\}$ converges weakly to u.

Let us interpret the solution u as the weak solution of a boundary value problem. Clearly as $u \in V = (H_0^1(\Omega))^2$ we have $u_1 = u_2 = 0$ on Γ where $u = (u_1, u_2)$. By taking $v = (v_1, 0)$ or $(0, v_2)$ with $v_1, v_2 \in H_0^1(\Omega)$ we get that

$$\int_\Omega \nabla u_2 \cdot \nabla v_1 + \int_\Omega u_1 v_1 = \int_\Omega f_1 v_1, \text{ for every } v_1 \in H_0^1(\Omega)$$

$$-\int_\Omega \nabla u_1 \cdot \nabla v_2 + \int_\Omega u_2 v_2 = \int_\Omega f_2 v_2, \text{ for every } v_2 \in H_0^1(\Omega).$$

Thus $(u_1, u_2) = u$ is the weak solution of the coupled system:

$$\left.\begin{aligned} -\Delta u_2 + u_1 &= f_1 \\ \Delta u_1 + u_2 &= f_2 \end{aligned}\right\} \text{ in } \Omega \qquad u_1 = u_2 = 0 \text{ on } \Gamma \tag{3.4.11}$$

Since $f_1, f_2, u_1, u_2 \in L^2(\Omega)$ we deduce from the regularity theorem that $u_1, u_2 \in H^2(\Omega)$. Thus the system (3.4.11) has a unique weak solution u belonging to

$$(H^2(\Omega) \cap H_0^1(\Omega))^2.$$

We shall see later in Chapter 4 that the system (3.4.11) is closely related to the solution of the *Schrodinger equation*.

The Galerkin method is a useful tool in proving the existence of solutions to non-linear problems. We shall see some examples in Chapter 5.

3.5 MAXIMUM PRINCIPLES

In this section we present some properties specially enjoyed by solutions of second order elliptic boundary value problems; they are called maximum principles.

Throughout we will asume that a_{ij}, $1 \leqslant i, j \leqslant n$ and a_0 are functions in $L^\infty(\Omega)$ such that $a_0 \geqslant 0$ and that $\{a_{ij}\}$ verifies the ellipticity condition (3.2.11).

Theorem 3.5.1 Let Ω be a bounded open set of $\mathbf{R}^n$ and with sufficiently smooth boundary Γ. Let $f \in L^2(\Omega)$ and $u \in H^1(\Omega) \cap C(\bar{\Omega})$ such that

$$\int_\Omega \sum_{i,j=1}^n a_{ij} \frac{\partial u}{\partial x_j} \frac{\partial v}{\partial x_i} + \int_\Omega a_0 uv = \int_\Omega fv, \text{ for every } v \in H_0^1(\Omega). \tag{3.5.1}$$

Then the following hold:

(i) $f \geqslant 0$ on Ω and $u \geqslant 0$ on Γ implies $u \geqslant 0$ on Ω.

(ii) if $a_0 \equiv 0$ and $f \geqslant 0$, then

$$u \geqslant \inf_\Gamma u, \text{ on } \Omega \tag{3.5.2}$$

(iii) if $f \equiv 0$ and $a_0 \equiv 0$, then

$$\inf_\Gamma u \leqslant u \leqslant \sup_\Gamma u. \tag{3.5.3}$$

Proof Since Γ is smooth enough, we have for $u \in H^1(\Omega)$, $|u| \in H^1(\Omega)$ (cf. Theorem 2.2.5, its corollary and Remark 2.2.1). Also it follows that u^+ and u^- are in $H^1(\Omega)$.

(i) If $u \geqslant 0$ on Γ then $u = |u|$ on Γ and so $u^- \in H_0^1(\Omega)$.

We can substitute it for v in (3.5.1). Now the supports of u^+ and u^- intersect only on the set $\{x \mid u = 0\}$. On this set u^+ and u^- also take the value zero and $\dfrac{\partial u^+}{\partial x_i}, \dfrac{\partial u^-}{\partial x_i}$ are zero a.e. on this set for $1 \leqslant i \leqslant n$. Hence we get, using $u = u^+ - u^-$, that

$$-\int_\Omega a_{ij} \frac{\partial u^-}{\partial x_i} \frac{\partial u^-}{\partial x_j} - \int_\Omega a_0 (u^-)^2 = \int_\Omega fu^-$$

The right hand side is $\geqslant 0$ since $f \geqslant 0$, $u^- \geqslant 0$. Hence from the ellipticity condition we get that

$$|u^-|_{1,\Omega}^2 \leqslant 0$$

and so (by Poincaré's inequality), $u^- = 0$. Hence $u = u^+ \geqslant 0$.

(ii) Let $a_0 \equiv 0$. Let

$$m = \inf_\Gamma u,$$

Since $a_0 \equiv 0$, the function $u - m$ also satisfies (3.5.1). Now as $f \geqslant 0$, we are in case (i) and so $u - m \geqslant 0$. Hence $u \geqslant m$.

(iii) If $f \equiv 0$ as well $-u$ also satisfies (3.5.1) and we now use case (ii). ∎

Remark 3.5.1 The above theorem is sometimes referred to as the weak maximum principle. ∎

Remark 3.5.2 The proof was based on the fact that if $u \in H^1(\Omega)$ then u^+

and $u^- \in H^1(\Omega)$ and this is a special property of the spaces $W^{1,p}(\Omega)$ only and not the higher order Sobolev spaces. Another important fact used more subtly in the proof was that when $u = 0$ on a set $E \subset \Omega$ then $\frac{\partial u}{\partial x_i} = 0$ a.e. on E. This result is a special case of a result of Stampacchia which states that $\nabla u = 0$ a.e. on the set $E = \{x | u(x) = \alpha\}$ where α is any fixed constant. It may happen that E itself is of measure zero and in this case we have no information on ∇u on this set. If meas $(E) > 0$ then $\nabla u = 0$ a.e. on this set. For a proof of this result, see Appendix 4. ∎

Remark 3.5.3 If $a_0 \geqslant 0$ the results of this theorem are also valid for the general second order elliptic equation of the form (3.2.15) with the variational formulation (3.2.16). For a proof, see Gilbarg and Trudinger [1]. Thus if $u = 0$ on Γ and $f = 0$ on Ω we will have from (3.5.2) that $u \geqslant 0$ and also $-u \geqslant 0$ (since $-u$ will also be a solution). Thus $u \equiv 0$ and this proves the uniqess of the solution. Then by Theorem 3.2.2, we have existence of a solution for all $f \in L^2(\Omega)$ (cf. Remark 3.2.4). ∎

Remark 3.5.4 The hypothesis $u \in C(\bar{\Omega})$ will be verified by a weak solution in certain cases by virtue of the regularity theorems and the Sobolev imbedding theorems. If, for instance, Γ is smooth enough, we know that $f \in L^2(\Omega)$ implies $u \in H^2(\Omega)$. If $n \leqslant 3$ then $H^2(\bar{\Omega}) \to C(\bar{\Omega})$ and thus $u \in C(\bar{\Omega})$ is automatically verified. ∎

We will now pass over to a stronger version of the maximum principle. We illustrate it in the case of the Laplace operator.

Let us denote by $\phi(\cdot)$ the following function:

$$\phi(\rho) = \begin{cases} \dfrac{1}{2\pi} \log \rho, & \text{if } n = 2 \\[2ex] \dfrac{-1}{(n-2)\omega_n} \rho^{2-n}, & \text{if } n \geqslant 3 \end{cases} \tag{3.5.4}$$

for $\rho \geqslant 0$, where ω_n is the surface measure of the unit sphere in $\mathbb{R}^n$. Notice that $\phi(|x - y|)$ is a fundamental solution at x of the Laplace operator (cf. Theorems 1.7.1 and 1.7.2).

Let Ω be a bounded domain in $\mathbb{R}^n$ with boundary Γ and let $x \in \Omega$. For $\epsilon > 0$ small enough, the ball $B(x; \epsilon)$ is contained in Ω. On the open set $\Omega_\epsilon = \Omega \setminus \overline{B(x; \epsilon)}$, the function $y \to \phi(|x - y|)$ denoted by $\phi_x(y)$, is a smooth function. Thus if $u \in C^2(\Omega) \cap C(\bar{\Omega})$, we have by Green's formula (cf. (2.7.2))

$$\int_{\Omega_\epsilon} (u\, \Delta\phi_x - \Delta u \cdot \phi_x) = \int_{\Gamma_\epsilon} \left(u \frac{\partial \phi_x}{\partial \nu} - \frac{\partial u}{\partial \nu} \phi_x \right)$$

where $\Gamma_\epsilon = \partial\Omega_\epsilon = \Gamma \cup \{y |\, |x - y| = \epsilon\}$. Letting $\epsilon \to 0$ and imitating the proof of Theorems 1.7.1 and 1.7.2, we get

$$u(x) = \int_\Omega \Delta u \cdot \phi_x - \int_\Gamma \left(\phi_x \frac{\partial u}{\partial \nu} - u \frac{\partial \phi_x}{\partial \nu} \right). \tag{3.5.5}$$

Now for $x \in \Omega$ apply the above result with Ω replaced by $B(x; \rho)$ where $\rho > 0$ such that $B(x; \rho) \subset \Omega$. Also the above equation is valid with ϕ_x replaced by $\phi_x - c$ for any constant c (since $\phi(|x - y|) - c$ is also a fundamental solution) and we choose $c = \phi(\rho)$. Then $\phi_x - c$ will vanish on the boundary of $B(x; \rho)$. Also

$$\frac{\partial}{\partial \nu}(\phi_x - \phi(\rho)) = \left.\frac{\partial \phi}{\partial r}\right|_{r=\rho} = \frac{1}{\omega_n} \rho^{1-n}, \; n \geqslant 2.$$

Substituting this in (3.5.5), we get

$$u(x) = \int_{B(x;\, \rho)} \Delta u(y)(\phi(|x - y|) - \phi(\rho))\, dy + \frac{1}{\omega_n \rho^{n-1}} \int_{|x-y|=\rho} u(y)\, dS(y) \tag{3.5.6}$$

If $\Delta u = 0$, we get the **Gauss Law of the Arithmetic Mean** for a harmonic function, viz.,

$$u(x) = \frac{1}{\omega_n \rho^{n-1}} \int_{|x-y|=\rho} u(y)\, dS(y). \tag{3.5.7}$$

If $\Delta u \leqslant 0$, then as $\phi(\cdot)$ is a monotonically increasing function we get that the integrand of the first integral of (3.5.6) is non-negative. Hence if $\Delta u \leqslant 0$, we have

$$u(x) \geqslant \frac{1}{\omega_n \rho^{n-1}} \int_{|x-y|=\rho} u(y)\, dS(y). \tag{3.5.8}$$

We can use this inequality to prove the following result, known as the **Strong Maximum Principle.**

Theorem 3.5.2 Let Ω be a bounded connected open set in $\mathbf{R}^n$ and $u \in C^2(\Omega) \cap C(\bar{\Omega})$ such that $\Delta u \leqslant 0$ in Ω. Then either u is a constant (and so $\Delta u \equiv 0$) or

$$u(x) > \inf_{\Omega} u, \quad \text{for every } x \in \Omega. \tag{3.5.9}$$

Proof Let $m = \inf u$ and write $\Omega = \Omega_1 \cup \Omega_2$ where

$$\Omega_1 = \{x \in \Omega \mid u(x) = m\}$$

and

$$\Omega_2 = \{x \in \Omega \mid u(x) > m\}.$$

Clearly as u is continuous, Ω_2 is an open set. Let $x \in \Omega_1$. If $\rho > 0$ is small enough so that $B(x; \rho) \subset \Omega$, we have by virtue of (3.5.8) that

$$0 \geqslant \int_{|x-y|=\rho} u(y)\, dS(y) - \omega_n \rho^{n-1} u(x)$$

$$= \int_{|x-y|=\rho} (u(y) - m)\, dS(y)$$

since $u(x) = m$. But then $u(y) - m \geqslant 0$ and so

$$\int_{|x-y|=\rho} (u(y) - m)\, dS(y) = 0$$

or $u(y) = m$ for all y such that $|x - y| = \rho$. The same result then will hold for all $0 \leqslant \rho_1 < \rho$. Thus $u(y) = m$ for all $y \in B(x; \rho)$ and thus Ω_1 is also open. Hence by the connectedness of Ω, we have $\Omega = \Omega_1$, in which case u is a constant or $\Omega = \Omega_2$ in which case (3.5.9) is true. ∎

Remark 3.5.5 The strong maximum principle is also available for operators of the form

$$- \sum_{i,j=1}^{n} \frac{\partial}{\partial x_i}\left(a_{ij}\frac{\partial}{\partial x_j}\right) + \sum_{i=1}^{n} a_i \frac{\partial}{\partial x_i} + a_0,$$

where $a_{ij} \in C^1(\bar{\Omega})$ and satisfy the ellipticity condition (3.2.11), $a_i\, a_0 \in C(\bar{\Omega})$, $a_0 \geqslant 0$. This is a result due to Hopf and for a proof, see Gilbarg and Trudinger [1] or Protter and Weinberger [1]. ∎

Corollary Let $u \in C^2(\Omega) \cap C(\bar{\Omega})$, $-\Delta u = f \geqslant 0$ in Ω and $u = 0$ on Γ. Then $u \equiv 0$ in Ω (so that $f \equiv 0$) or $u > 0$ in Ω. ∎

3.6 EIGENVALUE PROBLEMS

In this section we will consider the eigenvalue problems

$$\left.\begin{aligned} -\Delta w &= \lambda w, \quad \text{in } \Omega \\ w &= 0, \quad \text{on } \Gamma \end{aligned}\right\} \tag{3.6.1}$$

for a bounded open set $\Omega \subset \mathbb{R}^n$.

Theorem 3.6.1 There exists an orthonormal basis $\{w_m\}$ of $L^2(\Omega)$ and a sequence of positive real numbers $\{\lambda_m\}$ with $\lambda_m \to \infty$ as $m \to \infty$, such that

$$0 < \lambda_1 \leqslant \lambda_2 \leqslant \ldots \leqslant \lambda_m \leqslant \ldots,$$

$$\left.\begin{aligned} -\Delta w_m &= \lambda_m w_m \quad \text{on } \Omega \\ w_m &\in H_0^1(\Omega) \cap C^\infty(\Omega) \end{aligned}\right\} \tag{3.6.2}$$

Proof Given $f \in L^2(\Omega)$ we define $Gf \in H_0^1(\Omega)$ as the weak solution of the problem $-\Delta u = f$ in Ω, $u = 0$ on Γ. Thus for every $v \in H_0^1(\Omega)$ we have

$$\int_\Omega \nabla(Gf) \cdot \nabla v = \int_\Omega fv. \tag{3.6.3}$$

Now as Ω is bounded, $H_0^1(\Omega)$ is compactly embedded in $L^2(\Omega)$ by the Rellich-Kondrasov Theorem. Thus G considered as a map of $L^2(\Omega)$ into itself is compact. G is self-adjoint, for

$$\int_\Omega Gf \cdot g = \int_\Omega \nabla(Gf) \cdot \nabla(Gg) = \int_\Omega f \cdot Gg. \tag{3.6.4}$$

Also

$$\int_\Omega (Gf)f = \int_\Omega \nabla(Gf) \cdot \nabla(Gf) = |Gf|_{1,\Omega}^2 > 0, \text{ if } f \neq 0. \tag{3.6.5}$$

It follows then from the theory of compact, self-adjoint positive definite operators that there exists an orthonormal basis of eigenfunctions $\{w_m\}$ in $L^2(\Omega)$ and a sequence of positive eigenvalues $\{\mu_m\}$ decreasing to zero such that

$$Gw_m = \mu_m w_m. \tag{3.6.6}$$

Thus if we set $\lambda_m = \mu_m^{-1}$ then

$$w_m = G(\lambda_m w_m) \tag{3.6.7}$$

Since the range of G is in $H_0^1(\Omega)$, we have infact that $w_m \in H^1(\Omega)$ and by (3.6.3) and (3.6.7) we get that

$$\int_\Omega \nabla w_m \cdot \nabla v = \lambda_m \int_\Omega w_m v \tag{3.6.8}$$

for every $v \in H_0^1(\Omega)$. This implies that w_m satisfies (3.6.2) in the sense of distributions.

To show that $w_m \in C^\infty(\Omega)$, let $x \in \Omega$ and $B(x; r)$ a ball around x with radius $r > 0$ such that $B(x, r) \subset \Omega$. Now as $w_m \in L^2(B(x; r))$ and satisfies $-\Delta w_m = \lambda_m w_m$ in that domain, we have $w_m \in H^2(B(x; r))$, which in turn implies that $w_m \in H^4(B(x; r))$ and so on, by the interior regularity results (cf. Section 3.3). Thus $w_m \in H^k(B(x; r))$ for all k and so by the Sobolev imbedding theorems, $w_m \in C^\infty(B(x; r))$. Since $x \in \Omega$ was arbitrary, it follows that $w_m \in C^\infty(\Omega)$. ▮

Remark 3.6.1 If Ω is of class C^∞, we even have $w_m \in C^\infty(\overline{\Omega})$. ▮

Remark 3.6.2 If $H_0^1(\Omega)$ is provided with the inner-product

$$(u, v) = \int_\Omega \nabla u \cdot \nabla v,$$

then $\{\lambda_m^{-1/2} w_m\}$ is an orthonormal basis for $H_0^1(\Omega)$. Indeed

$$(\lambda_k^{-1/2} w_k, \lambda_m^{-1/2} w_m) = \lambda_k^{-1/2}\lambda_m^{-1/2} \int_\Omega \nabla w_k \, \nabla w_m = \lambda_k^{1/2}\lambda_m^{-1/2} \int_\Omega w_k w_m$$
$$= \delta_{km}.$$

Also if $u \in H_0^1(\Omega)$ such that $(u, w_m) = 0$ for all m, then

$$0 = \int_\Omega \nabla u \cdot \nabla w_m = \lambda_m \int_\Omega u w_m.$$

Thus $\int_\Omega u w_m = 0$ for all m and as $u \in L^2(\Omega)$ and $\{w_m\}$ is complete in $L^2(\Omega)$, we have $u = 0$. Thus $\{\lambda_m^{-1/2} w_m\}$ is an orthonormal basis for $H_0^1(\Omega)$. ▮

We now give a variational characterization of the eigenelements (w_m, λ_m) described above.

Theorem 3.6.2 Define, for $v \in H_0^1(\Omega)$, $v \neq 0$, the Rayleigh quotient

$$R(v) = \frac{|v|_{1,\Omega}^2}{|v|_{0,\Omega}^2} = \frac{\int_\Omega \nabla v \cdot \nabla v}{\int_\Omega v^2}. \tag{3.6.9}$$

Then

$$\lambda_1 = \min_{\substack{v \in H_0^1(\Omega) \\ v \neq 0}} R(v) = R(w_1). \tag{3.6.10}$$

$$\lambda_m = R(w_m) = \max_{\substack{v \in \text{span}\{w_1, \ldots, w_m\} \\ v \neq 0}} R(v) \tag{3.6.11}$$

$$\lambda_m = \min_{\substack{v \perp \{w_1, \ldots, w_{m-1}\} \\ v \neq 0}} R(v) \tag{3.6.12}$$

$$\lambda_m = \min_{\substack{W \subset H_0^1(\Omega) \\ \dim W = m}} \max_{v \in W} R(v) \tag{3.6.13}$$

for $m \geqslant 2$.

Proof: Let $W_k = \text{span}\{w_1, \ldots, w_k\}$. Clearly for all $m \geqslant 1$, $\lambda_m = R(w_m)$. If $v \perp \{w_1, \ldots, w_{m-1}\}$ then the Fourier expansion in $L^2(\Omega)$ of v is given by

$$v = \sum_{k=m}^{\infty} \alpha_k w_k, \quad \alpha_k = \int_\Omega v w_k$$

If we set $v_l = \sum_{k=m}^{l} \alpha_k w_k$, then $v_l \to v$ in $L^2(\Omega)$ and in $H_0^1(\Omega)$ as well, for, the Fourier expansion in $H_0^1(\Omega)$ will be

$$v = \sum_{k=m}^{\infty} \lambda_k^{-1/2} \beta_k w_k$$

with

$$\beta_k = \lambda_k^{-1/2} \int_\Omega \nabla v \cdot \nabla w_k = \lambda_k^{1/2} \int_\Omega v w_k = \lambda_k^{1/2} \alpha_k.$$

Thus

$$v = \sum_{k=m}^{\infty} \alpha_k w_k.$$

Thus $R(v_l) \to R(v)$. Now

$$R(v_l) = \frac{\sum_{k=m}^{l} \alpha_k^2 \int_\Omega \nabla w_k \cdot \nabla w_k}{\sum_{k=m}^{l} \alpha_k^2} = \frac{\sum_{k=m}^{l} \alpha_k^2 \lambda_k}{\sum_{k=m}^{l} \alpha_k^2}$$

$$\geqslant \lambda_m$$

Thus $R(v) \geqslant \lambda_m$ and (3.6.12) follows since the minimum is attained at w_m. The case $m = 1$ gives (3.6.10).

If $v \in W_m$ then $v = \sum_{k=1}^{m} \alpha_k w_k$ and thus

$$R(v) = \frac{\sum_{k=1}^{m} \lambda_k \alpha_k^2}{\sum_{k=1}^{m} \alpha_k^2} \leqslant \lambda_m$$

Thus the maximum of $R(v)$ over W_m is $\leqslant \lambda_m$ but it is attained at $w_m \in W_m$. Hence (3.6.11) follows.

Finally let W be a finite dimensional subspace of dimension m. Then there exists $w \in W$ such that $w \perp w_1, \ldots, w_{m-1}$ (since these only determine $m-1$ conditions of an m-dimensional vector). Hence $R(w) \geqslant \lambda_m$ and so

$$\max_{v \in W} R(v) \geqslant \lambda_m.$$

This is true for any m-dimensional subspace and so

$$\min_{\dim W = m} \max_{v \in W} R(v) \geqslant \lambda_m \tag{3.6.14}$$

and the maximum over the m-dimensional subspace W_m is exactly λ_m. Hence we have equality in (3.6.14) and so (3.6.13) holds. ∎

Remark 3.6.3 The characterization (3.6.13) of λ_m is intrinsic in the sense it is independent of the eigenfunctions chosen. ∎

We will now prove a very important and useful property of the eigenpair (w_1, λ_1).

Lemma 3.6.1 Let $w \neq 0$ in $H_0^1(\Omega)$ satisfy the relation

$$R(w) = \lambda_1. \tag{3.6.15}$$

Then w is an eigenfunction corresponding to λ_1.

Proof Let $v \in H_0^1(\Omega)$ be any element. Let $t > 0$. Then $w + tv \in H_0^1(\Omega)$ and by (3.6.10),

$$R(w + tv) \geqslant \lambda_1 = R(w).$$

By normalizing w, we can assume, without loss of generality, that $|w|_{0,\Omega}^2 = 1$ Thus

$$\frac{\int_\Omega \nabla(w + tv) \cdot \nabla(w + tv)}{\int_\Omega (w + tv)^2} \geqslant \int_\Omega \nabla w \cdot \nabla w = \lambda_1.$$

Cross multiplying by $\int_\Omega (w + tv)^2$ and simplifying we get

$$t^2 \int_\Omega \nabla v \cdot \nabla v + 2t \int_\Omega \nabla w \cdot \nabla v \geqslant \lambda_1 \left(2t \int_\Omega vw + t^2 \int_\Omega v^2 \right)$$

Divide throughout by $2t$ and then let $t \to 0$. Thus

$$\int_\Omega \nabla w . \nabla v = \lambda_1 \int_\Omega vw$$

and as $v \in H_0^1(\Omega)$ is arbitrary, the lemma is proved. ∎

Theorem 3.6.3 The first eigenvalue λ_1 is simple and the corresponding eigenfunction does not change sign in Ω. Thus we can choose w_1 such that $w_1 > 0$ in Ω (which is assumed to be connected).

Proof Let w be any eigenfunction corresponding to λ_1. We write $w = w^+ - w^-$. Now as $w^+, w^- \in H_0^1(\Omega)$ as well,

$$\int_\Omega \nabla w \cdot \nabla w^+ = \lambda_1 \int_\Omega ww^+$$

But as in Theorem 3.5.1 (see also Remark 3.5.2) this implies that

$$\int_\Omega \nabla w^+ \cdot \nabla w^+ = \lambda_1 \int_\Omega (w^+)^2 \tag{3.6.16}$$

and similarly

$$\int_\Omega \nabla w^- \cdot \nabla w^- = \lambda_1 \int_\Omega (w^-)^2. \tag{3.6.17}$$

If w does not have constant sign in Ω then both w^+ and w^- are not identically zero. Thus we deduce from (3.6.16) and (3.6.17) that

$$R(w^+) = \lambda_1 = R(w^-).$$

Hence by Lemma 3.6.1, both w^+ and w^- are eigenfunctions for λ_1, i.e. $-\Delta w^+ = \lambda_1 w^+$ and $-\Delta w^- = \lambda_1 w^-$. But as $w^+, w^- \geqslant 0$ by the strong maximum principle, $w^+ > 0$ and $w^- > 0$ (since neither is identically zero). But that is impossible since w^+ must vanish when $w^- \neq 0$ and vice versa. Thus w has constant sign and is non-zero inside Ω. Hence we can choose a $w_1 > 0$ in Ω which is an eigenfunction corresponding to λ_1.

If λ_1 were not simple, then there exists another eigenfunction orthogonal to w_1, say, $\widetilde{w}_1$. But then $\widetilde{w}_1$ must also have constant sign and then it is impossible for the integral

$$\int_\Omega w_1 \widetilde{w}_1$$

to vanish. Thus λ_1 is a simple eigenvalue. ∎

Remark 3.6.4 This theorem is usually proved as a consequence of the Krein-Rutman Theorem. In finite dimensional spaces we have the analogous result known as the Perron-Frobenius Theorem which states that for a non-negative irreducible matrix the spectral radius is a simple eigenvalue and that there exists a strictly positive eigenvector. Here the operator G defined in Theorem 3.6.1 can be shown to have analogous infinite dimensional properties and so its spectral radius μ_1 is a simple eigenvalue

with eigenfunction of constant sign. The proof given above was communicated by P. Rabier [1] and modified by the author. The advantage of this proof is that it works for any elliptic eigenvalue problem of the form: $u \in H_0^1(\Omega)$

$$\int_\Omega \sum_{i,j=1}^{n} a_{ij} \frac{\partial u}{\partial x_j} \frac{\partial v}{\partial x_i} + \int_\Omega a_0 uv = \lambda \int_\Omega uv, \tag{3.6.18}$$

for every $v \in H_0^1(\Omega)$, with $\{a_{ij}\}$ symmetric and $a_0 \geqslant 0$ so that the corresponding Rayleigh quotient characterization and the strong maximum principle work.

Remark 3.6.5 To apply the strong maximum principle we also need $u \in C(\bar{\Omega})$. This comes for even Lipschitzian domains Ω when $n \leqslant 3$ since $u \in H^2(\Omega)$ by regularity and $H^2(\Omega) \subset C(\bar{\Omega})$ when $n \leqslant 3$. Otherwise we need to assume more smoothness on the domain so that we have successively higher and higher regularity of u to get $u \in C(\bar{\Omega})$ by Sobolev inclusion. ∎

Remark 3.6.6 Just as the problem $-\Delta u = f$ in Ω, $u = 0$ on Γ modelled the displacement of a membrane fixed along Γ and acted on by a force f (cf. Example 1.1.2) the eigenvalue problems describes the vibration of a membrane fixed along Γ ($\Omega \subset \mathbb{R}^2$). The harmonics of a membrane fixed along Γ are given by $w_m(x) \sin \sqrt{\lambda_m} t$. In this connection we should mention the problem of deducing geometric properties of a domain when the eigenvalues are known for the Laplace operator with Dirichlet boundary conditions. In particular if two domains Ω_1 and Ω_2 are such that they have the same eigenvalues, then are they isometric to each other? This is a famous open problem and M. Kac [1] has named it "can one hear the shape of a drum" in keeping with the physical interpretation given above. ∎

3.7 INTRODUCTION TO THE FINITE ELEMENT METHOD

Let V be a real Hilbert space and let $a : V \times V \to \mathbb{R}$ be a V-elliptic and continuous bilinear form. Let $f \in V$. Then by the Lax-Milgram Theorem there exists a unique $u \in V$ such that

$$a(u, v) = (f, v), \quad \text{for ever } v \in V. \tag{3.7.1}$$

We have seen in Section 3.2 several examples of this set-up. We now turn to the approximation of the solution u. The Galerkin method described in Section 3.4 gives us an idea as to how to proceed to do this.

Let $h > 0$ be a parameter. To it we associate V_h which is a *finite dimensional subspace* of V. Now consider the following problem: find $u_h \in V_h$ such that

$$a(u_h, v_h) = (f, v_h), \quad \text{for every } v_h \in V_h. \tag{3.7.2}$$

Since $a(\cdot, \cdot)$ is continuous and V-elliptic, it is so on V_h as well and by the Lax-Milgram Theorem, u_h exists uniquely.

The calculation of u_h reduces to the solution of a system of linear equations. Let $N = N(h)$ be the dimension of V_h. Let $\{w_h^1, \ldots, w_h^N\}$ be a basis for V_h. Then (3.7.2) is equivalent to

$$a(u_h, w_h^i) = (f, w_h^i), \quad 1 \leqslant i \leqslant N. \tag{3.7.3}$$

Now write $u_h = \sum_{j=1}^{N} u_h^j w_h^j$. Substituting this into (3.7.3) we get

$$\sum_{j=1}^{N} a(w_h^j, w_h^i) u_h^j = (f, w_h^i), \quad 1 \leqslant i \leqslant N.$$

This is a system of linear equations in the N unknowns $(u_h^1, \ldots, u_h^N)$ and can be put in the matrix form

$$A_h U_h = F_h \tag{3.7.4}$$

where $A_h = (a_{ij}^h)$, $1 \leqslant i, j \leqslant N$ is an $N \times N$ matrix, given by

$$a_{ij}^h = a(w_h^j, w_h^i) \tag{3.7.5}$$

and U_h has ith component u_h^i and F_h has ith component (f, w_h^i).

It is easy to see that if $a(\cdot\,, \cdot)$ is symmetric so is A_h and the ellipticity of $a(\cdot\,, \cdot)$ translates into the positive definiteness of A_h : if $\xi \in \mathbb{R}^N$ then

$$\xi^T A_h \xi = \sum_{i,j=1}^{N} a_{ij}^h \xi_i \xi_j = \sum_{i,j=1}^{N} a(w_h^j, w_h^i) \xi_j \xi_j$$
$$= a(v, v) > 0 \quad \text{if } \xi \neq 0,$$

where $v = \sum_{i=1}^{N} \xi_i w_h^i$.

Thus the linear system can be set up once the basis of V_h is known and can be solved by any of the standard methods of solving linear systems.

Let us look at the error analysis.

Theorem 3.7.1 (*Céa*) There exists a constant $C > 0$, which does not depend on h, such that

$$\|u - u_h\| \leqslant C \inf_{v_h \in V_h} \|u - v_h\|. \tag{3.7.6}$$

Proof We have

$$\alpha \|u - u_h\|^2 \leqslant a(u - u_h, u - u_h)$$
$$= a(u - u_h, u - v_h) + a(u - u_h, v_h - u_h)$$

for any $v_h \in V_h$. But from (3.7.1) and (3.7.2) it follows that

$$a(u - u_h, v_h - u_h) = 0.$$

Thus by the continuity of $a(\cdot\,, \cdot)$,

$$\alpha \|u - u_h\|^2 \leqslant M \|u - u_h\| \, \|u - v_h\|$$

from which, on taking the infinum over $v_h \in V_h$, (3.7.6) will follow with $C = M/\alpha$. ∎

The importance of the above result is that it reduces the study of the error in the solution to the study of the *interpolation error* between the spaces V and V_h, which is a problem in approximation theory. In particular given $u \in V$ if we can prove there exists $\Pi_h u \in V_h$ an 'interpolated function' such that

$$\|u - \Pi_h u\| \leqslant \phi(h) \tag{3.7.7}$$

where $\phi(h) \to 0$ as $h \to 0$ then we will have by the above theorem,

$$\|u - u_h\| \leqslant C\|u - \Pi_h u\| \leqslant C\phi(h)$$

and so we can prove convergence of the approximate solution u_h to u as $h \to 0$, and infact have an order of convergence as well.

The finite element method is a systematic method to construct subspaces (V_h) of the various Sobolev spaces (V) so that (i) $V_h \subset V$ and (ii) study error estimates via a rigorous interpolation theory. The finite element analysis of a problem will proceed along the following lines:

Step 1 Given an elliptic boundary value problem, put it in weak form and prove the existence and uniqueness of the weak solution by the Lax-Milgram Theorem.

Step 2 Once the space V, the bilinear form $a(\cdot\,,\cdot)$ and $f \in V$ are identified, choose the spaces $V_h \subset V$.

Step 3 Identify a basis $\{w_h^i\}$ of V_h such that the matrix A_h is as sparse as possible, i.e. $a_{ij}^h = 0$ for as many entries as possible. This is only for the convenience of implementation on the computer. A banded matrix can be easily stored even when its order is high. Note that as $h \to 0$ we expect V_h to be 'closer and closer' to V (cf (3.7.7)) and so $N = N(h)$ will tend to infinity. Thus A_h will be a very large matrix.

Step 4 Error analysis. First study the interpolation error between V and V_h. Usually it will be possible to obtain an inequality like (3.7.7) only when $u \in V$ and is sufficiently regular (for instance if $V = H_0^1(\Omega)$, we will need u to be atleast in $H^2(\Omega)$). Thus we will be called upon to use the regularity theorems available for weak solutions. ∎

In Section 3.2, we have seen several examples of the programme set out in Step 1 above. Thus we will have a bounded domain Ω and a problem set out in a space V which will usually be a subspace of $H^1(\Omega)$ (if the problem is of the second order) $H^2(\Omega)$ (for the fourth order problems).

The basic idea to construct the spaces V_h is to partition the domain into smaller entities, say, triangles. Assume, for simplicity, that Ω is a polygonal domain. A triangulation $\mathcal{T}_h$ of Ω is a partition of Ω into triangles (i.e., 'n'-simplices in $\mathbb{R}^n$) such that each 'side' is either part of the boundary Γ or is a 'side' of an adjacent triangle. Thus we do not allow triangles as shown in Fig. 13.

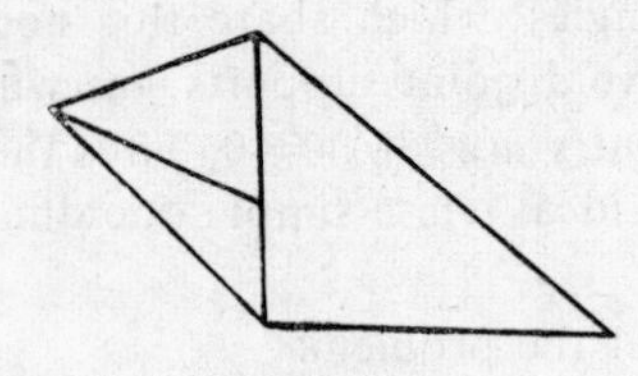

Fig. 13

Thus we have a triangulation as follows:

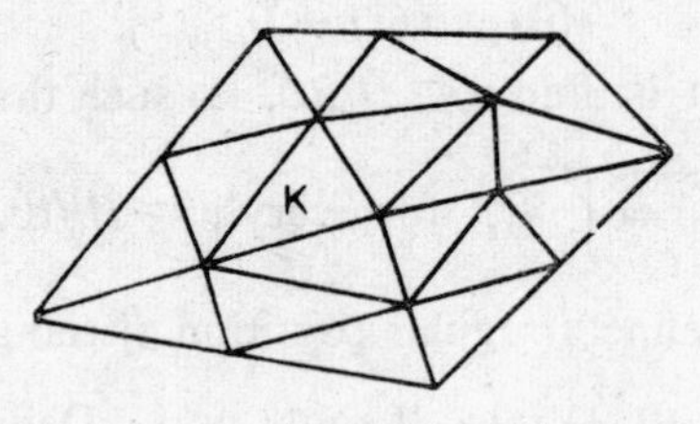

Fig. 14

We denote by K any generic triangle in $\mathcal{T}_h$. The 'h' now stands for

$$h = \max_{K \in \mathcal{T}_h} \operatorname{diam}(K)$$

so that as $h \to 0$ the triangles are smaller and smaller and we get a fine mesh on Ω.

Now we say that V_h must be a space of piecewise polynomials with respect to $\mathcal{T}_h$. That is we fix an integer k and if P_k stands for polynomials of degree $\leqslant k$ in n-variables, then we wish that if $v \in V_h$ then $v|_K \in P_k$ for every $K \in \mathcal{T}_h$. Thus V_h is clearly finite dimensional and once k is fixed, the dimension of V_h can be increased by just refining the given triangulation, Thus as $h \to 0$, $N(h) \to \infty$.

When do we have $V_h \subset V$? If $V \subset H^1(\Omega)$ then $V_h \subset V$ if and only if $V_h \subset C(\bar{\Omega})$ (cf. Exercise 2.15). Thus if $V = H_0^1(\Omega)$, then we need $V_h \subset C(\bar{\Omega})$ and $v_h = 0$ on $\partial\Omega = \Gamma$ for all $v_h \in V_h$. Similarly it can be proved that if $V \subset H^2(\Omega)$ then it is necessary and sufficient that $V_h \subset C^1(\bar{\Omega})$. Thus we need to verify that elements of V_h satisfy some inter-element continuity conditions; i.e., if K_1 and K_2 are adjacent triangles, then $v_h|_{K_1}$ and $v_h|_{K_2}$ must satisfy continuity conditions on their common side.

Once V_h (and thus the triangulation and the polynomial degree k) are fixed we proceed to construct a basis for V_h. For this purpose we identify a certain number of degrees of freedom of a polynomial of degree k, i.e. the polynomial is uniquely fixed once these values are known. It could be value of the polynomial at certain nodes or values of some of its derivatives as well. Once the degrees of freedom and the nodes where they are located are fixed, we associate to each node a basis function which has exactly one degree of freedom (at that node) equal to 1 and all other degrees of freedom zero. Thus dim $V_h = N(h)$, the number of nodes of the degrees of freedom. The advantage of this is that the support of a basis function

will be only those triangles which share this node and thus if two basis functions w_h^i and w_h^j have disjoint supports—as will often be the case—the corresponding matrix entry $a(w_h^j, w_h^i) = 0$. Thus the matrix will be sparse.

Let us illustrate these ideas via a simple one-dimensional example.

Example 3.7.1 Consider the problem:

$$\left.\begin{aligned} -\frac{d^2u}{dx^2} &= f \quad \text{in } (0, 1) \\ u(0) = u(1) &= 0 \end{aligned}\right\} \tag{3.7.8}$$

The weak formulation is: find $u \in H_0^1((0, 1))$ such that

$$\int_0^1 u'v' = \int_0^1 fv \quad \text{for every } v \in H_0^1((0, 1)). \tag{3.7.9}$$

Set $\Omega = (0, 1)$, and define a regular partition ('triangulation') as follows: Let $h = \dfrac{1}{N}$ and divide (0, 1) into N equal parts. Define

$$x_i = ih, \quad 0 \leqslant i \leqslant N. \tag{3.7.10}$$

We approximate $H_0^1(\Omega) = V$ by V_h, made up of piecewise linear functions. Thus if $v_h \in V_h$ then $v_h|_{[x_{i-1}, x_i]}$ must be linear for all $1 \leqslant i \leqslant N$ and since $V \subset H^1(\Omega)$, v_h must be continuous, i.e.

$$v_h\Big|_{[x_{i-1}, x_i]}(x_i) = v_h\Big|_{[x_i, x_{i+1}]}(x_i), \quad \text{for all } 1 \leqslant i \leqslant N-1. \tag{3.7.11}$$

Further as $V = H_0^1(\Omega)$, $v_h(x_0) = v_h(0) = 0$ and $v_h(x_N) = v_h(1) = 0$. A typical element of v_h will have a graph as shown in Fig. 15 below.

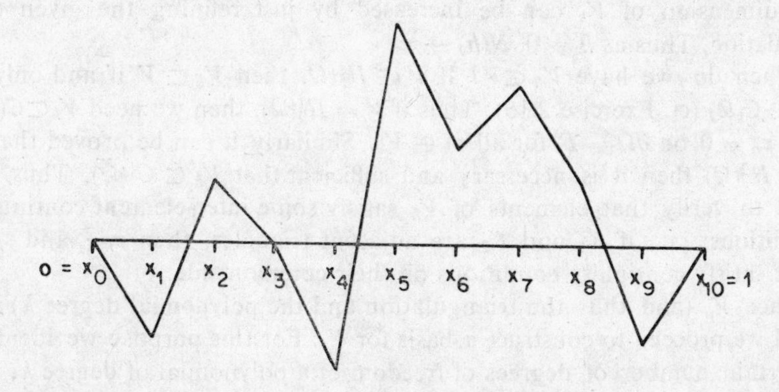

Fig. 15

Since the function is linear in each interval $[x_{i-1}, x_i]$, it is completely determined by its values at x_{i-1} and x_i in this interval. Thus the degrees of freedom are located at the nodes $\{x_i\}$ $1 \leqslant i \leqslant N-1$ (since x_0 and x_N have fixed values, $=0$, for $v_h \in V_h$) and we can now define the basis $\{w_h^i\}$, $1 \leqslant i \leqslant N-1$ as follows: w_h^i has value 1 at x_i and 0 at all x_j, $j \neq i$.

Thus

$$w_h^i(x) = \begin{cases} \frac{1}{h}(x - x_{i-1}), & \text{if } x \in [x_{i-1}, x_i] \\ \frac{1}{h}(x_{i+1} - x), & \text{if } x \in [x_i, x_{i+1}] \\ 0, & \text{otherwise.} \end{cases} \tag{3.7.12}$$

i.e. w_h^i is the 'roof function' shown below:

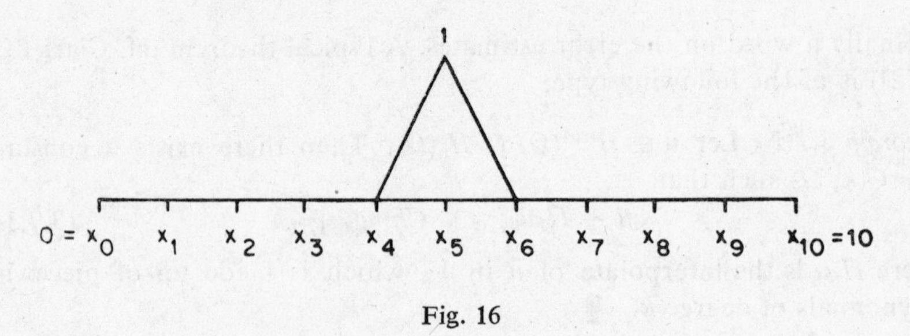

Fig. 16

Thus the support of w_h^i is $[x_{i-1}, x_{i+1}]$ and as

$$a(u, v) = \int_0^1 u'v'.$$

We have $a(w_h^i, w_h^j) = 0$ if $j \neq i-1, i, i+1$. Thus the matrix is *tridiagonal* in this case. We also have

$$a(w_h^i, w_h^i) = \frac{1}{h^2}\int_{x_{i-1}}^{x_i} dx + \frac{1}{h^2}\int_{x_i}^{x_{i+1}} dx = \frac{2}{h}.$$

$$a(w_h^i, w_h^{i+1}) = -\frac{1}{h^2}\int_{x_i}^{x_{i+1}} dx = -\frac{1}{h} = a(w_h^i, w_h^{i-1}).$$

Thus we get the linear system:

$$\frac{1}{h}\begin{bmatrix} 2 & -1 & & & \bigcirc \\ -1 & 2 & -1 & & \\ & \cdot & \cdot & \cdot & \\ & & \cdot & \cdot & \cdot \\ \bigcirc & & -1 & 2 & -1 \end{bmatrix} \begin{bmatrix} u_h(x_1) \\ u_h(x_2) \\ \vdots \\ u_h(x_{N-1}) \end{bmatrix} = \begin{bmatrix} \int_0^1 f w_h^1 \\ \int_0^1 f w_h^2 \\ \vdots \\ \int_0^1 f w_h^{N-1} \end{bmatrix} \tag{3.7.13}$$

The matrix above is symmetric, positive definite and tridiagonal. It can also be shown to be monotone, i.e. its inverse has all entries positive. Thus if $f \geqslant 0$, then as $w_h^i \geqslant 0$ for all i, the right hand side vector will have all components positive and so $u_h(x_i) > 0$ for all i. Thus the approximate solution also satisfies the strong maximum principle. ∎

Remark 3.7.1 Those who have worked with finite differences will realize

that the linear system (3.7.13) is essentially the same as that obtained by the finite difference method. Only the matrix will have a factor $1/h^2$ instead of $1/h$. This is balanced by the fact that the right hand side components will be of the form $f(x_i)$ instead of $\int_0^1 fw_h^i$. In fact in one-dimensional problems the finite element method has no special advantage. It is only in higher dimensions that it is advantageous as it can tackle in a systematic manner different geometries of the domain as well as different boundary conditions. ∎

Finally a word on the error estimates. A typical theorem (cf. Ciarlet [1] or [2]) is of the following type:

Theorem 3.7.2 Let $u \in H^{k+1}(\Omega) \cap H_0^1(\Omega)$. Then there exists a constant $C = C(k, \Omega)$ such that

$$\|u - \Pi_h u\|_{1,\Omega} \leqslant Ch^k|u|_{k+1,\Omega}, \tag{3.7.14}$$

where $\Pi_h u$ is the interpolate of u in V_h which is made up of piecewise polynomials of degree k. ∎

To apply the above theorem we need to use the regularity theorems. If $f \in L^2(\Omega)$ and Ω is a polygon, we only know that $u \in H^2(\Omega)$, u the solution of an elliptic problem with data f and Dirichlet boundary conditions. Thus it suffices to take $k = 1$, i.e. piecewise linear functions to get

$$\|u - u_h\|_{1,\Omega} \leqslant Ch|u|_{2,\Omega} \leqslant Ch|f|_{0,\Omega}. \tag{3.7.15}$$

If $u \notin H^{k+1}$ for a $k > 1$ then we cannot expect the order of convergence h^k by using polynomials of degree k. Thus the solution is not necessarily more accurate than when we use piecewise linear functions. The regularity of the solution thus dictates the optimal choice of the space V_h.

We will now illustrate the error analysis in the case of Example 3.7.1. The analysis in higher dimensions and for more general situations follows the same lines of argument but each step will be more technical.

Example 3.7.1 (contd.) Let $f \in L^2(\Omega)$ so that the solution u of (3.7.8) is in $H^2(\Omega)$, $\Omega = (0, 1)$. For a given partition of step size h, let u_h be the corresponding solution of the approximate problem. Now as $H^2(\Omega) \subset C^1(\bar{\Omega})$, $(n = 1)$, we can define the interpolate of u as follows:

$$\Pi_h u(x_i) = u(x_i),\ \Pi_h u \in V_h.$$

(cf. Fig. 17).

On any subinterval $[x_{i-1}, x_i]$, notice that

$$(\Pi_h u)'(x) = \frac{u(x_i) - u(x_{i-1})}{h} = \frac{1}{h}\int_{x_{i-1}}^{x_i} u'(t)\,dt = \bar{u}'_i,$$

where $\bar{u}'_i$ is the average of u' on that interval. Now we estimate $u - \Pi_h u$ in $H_0^1(\Omega)$. For this

$$|u' - (\Pi_h u)'|_{0,\Omega}^2 = \sum_{i=1}^{N}\int_{x_{i-1}}^{x_i} |u' - (\Pi_h u)'|^2\,dx \tag{3.7.16}$$

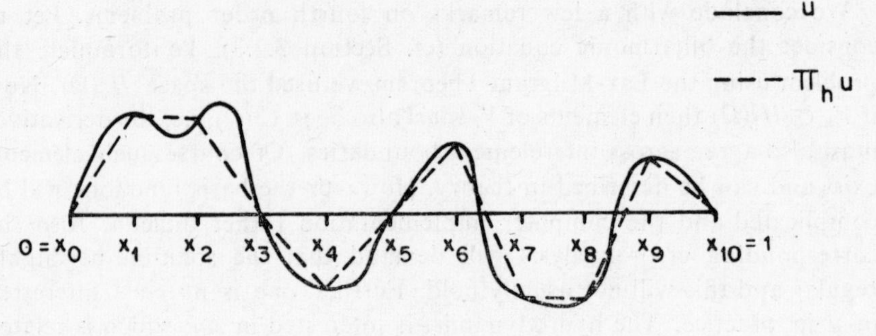

Fig. 17

Now,

$$\int_{x_{i-1}}^{x_i} |u' - (\Pi_h u)'|^2 \, dx = \int_{x_{i-1}}^{x_i} |u' - \bar{u}_i'|^2 \, dx.$$

Let v be any function in $H^1((x_{i-1}, x_i))$ and $\bar{v}$ its average. Set $y = (x - x_{i-1})/h$ and then $y \to x$ is a bijection of $[0, 1]$ onto $[x_{i-1}, x_i]$. Let

$$\hat{v}(y) = v(x) = v(x_{i-1} + yh)$$

Then $\hat{v} \in H^1((0, 1))$ and $\bar{\hat{v}} = \bar{v}$ as is easy to see from the change of variable. Thus we can write

$$\int_{x_{i-1}}^{x_i} |v - \bar{v}|^2 \, dx = h \int_0^1 |\hat{v} - \bar{\hat{v}}|^2 \, dy$$

$$\leqslant Ch \int_0^1 |\hat{v}'|^2 \, dy$$

where $C > 0$ depends only on $[0, 1]$, by the Poincaré-Wirtinger inequality (cf. Theorem 2.5.5). Again,

$$\frac{d\hat{v}}{dy}(y) = h \frac{dv}{dx}(x)$$

and so changing back to x, we get

$$\int_{x_{i-1}}^{x_i} |v - \bar{v}|^2 \, dx \leqslant \frac{Ch^3}{h} \int_{x_{i-1}}^{x_i} |v'|^2 \, dx = Ch^2 \int_{x_{i-1}}^{x_i} |v'|^2 \, dx.$$

Applying this to u' and substituting in (3.7.16) we get

$$|u' - (\Pi_h u)'|_{0,\Omega}^2 \leqslant Ch^2 \sum_{i=1}^{N} \int_{x_{i-1}}^{x_i} |u''|^2$$

$$= Ch^2 \int_0^1 |u''|^2 = Ch^2 |f|_{0,\Omega}^2.$$

Now by Céa's Theorem (Theorem 3.7.1) we get

$$|u - u_h|_{1,\Omega} \leqslant C|u - \Pi_h u|_{1,\Omega} \leqslant Ch|f|_{0,\Omega},$$

which gives the order of convergence. Notice that we only estimated $|u - u_h|_{1,\Omega}$ but the same estimate is valid for $\|u - u_h\|_{1,\Omega}$ by virtue of Poincaré's inequality.

We conclude with a few remarks on fourth order problems. Let us consider the biharmonic equation (cf. Section 3.2.3). To formulate the problem using the Lax-Milgram Theorem we used the space $H_0^2(\Omega)$. Now if $V_h \subset H_0^2(\Omega)$ then elements of V_h must also be in $C^1(\bar{\Omega})$, i.e. the derivatives must also agree across interelement boundaries. Of course such elements exist and can be described in theory. However the basis functions will be complicated and the computer implementation rather difficult. Also the corresponding error analysis will demand that the solution be highly regular and this will not usually hold. Further one is not just interested in u in practice. The hydrodynamist is interested in Δu (which is related to the vorticity) and structural analyst in all the second derivatives of u (which is related to the stress tensor). For all these reasons, a mixed formulation is preferred. For instance the formulation (3.2.40)-(3.2.41) gives both u and Δu simultaneously. Also the spaces involved are only $\tilde{\Sigma} = H^1(\Omega)$ and $\tilde{V} = H_0^1(\Omega)$. The construction of $\Sigma_h \subset \tilde{\Sigma}$ and $V_h \subset \tilde{V}$ now just involves piecewise polynomials which are in $C(\bar{\Omega})$ and so is much easier to do. It is possible to develop a detailed error analysis for such systems as well. For a complete description, see Girault and Raviart [1]. Such methods are called *mixed finite element methods.*

COMMENTS

1. *Variational Inequalities* The starting point for the study of variational inequalities is the theorem of Stampacchia (cf. Theorem 3.1.3). As already explained, when a quadratic functional is minimized over whole space we get equality conditions satisfied by the solutions. These are called the Euler equations in mechanics. When we minimize over a convex set we get inequalities. For a complete study of existence and regularity of solutions of variational inequalities, see Kinderlehrer and Stampacchia [1]. Variational inequalities have a lot of applications in mechanics and physics. They are useful in the study of *free boundary problems* (eg. *Stefan problems*). For a description of these applications, see Duvaut and Lions [1].

2. *Other Existence and Regularity Theories* In this chapter we have only presented the Hilbert space (or L^2) theory of elliptic boundary value problems. There is also an L^p-theory of existence and regularity for $1 < p < \infty$ and the principal reference for this is the work of Agmon, Douglis and Nirenberg [1]. The existence and regularity theory in classical spaces of Holder continuous functions is due to Schauder (cf. Gilbarg and Trudinger [1]). Most of these results extensively use the theory of singular integrals. These results can also be generalized to elliptic operators of order $2m$ and to elliptic systems.

Existence of solutions can be proved by a variety of ways when the variational method fails. For instance there is the method of sub- and super-solutions due to Perron (cf. F. John [1]). There is also the duality method for data in $L^1(\Omega)$ or even measures. In case the data is not regular

enough one can also use a density method wherein the data is approximated by smooth functions and for these problems which can be solved, *a proiri* estimates can be used and a solution to the original problem obtained by passing to the limit. For examples of these situations, see Brezis [1].

3. *Regularity Theorems* The proof of the regularity result for $\mathbb{R}^n_+$ presented in this chapter seems simpler than the usual proofs, but works well only for the Laplace operator. The usual proofs are based on the method of difference quotients due to Nirenberg (cf. Brezis [1]). This uses the characterization of Sobolev spaces via difference quotients (cf. Comment No. 6 of Chapter 2 and also Exercise 2.8). See also Agmon [1] and Friedman [1] for an approach using Garding's inequality.

We have stated regularity results only for the homogeneous boundary value problems. We can also state regularity results for inhomogeneous problems. For instance if $-\Delta u = f$ in Ω and $u = g$ on Γ with $f \in L^2(\Omega)$ and $g \in H^{3/2}(\Gamma)$, we can have $u \in H^2(\Omega)$ for suitable domains. Note that for $u \in H^2(\Omega)$, its trace on Γ must necessarily lie in $H^{3/2}(\Gamma)$. Similarly we can have higher order regularity for smooth data and smooth domains.

For regularity of solutions in domains with corners or angular points, see the work of Grisvard [1] or Kondrat'ev [1].

4. *Maximum Principles* Maximum principles are specially true for second order problems. However in some very special cases some forms of maximum principles are available for higher order equations. A standard reference for maximum principles is the book by Protter and Weinberger [1]. See also Gilbarg and Trudinger [1].

5. *The Finite Element Method* The first recorded use of the finite element method goes back to Courant in the 1940's when he suggested the use of piecewise linear functions to approximate the solution of variational problems. The major advances of this method—both in theory and practice—were seen in the late 60's and early 70's with the advent of high speed computers. At first it was only used for problems in structural engineering and elasticity. But with the discovery of mixed formulations the method has been successfully used to solve fluid flow problems.

The main advantages of this method are that it gives a unified mathematical basis to deal with different types of boundary conditions and can tackle any type of geometry while the finite difference method is best suited to rectangular domains. There is a rigourous error analysis and convergence theory.

We have given a very rudimentary treatment of this subject in this chapter. For a thorough mathematical treatment of direct methods (i.e. based on the Lax-Milgram Theorem) see Ciarlet [1] or [2]. For mixed methods see Girault and Raviart [1], where the emphasis is on fluid flow problems.

When the domain is not a polygon, we use curved triangles (or quadrilaterals) and these are called iso-parametric finite elements. Then the computational domain Ω_h differs from Ω. Even if $\Omega_h = \Omega$, we can use a $V_h \not\subset V$. Such methods are called non-conforming. Even when $V_h \subset V$, by virtue of numerical integration or other methods, it is possible that we replace $a(\cdot, \cdot)$ by another approximate bilinear form $a_h(\cdot, \cdot)$ or $(f, \cdot)$ by $(f_h, \cdot)$. These are called 'Variational Crimes' by G. Strang (cf. Strang and Fix [1]). They work well for a large class of problems (cf. Ciarlet [1] or [2]).

6. *Bibliography* The main reference for all aspects of second order elliptic problems is the book by Gilbarg and Trudinger [1]. See also Agmon [1] and Lions and Magenes [1]. For maximum principles see Protter and Weinberger [1]. Another important work both from the point of view of theory and applications is the book by Courant and Hilbert [1], especially for eigenvalue problems. For the finite element method see Ciarlet [1] and [2], Girault and Raviart [1] and Strang and Fix [1].

EXERCISES 3

3.1 (Robin Boundary Condition). Let $\Omega \subset \mathbb{R}^n$ be a bounded open set with boundary Γ. Let $\alpha > 0$. Define

$$a(u, v) = \int_\Omega \nabla u \cdot \nabla v + \int_\Omega uv + \alpha \int_\Gamma uv, \quad u, v \in H^1(\Omega).$$

Let $f \in L^2(\Omega)$. Show that there exists a unique $u \in H^1(\Omega)$ such that

$$a(u, v) = \int_\Omega fv, \quad \text{for every } v \in H^1(\Omega).$$

If u is smooth enough interpret u as the solution of a boundary value problem.

3.2 (a) Show that the bilinear form

$$a(u, v) = \int_\Omega \nabla u \cdot \nabla v + \left(\int_\Omega u\right)\left(\int_\Omega v\right), \quad u, v \in H^1(\Omega)$$

is $H^1(\Omega)$-elliptic, where $\Omega \subset \mathbb{R}^n$ is a bounded and connected open set.

(b) Deduce that if $f \in L^2(\Omega)$, there exists a unique $u \in H^1(\Omega)$ such that

$$a(u, v) = \int_\Omega fv, \quad \text{for every } v \in H^1(\Omega)$$

and that

$$\int_\Omega u = \frac{1}{\text{meas}\,(\Omega)} \int_\Omega f.$$

(c) If $\int_\Omega f = 0$, what is the boundary value problem solved by u?

3.3 (a) Consider the boundary value problem

$$-\frac{d^2u}{dx^2} + u = f \text{ in } (0, 1)$$

$$u(0) = u(1)$$

$$u'(0) = u'(1).$$

If $f \in L^2((0, 1))$ give a formulation of this problem in a suitable subspace of $H^1(\Omega)$ and prove the existence and uniqueness of the weak solution.

(b) If $f \in H^1((0, 1))$, show that the above weak solution is a classical solution.

3.4 (a) (The Obstacle Problem). Let $\Omega \subset \mathbb{R}^2$ be a bounded open set with boundary Γ. Let $\chi \in H^2(\Omega)$ with $\chi \leqslant 0$ on Γ. Define

$$K = \{v \in H_0^1(\Omega) \mid v \geqslant \chi \quad \text{a.e. in } \Omega\}.$$

Show that there exists a unique $u \in K$ such that

$$J(u) = \min_{v \in K} J(v)$$

where

$$J(v) = \tfrac{1}{2} \int_\Omega \nabla v \cdot \nabla v - \int_\Omega fv, \quad f \in L^2(\Omega).$$

(b) If u is smooth enough, show that it satisfies the conditions:

$$\left.\begin{aligned} u &\geqslant \chi \quad \text{in } \Omega \\ -\Delta u &= f \quad \text{on the set } \{x \mid u(x) > \chi(x)\} \\ u &= 0 \quad \text{on } \Gamma \end{aligned}\right\}.$$

(Note: This problem models the behaviour of a membrane fixed along Γ and stretched over an "obstacle" χ. The region of contact $C = \{x \in \Omega \mid u(x) = \chi(x)\}$ is unknown and on the (unknown) domain $\Omega \setminus C$, u satisfies a differential equation. This is an example of a *free boundary problem.*)

3.5 Let $\Omega \subset \mathbb{R}^n$ be a bounded open set and $x \in \Omega$. Show that there exists a function G_x such that

$$\left.\begin{aligned} -\Delta G_x &= \delta_x \quad (\text{in } \mathscr{D}'(\Omega)) \\ G_x &= 0 \quad \text{on } \Gamma \end{aligned}\right\}.$$

3.6 Let ϕ_x stand for the fundamental solution of the Laplacian given by

$$\phi_x(y) = \frac{1}{2\pi} \log |x - y|, \quad \text{when } n = 2$$

and

$$\phi_x(y) = \frac{-1}{(n-2)\omega_n |x - y|^{n-2}}, \quad \text{when } n \geqslant 3,$$

where ω_n is the surface measure of the unit sphere in $\mathbb{R}^n$.
Let $\Gamma = \{x \in \mathbb{R}^n | \ |x| = R\}$. Compute

$$\int_\Gamma \phi_x(y)\, dS(y)$$

for all x such that $|x| \leqslant R$.

3.7 Let $\Omega \subset \mathbb{R}^n$ be a bounded open set of class C^2. Show that the map $u \to |\Delta u|_{0,\Omega}$ defines a norm on $H^2(\Omega) \cap H_0^1(\Omega)$ equivalent to the norm $\|\cdot\|_{2,\Omega}$ (cf. Exercise 2.17).

3.8 Let V be a separable Hilbert space. Prove the Lax-Milgram Theorem for V by the Galerkin method.

3.9 (a) Let $\Omega \subset \mathbb{R}^n$ a bounded open set, $f \in L^2(\Omega)$ and $u \in H^1(\Omega) \cap C(\bar{\Omega})$ such that

$$\int_\Omega \nabla u \cdot \nabla v + \int_\Omega uv = \int_\Omega fv, \quad \text{for every } v \in H_0^1(\Omega).$$

Show that

$$\min \{\inf_\Gamma u, \inf_\Omega f\} \leqslant u(x) \leqslant \max \{\sup_\Gamma u, \sup_\Omega f\},$$

for every $x \in \Omega$.

(b) If $f = 0$ show that

$$|u|_{0,\infty,\Omega} \leqslant |u|_{0,\infty,\Gamma}.$$

3.10 Let $\Omega = \{x \in \mathbb{R}^n | \ |x| > 1\}$ and let $-\Delta u + u = 0$ on Ω, $u \in H^1(\Omega)$. Show that

$$|u|_{0,\infty,\Omega} \leqslant |u|_{0,\infty,\Gamma},$$

where $\Gamma = \{x | \ |x| = 1\}$.

3.11 (Maximum Principle for the Neumann Problem.) Let $\Omega \subset \mathbb{R}^n$ a bounded open set, $f \in L^2(\Omega)$ and $u \in H^1(\Omega)$ such that

$$\int_\Omega \nabla u \cdot \nabla v + \int_\Omega uv = \int_\Omega fv, \quad \text{for every } v \in H^1(\Omega).$$

Show that

$$\inf_\Omega f \leqslant u(x) \leqslant \sup_\Omega f \quad \text{a.e. } x \in \Omega.$$

3.12 Show that the family $\{\sqrt{2} \sin n\pi x)$, $n = 1, 2, \ldots$ is an orthonormal basis for $L^2((0, 1))$. Using the fact that the function $f(x) = x$ is in $L^2((0, 1))$, compute

$$\sum_{n=1}^{\infty} \frac{1}{n^2}.$$

3.13 Consider Poincaré's inequality in the form:

$$|u|_{0,\Omega} \leqslant C|u|_{1,\Omega} \quad \text{for every } u \in H_0^1(\Omega),$$

where $\Omega \subset \mathbb{R}^n$ is a bounded open set. Find the best possible value for C.

3.14 (a) Let V and W be Hilbert spaces and $a(\cdot, \cdot)\colon V \times W \to \mathbb{R}$ a bilinear form such that there exist positive constants M, β, γ

$$|a(v, w)| \leqslant M\|v\|_{\hat{V}}\,\|w\|_{\hat{W}}, \quad \text{for every } v \in V,\ w \in W,$$

$$\sup_{\substack{v \in V \\ v \neq 0}} \frac{a(v, w)}{\|v\|_V} \geqslant \beta\|w\|_W \quad \text{for every } w \in W,$$

$$\sup_{\substack{w \in W \\ w \neq 0}} \frac{a(v, w)}{\|w\|_W} \geqslant \gamma\|v\|_V, \quad \text{for every } v \in V.$$

If $f \in V$ and $g \in W$ show that there exist unique $u \in V$ and $z \in W$ such that

$$a(u, w) = (g, w)_W, \quad \text{for every } w \in W$$
$$a(v, z) = (f, v)_V, \quad \text{for every } v \in V.$$

(b) Let $V = W$ and let $V_h \subset V$ be a finite dimensional space such that there exists $\beta_h > 0$ and

$$\sup_{\substack{w_h \in V_h \\ w_h \neq 0}} \frac{a(v_h, w_h)}{\|w_h\|} \geqslant \beta_h\|v_h\|, \quad \text{for every } v_h \in V_h.$$

Show that there exists a unique $u_h \in V_h$ such that

$$a(u_h, v_h) = (g, v_h), \quad \text{for every } v_h \in V_h.$$

Show also that

$$\|u - u_h\| \leqslant \left(1 + \frac{M}{\beta_h}\right) \inf_{v_h \in V_h} \|u - v_h\|.$$

3.15 (a) (Aubin-Nitsche Theorem). Let V, H be Hilbert spaces such that $V \subset H$ and V dense in H. Let $\|\cdot\|$ be the norm in V, $|\cdot|$ the norm in H and $(\cdot, \cdot)$ the inner-product in H. Let $a(\cdot, \cdot)\colon V \times V \to \mathbb{R}$ be V-elliptic and such that

$$|a(u, v)| \leqslant M\|u\|\,\|v\|, \quad \text{for every } u, v \in V.$$

Let $f \in H$. Let $V_h \subset V$ be a finite dimensional subspace. If $u \in V$ and $u_h \in V_h$ are the unique solutions of

$$a(u, v) = (f, v) \quad \text{for every } v \in V$$
$$a(u_h, v_h) = (f, v_h) \quad \text{for every } v_h \in V_h$$

show that

$$|u - u_h| \leqslant M\|u - u_h\| \sup_{g \in H} \left\{\frac{1}{|g|} \inf_{\phi_h \in V_h} \|\phi - \phi_h\|\right\}$$

where ϕ is the unique solution of

$$a(v, \phi) = (g, v), \quad \text{for every } v \in V.$$

(b) Deduce that in the case of Example 3.7.1,

$$|u - u_h|_{0,\Omega} \leqslant Ch^2|f|_{0,\Omega}.$$

FOUR

Introduction to Semigroups and Applications

4.1 UNBOUNDED LINEAR OPERATORS

The theory of semigroups of linear operators plays an important role in the study of evolution equations in an abstract framework. One of the essential ingredients of this theory is the notion of an unbounded linear operator.

Let V and W be Banach spaces. By a linear operator on V with values in W, we mean a linear map A defined on a subspace $D(A)$ (called the **domain** of A) into W. Its **range** will be a subspace of W denoted by $\mathcal{R}(A)$ and its **null-space** (or kernel) a subspace of $D(A)$ denoted by $N(A)$.

Definition 4.1.1 A linear operator $A : D(A) \subset V \to W$ is said to be **bounded** if there exists $C > 0$ such that

$$\|Au\|_W \leqslant C\|u\|_V, \quad \text{for every } u \in D(A). \tag{4.1.1}$$

Otherwise it is said to **unbounded.** It is said to be **densely defined** if $\overline{D(A)} = V$. It is said to be **closed** if the **graph**

$$G(A) = \{(u, Au) \mid u \in D(A)\} \subset V \times W \tag{4.1.2}$$

is closed as a subspace of $V \times W$. ∎

Remark 4.1.1 A bounded linear operator whose domain is all of V is a continuous linear operator as we have understood upto now. In general a bounded linear operator extends uniquely as a continuous liner operator to $\overline{D(A)}$. If A is a closed operator with $D(A) = V$, then by the Closed Graph Theorem, A is a continuous linear operator. If A is a closed operator, it follows that $N(A)$ is a closed subspace. ∎

Most of the unbounded operators we will consider will be densely defined.

Example 4.1.1 Let $\Omega = (0, 1) \subset \mathbb{R}$ and let $V = W = L^2(\Omega)$. Let A be

defined as follows:

$$\left.\begin{aligned} D(A) &= H_0^1(\Omega) \\ Au &= u', \quad u \in D(A). \end{aligned}\right\} \tag{4.1.3}$$

Then A is an unbounded, densely defined operator. Clearly $H_0^1(\Omega)$ is dense in $L^2(\Omega)$, since $\mathscr{D}(\Omega) \subset H_0^1(\Omega)$. Also consider the sequence $\left\{\frac{\sqrt{2}}{n\pi} \sin(n\pi x)\right\}$, $n = 1, 2, \ldots$. If $u_n = \frac{\sqrt{2}}{n\pi} \sin(n\pi x)$ then $|u_n|_{0,\Omega} = 1/n\pi \to 0$ as $n \to \infty$. But $u_n' = \sqrt{2} \cos(n\pi x)$ and $|u_n'|_{0,\Omega} = 1$ for all n. Thus A is not bounded. ∎

Example 4.1.2 Let Ω be a bounded open set in $\mathbb{R}^n$ and let $V=W=L^2(\Omega)$. Let

$$\left.\begin{aligned} D(A) &= H^2(\Omega) \cap H_0^1(\Omega) \\ Au &= \Delta u, \; u \in D(A). \end{aligned}\right\} \tag{4.1.4}$$

Again A is densely defined. It is unbounded for if we consider $\{w_m\}$ the sequence of eigenfunctions of Δ, i.e., $w_m \in H^2(\Omega) \cap H_0^1(\Omega)$, $-\Delta w_m = \lambda_m w_m$ (cf. Section 3.6) then $|w_m|_{0,\Omega} = 1$ while $|Aw_m|_{0,\Omega} = \lambda_m \to \infty$ as $m \to \infty$. Thus A is not bounded. ∎

Let A be a densely defined operator on V with values in W. We now define the space

$$D(A^*) = \{f \in W^* \mid \text{There exists } C > 0 \text{ such that } |f(Au)| \leqslant C\|u\|_V, \quad \text{for every } u \in D(A)\}. \tag{4.1.5}$$

Clearly $D(A^*)$ is a subspace of W^*. Now if $f \in D(A^*)$ we define a linear functional g by

$$g(u) = f(Au), \; u \in D(A).$$

By definition of $D(A^*)$, g is a continuous linear functional on $D(A)$ and so by density extends uniquely as a continuous linear functional on V. We denote this unique extension as $A^*f \in V^*$. Thus $A^*: D(A^*) \subset W^* \to V^*$ is a linear operator and is called the **adjoint** of A. An important relation connecting A and A^* is

$$(A^*f)(u) = f(Au) \quad \text{for all } f \in D(A^*), \, u \in D(A). \tag{4.1.6}$$

Theorem 4.1.1 If A: $D(A) \subset V \to W$ is a densely defined linear operator then A^*: $D(A^*) \subset W^* \to V^*$ is always closed.

Proof We need to show that $G(A^*)$ is closed in $W^* \times V^*$. To do this, let $f_n \to f$ in W^* and $A^*f_n \to g$ in V^*. We must now show that this implies that $f \in D(A^*)$ and that $A^*f = g$.

Let $u \in D(A)$. Then

$$(A^*f_n)(u) = f_n(Au)$$

and passing to the limit as $n \to \infty$,

$$g(u) = f(Au) \tag{4.1.7}$$

and so $|f(Au)| \leqslant \|g\|_{V^*} \|u\|_V$ showing that $f \in D(A^*)$. By the density of $D(A)$ in V, it follows from (4.1.7) that $g = A^*f$. ∎

We have the following result for bounded linear operators.

Theorem 4.1.2 Let $A: D(A) \subset V \to W$ be a densely defined and closed linear operator. Then if $D(A) = V$, A is bounded and A^* is also bounded. Further, in this case

$$\|A\| = \|A^*\|. \tag{4.1.8}$$

Proof If $D(A) = V$ then by the Closed Graph Theorem, A is bounded. Hence if $f \in W^*$,

$$|f(Au)| \leqslant \|f\|_{W^*} \|Au\|_W \leqslant \|A\| \, \|f\|_{W^*} \|u\|_V. \tag{4.1.9}$$

Thus $D(A^*) = W^*$ and so again by the Closed Graph Theorem (since A^* is always closed), A^* is bounded. Now for every $u \in V$ and every $f \in W^*$,

$$f(Au) = (A^*f)(u).$$

It follows from the Hahn-Banach Theorem that

$$\|Au\|_W \leqslant \sup_{\substack{f \in W^* \\ \|f\|_{W^*}=1}} |f(Au)| \leqslant \|A^*\| \, \|u\|.$$

Thus $\|A\| \leqslant \|A^*\|$. On the other hand,

$$\|A^*f\|_{V^*} = \sup_{\substack{\|u\|_V \leqslant 1 \\ u \in V}} |(A^*f)(u)| \leqslant \|A\| \, \|f\|_{W^*} \quad \text{(cf. (4.1.9))}.$$

Thus $\|A^*\| \leqslant \|A\|$ and (4.1.8) is proved. ∎

Remark 4.1.2 The consequence of the Hahn-Banach Theorem mentioned in the above proof is the one which proves that the canonical imbedding of V into V^{**} is an isometry: "Given $w \in W$, there exists $f \in W^*$ such that $\|f\|_{W^*} = 1$ and $f(w) = \|w\|$". ∎

If $V = W$ and is a Hilbert space, we can, by the Riesz Representation Theorem, identify V^* with V and the adjoint may again be considered as a linear operator in V.

Definition 4.1.2 Let V be a Hilbert space and $A : D(A) \subset V \to V$ be a densely defined linear operator. A is said to be **symmetric** if for every $u, v \in D(A)$,

$$(Au, v) = (u, Av). \tag{4.1.10}$$

It is said to be **self-adjoint** if $A^* = A$. ∎

Remark 4.1.3 When $D(A) = V$ and A is bounded, there is no distinction between symmetric and self-adjoint operators. However if A is an unbounded densely defined operator, A being symmetric only means that $D(A) \subset D(A^*)$ and that $A^*|_{D(A)} = A$. If however A is self-adjoint, we have $D(A) = D(A^*)$ and $A^* = A$. ∎

Example 4.1.3 Let $V = L^2(\Omega)$, $\Omega = (0, 1)$. Let A be as in Example 4.1.1. Now $D(A^*) = \left\{ v \in L^2(\Omega) \,\Big|\, \left|\int_0^1 vu'\right| \leqslant C|u|_{0,\Omega} \text{ for every } u \in H_0^1(\Omega) \right\}$.

In particular the inequality in the above definition must be valid for every $\phi \in \mathscr{D}(\Omega)$. This then implies that $v \in H^1(\Omega)$ (cf. Exercise 2.7). Conversely if $v \in H^1(\Omega)$ then by integration by parts,

$$\int_0^1 vu' = -\int_0^1 v'u \tag{4.1.11}$$

and so $v \in D(A^*)$. Thus $D(A^*) = H^1(\Omega)$ and by (4.1.11) it follows that

$$A^*v = -v'. \quad ∎ \tag{4.1.12}$$

Example 4.1.4 Let $\Omega \subset \mathbb{R}^n$ be a bounded open set of class C^2 and let $V = L^2(\Omega)$. Let A be as in Example 4.1.2. Let $u \in D(A)$. Then if $v \in D(A)$,

$$\int_\Omega u\, Av = \int_\Omega u\, \Delta v = -\int_\Omega \nabla u \cdot \nabla v = \int_\Omega \Delta u \cdot v = \int_\Omega Au \cdot v, \tag{4.1.13}$$

Thus $\left|\int_\Omega u\, Av\right| \leqslant C|v|_{0,\Omega}$ and so $u \in D(A^*)$ and A is symmetric. We now show that A is self-adjoint. To to this, we need to show that $D(A^*) = D(A)$. Since $D(A) \subset D(A^*)$, we need only to prove the reverse inclusion.

Define $G : L^2(\Omega) \to L^2(\Omega)$ by $u = Gf$ where

$$-\Delta u + u = f \text{ in } \Omega, \quad u \in H_0^1(\Omega). \tag{4.1.14}$$

Such a u always exists and by a regularity theorem (cf. Section 3.3) we have that $u \in H^2(\Omega) \cap H_0^1(\Omega) = D(A)$. Thus we have that $\mathscr{R}(I - A) = V$ and $G = (1 - A)^{-1}$. Again it is easy to see that G is self-adjoint. For,

$$\int_\Omega Gu \cdot v = \int_\Omega \nabla(Gv) \cdot \nabla(Gu) + \int_\Omega Gv \cdot Gu = \int_\Omega Gv \cdot u,$$

which shows that G is symmetric and hence self-adjoint since G is a bounded linear operator. Now given any $w \in V$, let $v \in D(A)$ be such that $v = Gw$. If $u \in D(A^*)$ then set $f = u + A^*u$. Then

$$\int_\Omega fv = \int_\Omega (u + A^*u)v = \int_\Omega u(v + Av)$$

$$= \int_\Omega u(v - \Delta v) = \int_\Omega uw.$$

Thus

$$\int_\Omega Gf\cdot w = \int_\Omega fGw = \int_\Omega fv = \int_\Omega uw$$

or $u = Gf$ and so in particular $u \in H^2(\Omega) \cap H_0^1(\Omega) = D(A)$. Thus $D(A^*) \subset D(A)$ and so A is self-adjoint. ∎

Remark 4.1.4 An operator $A : D(A) \subset V \to V$, V a Hilbert space, is said to be **monotone** if $(Au, u) \geqslant 0$ for every $u \in D(A)$ and **maximal monotone** if it is monotone and if $\mathcal{R}(I + A) = V$. It is easy to see that in the above example, $-A$ is maximal monotone. The same proof can be adapted to show that any symmetric maximal monotone operator in a Hilbert space is self-adjoint. ∎

If A is a closed operator then we can consider $D(A)$ with the **graph norm.**

$$\|u\|_{D(A)} = \|u\|_V + \|Au\|_W. \tag{4.1.15}$$

and then $D(A)$ will be a Banach space; for if $\{u_n\}$ is a Cauchy sequence in $D(A)$ then $\{u_n\}$ is Cauchy in V and $\{Au_n\}$ will be in Cauchy W. Thus $u_n \to u$ in V and $Au_n \to w$ in W and by the closedness of A, $u \in D(A)$, $Au = w$ and further $u_n \to u$ in $D(A)$. In future when we refer to $D(A)$ as an independent space we shall imply this structure on it. If V is Hilbert space and $A : D(A) \subset V \to V$, then $D(A)$ is a Hilbert space for the innerproduct

$$(u, v)_{D(A)} = (u, v)_V + (Au, Av)_V. \tag{4.1.16}$$

4.2 THE EXPONENTIAL MAP

Let V be a Banach space and $A : V \to V$ a bounded linear operator. Let us consider the ordinary differential equation

$$\left.\begin{aligned} \frac{du}{dt}(t) &= Au(t),\ t \geqslant 0 \\ u(0) &= u_0 \in V \end{aligned}\right\} \tag{4.2.1}$$

The above initial value problem has a unique solution for all time t (for instance by the famous theorem of Picard on Lipschitz continuous functions). This solution can be described explicitly. Consider the linear map defined by the following series:

$$\exp(A) = e^A = I + A + \frac{A^2}{2!} + \frac{A^3}{3!} + \ldots = I + \sum_{k=1}^{\infty} \frac{A^k}{k!}. \tag{4.2.2}$$

The series is absolutely convergent and so defines a bounded linear operator on V. We now define

$$u(t) = e^{tA}u_0. \tag{4.2.3}$$

We verify that this indeed defines the solution of (4.2.1). First of all,

clearly $u(0) = u_0$. Now

$$\frac{e^{tA}u_0 - u_0}{t} = \sum_{k=1}^{\infty} \frac{t^{k-1}}{k!} A^k u_0 = Au_0 + \sum_{k=2}^{\infty} \frac{t^{k-1}}{k!} A^k u_0.$$

Now,

$$\left\| \sum_{k=2}^{\infty} \frac{t^{k-1}}{k!} A^k u_0 \right\| \leqslant \sum_{k=2}^{\infty} \frac{|t|^{k-1}}{k!} \|A\|^k \|u_0\| \leqslant |t| \, \|A\|^2 \|u_0\| \sum_{k=0}^{\infty} \frac{|t|^k}{k!} \|A\|^k$$

$$= |t| \, \|A\|^2 \|u_0\| e^{|t| \, \|A\|} \to 0 \text{ as } |t| \to 0.$$

Again it is a simple exercise to check that for $t, s \in \mathbb{R}$.

$$e^{(t+s)A} = e^{tA} e^{sA} \tag{4.2.4}$$

and so

$$\begin{aligned} \frac{du(t)}{dt} &= \lim_{h \to 0} \frac{u(t+h) - u(t)}{h} \\ &= \lim_{h \to 0} e^{tA} \frac{(e^{hA}u_0 - u_0)}{h} = e^{tA} A u_0 \\ &= A e^{tA} u_0 = Au(t). \end{aligned}$$

Thus we have shown that the solution of the initial value problem (4.2.1) admits an explicit solution of the form (4.2.3). There are however several partial differential equations of evolution type which can be put in the form (4.2.1) *but with A unbounded.* We then wish to investigate when such problems are solvable. In particular we wish to know if the analogue of the exponential map e^A exists. This we shall investigate in the following sections.

To conclude this section, we derive an estimate for $\frac{du}{dt}(t)$ when $t > 0$ in the case when V is a Hilbert space and $A : V \to V$ a bounded linear operator which is self-adjoint and monotone.

Theorem 4.2.1 Let V be a Hilbert space and $A : V \to V$ a self-adjoint bounded linear operator. Let $-A$ be monotone. If $u(t) = e^{tA}u_0$, then

$$\left\| \frac{du}{dt}(t) \right\| \leqslant \frac{1}{t} \|u_0\|. \tag{4.2.5}$$

Proof *Step 1* We have

$$\frac{d}{dt}(u(t)) = Au(t), \; u(0) = u_0. \tag{4.2.6}$$

Let $t > 0$ be fixed. By taking the scalar product on both sides of the above equation with $u(t)$ and integrating on $[0, t]$ we get

$$\int_0^t \left(\frac{du}{dt}(\tau), u(\tau) \right) d\tau = \int_0^t (Au(\tau), u(\tau)) \, d\tau$$

or,

$$\tfrac{1}{2}\|u(t)\|^2 - \int_0^t (Au(\tau), u(\tau)) \, d\tau = \tfrac{1}{2}\|u_0\|^2. \tag{4.2.7}$$

Again by taking the scalar product of (4.2.6) with $t \frac{du}{dt}(t)$ and integrating over $[0, t]$, we get

$$\int_0^t \tau \left\| \frac{du}{dt}(\tau) \right\|^2 d\tau - \int_0^t \left(Au(\tau), \frac{du}{dt}(\tau) \right) \tau \, d\tau = 0. \tag{4.2.8}$$

Now,

$$\frac{d}{dt}(Au(t), u(t)) = \left(A\frac{du}{dt}(t), u(t) \right) + \left(Au(t), \frac{du}{dt}(t) \right)$$
$$= 2\left(Au(t), \frac{du}{dt}(t) \right)$$

using the self-adjointness of A. Thus

$$\int_0^t \tau \left(Au(\tau), \frac{du}{dt}(\tau) \right) d\tau = \tfrac{1}{2} \int_0^t \tau \frac{d}{d\tau}(Au(\tau), u(\tau)) \, d\tau$$
$$= \frac{t}{2}(Au(t), u(t)) - \tfrac{1}{2} \int_0^t (Au(\tau), u(\tau)) \, d\tau \tag{4.2.9}$$

by integration by parts. Combining (4.2.7), (4.2.8) and (4.2.9) we get

$$\tfrac{1}{2}\|u(t)\|^2 + 2 \int_0^t \tau \left\| \frac{du}{dt}(\tau) \right\|^2 dt - t(Au(t), u(t)) = \tfrac{1}{2}\|u_0\|^2. \tag{4.2.10}$$

Step 2 We also have from (4.2.7) and the fact that $-A$ is monotone, that $\|u(t)\| \leqslant \|u_0\|$ for any $t > 0$, and in fact that $\|u(t)\|$ decreases as t increases. Now the function $v(t) = \frac{du}{dt}(t)$ satisfies the equation

$$\frac{dv(t)}{dt} = Av(t), \; v(0) = Au_0. \tag{4.2.11}$$

Thus we get that $\|v(t)\| \leqslant \|Au_0\|$ and infact that $\|v(t)\|$ is a decreasing function of t. Thus we deduce that

$$\int_0^t \tau \left\| \frac{du(\tau)}{d\tau} \right\|^2 d\tau \geqslant \left\| \frac{du(t)}{dt} \right\|^2 \int_0^t \tau \, d\tau = \left\| \frac{du(t)}{dt} \right\|^2 \frac{t^2}{2}.$$

Using this and the monotonicity of $-A$ in (4.2.10) we deduce that

$$\tfrac{1}{2}\|u(0)\|^2 \geqslant t^2 \left\| \frac{du}{dt}(dt) \right\|^2$$

from which the estimate (4.2.5) follows. ∎

4.3 C_0-SEMIGROUPS

Let V be a Banach space and let $A : V \to V$ be a bounded linear operator. Given $u_0 \in V$ we saw that the function $u(t) = e^{tA}u_0$ was the unique solution of the initial value problem (4.2.1). Let us briefly look at the properties of $u(t)$. For fixed t, the map $u_0 \to u(t)$ is a linear map on V.

Further $\|u(t)\| \leqslant e^{t\|A\|}\|u_0\|$ and so the above linear map is bounded. Further as $t \downarrow 0$, we have $u(t) \to u_0$ and also $u(0) = u_0$. Finally by the uniqueness of the solution to the problem if we start with initial data $u(t_0)$ and allow the solution to evolve for time t, we must get $u(t + t_0)$. This can also be seen from the relation (4.2.4). We generalize all these properties to formulate the following definition:

Definition 4.3.1 Let V be a Banach space and $\{S(t)\}_{t \geqslant 0}$ a family of bounded linear operators on V. It is said to be a C_0-**Semigroup** if the following are true:

(i) $S(0) = I$, the identity operator on V.
(ii) $S(t + s) = S(t)S(s)$, for all $t, s \geqslant 0$.
(iii) For every $u \in V$,

$$S(t)u \to u \quad \text{as} \quad t \downarrow 0. \quad \blacksquare$$

Remark 4.3.1 The property (ii) above is called the semigroup property. Property (iii) is connected with the continuity with respect to t. ▮

Example 4.3.1 If A is a bounded linear operator on V, then we saw in Section 4.2 that $S(t) = e^{tA}$ defines a C_0-semigroup. ▮

Example 4.3.2 Let V be the space of bounded uniformly continuous functions on $\mathbb{R}$ with the sup-norm. Define

$$S(t)f(s) = f(t + s) \tag{4.3.1}$$

Then $\{S(t)\}_{t \geqslant 0}$ defines a C_0-semigroup on V. ▮

We will now study some elementary properties of semigroups.

Theorem 4.3.1 Let $\{S(t)\}_{t \geqslant 0}$ be a C_0-semigroup on V. Then there exist $M \geqslant 1$ and ω such that

$$\|S(t)\| \leqslant Me^{\omega t}, \text{ for all } t \geqslant 0. \tag{4.3.2}$$

Proof There exist $M \geqslant 1$ and $\delta > 0$ such that for $0 \leqslant t \leqslant \delta$,

$$\|S(t)\| \leqslant M.$$

If not, then there exist $t_n \to 0$ such that $\{\|S(t_n)\|\}$ is unbounded. But since $S(t_n)u \to u$ as $n \to \infty$, $\{\|S(t_n)u\|\}$ is bounded for every $u \in V$ and by the Banach-Steinhaus Theorem $\{\|S(t_n)\|\}$ has to be bounded.

Now define $\omega = \delta^{-1} \log M \geqslant 0$. Given any $t \geqslant 0$ we can find an integer $n \geqslant 0$ and $0 \leqslant \eta < \delta$ such that

$$t = n\delta + \eta.$$

Then

$$S(t) = (S(\delta))^n S(\eta)$$

and so

$$\|S(t)\| \leqslant \|S(\delta)\|^n \|S(\eta)\| \leqslant M^n \cdot M \leqslant Me^{\omega t}$$

since $\log (M^n) = n \log M \leqslant n\omega\delta \leqslant \omega t$. ∎

Corollary For every $u \in V$, the mapping $t \to S(t)u$ is a continuous map from $[0, \infty)$ into V.

Proof Let $h \geqslant 0$. Then

$$\begin{aligned} \|S(t+h)u - S(t)u\| &\leqslant \|S(t)\| \, \|S(h)u - u\| \\ &\leqslant Me^{\omega t}\|S(h)u - u\| \end{aligned}$$

which tends to zero as $h \downarrow 0$. Again

$$\begin{aligned} \|S(t) - S(t-h)u\| &\leqslant \|S(t-h)\| \, \|S(h)u - u\| \\ &\leqslant Me^{\omega t}\|S(h)u - u\| \end{aligned}$$

which tends to zero as $h \downarrow 0$. Hence the result. ∎

Remark 4.3.2 We write that for $u \in V$ fixed,

$$S(\cdot)u \in C([0, \infty); V). \quad ∎$$

Definition 4.3.2 If $M = 1$ and $\omega = 0$, so that $\|S(t)\| \leqslant 1$ for all $t \geqslant 0$, we say that $\{S(t)\}$ is a **contraction semigroup.** ∎

For $v \in V$, since $t \to S(t)u$ is continuous, we know how to integrate this function (cf. Appendix 3). Thus we can prove the following result:

Lemma 4.3.1

$$\lim_{h \downarrow 0} \frac{1}{h} \int_t^{t+h} S(\tau)u \, d\tau = S(t)u$$

for every $u \in V$.

Proof

$$\begin{aligned} \left\| \frac{1}{h} \int_t^{t+h} S(\tau)u \, d\tau - S(t)u \right\| &= \left\| \frac{1}{h} \int_t^{t+h} (S(\tau) - S(t))u \, d\tau \right\| \\ &\leqslant \frac{1}{h} \int_t^{t+h} \|S(\tau)u - S(t)u\| \, d\tau \end{aligned}$$

Choosing h small enough so that $\|S(t)u - S(\tau)u\| < \epsilon$ for $|t - \tau| < h$ we get that

$$\left\| \frac{1}{h} \int_t^{t+h} S(\tau)u \, d\tau - S(t)u \right\| < \epsilon$$

and the result follows. ∎

We now introduce an important concept in the theory of semigroups.

Definition 4.3.3 Let $\{S(t)\}_{t\geq 0}$ be a C_0-semigroup on V. The **infinitesimal generator** of the semigroup is a linear operator A given by

$$\left.\begin{aligned} D(A) &= \left\{u \in V \mid \lim_{t\downarrow 0} \frac{S(t)u - u}{t} \text{ exists}\right\} \\ Au &= \lim_{t\downarrow 0} \frac{S(t)u - u}{t}, \; u \in D(A). \end{aligned}\right\} \tag{4.3.3}$$

■

Example 4.3.3 If A is a bounded linear operator on V and $S(t) = e^{tA}$, then we proved in Section 4.2 that for any $u_0 \in V$,

$$\lim_{t\downarrow 0} \frac{e^{tA}u_0 - u_0}{t} = Au_0.$$

Thus the infinitesimal generator of e^{tA} is A. ■

Example 4.3.4 If V and $S(t)$ are as in Example 4.3.2, then

$$\lim_{t\downarrow 0} \frac{(S(t)f - f)(s)}{t} = D^+f(s)$$

if it exists. Thus we see that in this case if $f \in D(A)$ then D^+f exists at all points and is bounded and uniformly continuous. But then

$$\frac{f(s) - f(s-t)}{t} = D^+f(s-t) + \frac{o(t)}{t} \to D^+f(s), \text{ as } t \downarrow 0$$

by the continuity of D^+f. Hence D^-f exists everywhere and $D^-f = D^+f$. Hence f is differentiable everywhere. So

$$\left.\begin{aligned} D(A) &= \{f \in V \mid f' \text{ exists everywhere and } f' \in V\} \\ Af &= f', \quad f \in D(A). \end{aligned}\right\} \tag{4.3.4}$$

■

Theorem 4.3.2 Let $\{S(t)\}_{t\geq 0}$ be a C_0-semigroup and let A be its infinitesimal generator. Let $u \in D(A)$. Then

$$S(t)u \in C^1([0, \infty); V) \cap C([0, \infty); D(A))$$

and

$$\frac{d}{dt}(S(t)u) = AS(t)u = S(t)Au. \tag{4.3.5}$$

Proof Let $u \in D(A)$. Then

$$\left(\frac{S(h) - I}{h}\right) S(t)u = S(t) \frac{(S(h)u - u)}{h} \to S(t)Au$$

as $h \downarrow 0$, by definition of A. Thus $S(t)u \in D(A)$ and

$$AS(t)u = S(t)Au = D^+S(t)u.$$

Next consider

$$\frac{S(t)u - S(t-h)u}{h} = S(t-h) \frac{(S(h)u - u)}{h}.$$

Hence

$$\frac{S(t)u - S(t-h)u}{h} - S(t)Au = S(t-h)\left(\frac{S(h)u - u}{h} - Au\right) + (S(t-h) - S(t))Au.$$

But

$$\left\|S(t-h)\left(\frac{S(h)u-u}{h} - Au\right)\right\| \leqslant Me^{\omega t}\left\|\frac{S(h)u-u}{h} - Au\right\| \to 0 \quad \text{as } h \downarrow 0$$

and

$$\|(S(t-h) - S(t))Au\| \to 0 \quad \text{as } h \downarrow 0$$

by Corollary above. Thus

$$D^{-}S(t)u = S(t)Au = D^{+}S(t)u.$$

and thus

$$\frac{d}{dt}(S(t)u) = S(t)Au = AS(t)u.$$

This proves (4.3.5). We also now have from Corollary above that $t \to S(t)Au$ is continuous and so $S(t)u$ is thus continuously differentiable from $[0, \infty)$ with values in V. Also $S(t)u \in D(A)$ as proved and so it has values in $D(A)$ as well. Further the continuous differentiability into V also proves the continuity into $D(A)$ (with the graph norm). Hence the theorem is completely proved. ∎

Remark 4.3.3 If A is the infinitesimal generator of a C_0-semigroup $\{S(t)\}$ then we know by the above theorem that

$$u(t) = S(t)u_0$$

defines the unique solution of the initial value problem

$$\left.\begin{aligned} \frac{du(t)}{dt} &= Au(t),\ t \geqslant 0 \\ u(0) &= u_0 \end{aligned}\right\} \tag{4.3.6}$$

provided the initial data is in $D(A)$. (If $u_0 \notin D(A)$, then $S(t)u_0$ is not differentiable with respect to t. We can however consider $u(t) = S(t)u_0$ as a *generalized solution* of (4.3.6)). For the uniqueness, if $v(t)$ is any other solution, define $w(s) = S(t-s)v(s)$. Then

$$\frac{dw}{ds} = -AS(t-s)v(s) + S(t-s)Av(s) = 0.$$

Thus $w(t) = w(0)$ or $v(t) = S(t)u_0 = u(t)$. ∎

We list some further properties of the infinitesimal generator of a C_0-semigroup.

Theorem 4.3.3 Let A be the infinitesimal generator of a C_0-semigroup $\{S(t)\}$ on V. Then for any $u \in V$, $\int_0^t S(\tau)u \, d\tau \in D(A)$ and

$$A\left(\int_0^t S(\tau)u \, d\tau\right) = S(t)u - u. \tag{4.3.7}$$

Proof Let $h > 0$. Consider

$$\begin{aligned}\left(\frac{S(h) - I}{h}\right)\int_0^t S(\tau)u \, d\tau &= \frac{1}{h}\int_0^t (S(\tau + h)u - S(\tau)u) \, d\tau \\ &= \frac{1}{h}\left[\int_t^{t+h} S(\tau)u \, d\tau - \int_0^h S(\tau)u \, d\tau\right] \\ &\to S(t)u - u\end{aligned}$$

by Lemma 4.3.1. Thus $\left(\int_0^t S(\tau)u \, d\tau\right) \in D(A)$ and (4.3.7) follows by the definition of A. ∎

Corollary If A is the infinitesimal generator of a C_0-semigroup $\{S(t)\}$ then A has to be closed and densely defined.

Proof Let $u \in V$. Then $\int_0^h S(\tau)u \, d\tau \in D(A)$ and by Lemma 4.3.1, $\frac{1}{h}\int_0^h S(\tau)u \, d\tau \to S(0)u = u$. Thus $D(A)$ is dense in V. If $u_n \in D(A)$ and $u_n \to u$ in V and $Au_n \to v$ in V we need to show that $u \in D(A)$ and $Au = v$, in order to prove that A is closed. But

$$\frac{S(h)u - u}{h} = \lim_{n\to\infty} \frac{S(h)u_n - u_n}{h} = \lim_{n\to\infty} \frac{1}{h}\int_0^h S(\tau)Au_n \, d\tau$$

since for any $w \in D(A)$ and any $h_1, h_2 \geqslant 0$, we have on integrating (4.3.5) between h_1 and h_2,

$$S(h_2)w - S(h_1)w = \int_{h_1}^{h_2} S(\tau)Aw \, d\tau = \int_{h_1}^{h_2} AS(\tau)w \, d\tau.$$

But $Au_n \to v$ and so

$$\frac{S(h)u - u}{h} = \frac{1}{h}\int_0^h S(\tau)v \, d\tau \to v \quad \text{as } h \to 0$$

by Lemma 4.3.1. Thus $u \in D(A)$ and $Au = v$. ∎

The above corollary shows that in order that an unbounded operator generate a C_0-semigroup, it must necessarily be closed and densely defined. In the next section we will investigate further necessary and sufficient conditions on A.

Theorem 4.3.4 If two C_0-semigroups $\{S_1(t)\}$ and $\{S_2(t)\}$ have the same infinitesimal generator, A, then they are identical.

Proof Let $F(s) = S_1(t-s)S_2(s)u$ for $u \in D(A)$. Then

$$\frac{dF(s)}{ds} = -AS_1(t-s)S_2(s)u + S_1(t-s)AS_2(s)u$$

$$= 0$$

since A commutes with $S_1(t-s)$. Thus $F(s) =$ constant and so $F(t) = F(0)$, i.e. $S_2(t)u = S_1(t)u$ for all $t \geqslant 0$. But $D(A)$ is dense in V and so $S_2(t)u = S_1(t)u$ for all $u \in V$ and for all $t \geqslant 0$. ∎

Corollary If $\{S(t)\}$ is a C_0-semigroup whose infinitesimal generator A is bounded then $S(t) = e^{tA}$. ∎

4.4 THE HILLE-YOSIDA THEOREM

It was seen in the previous section that if A is the infinitesimal generator of a C_0-semigroup then the initial value problem (4.3.6) always has a unique solution provided $u_0 \in D(A)$. We also saw that if A is the infinitesimal generator of a C_0-semigroup then, necessarily, A is closed and densely defined.

Let us now assume that A generates a *contraction semigroup* $\{S(t)\}$, i.e. $\|S(t)\| \leqslant 1$ for all t. Let $u \in V$ and $\lambda > 0$. Then consider the integral

$$\int_s^t e^{-\lambda\tau} S(\tau)u \, d\tau.$$

This integral is well-defined and as $\|S(\tau)u\| \leqslant \|u\|$ for all τ, we deduce that this integral tends to zero as $t, s \to \infty$. Hence the integral

$$\int_0^\infty e^{-\lambda\tau} S(\tau)u \, d\tau$$

exists as an improper Riemann integral and we define

$$R(\lambda)u = \int_0^\infty e^{-\lambda\tau} S(\tau)u \, d\tau. \tag{4.4.1}$$

Clearly $u \to R(\lambda)u$ defines a linear operator on V and

$$\|R(\lambda)u\| \leqslant \|u\| \int_0^\infty e^{-\lambda\tau} \, d\tau = \frac{1}{\lambda} \|u\|.$$

Thus $R(\lambda)$ is a bounded linear operator and

$$\|R(\lambda)\| \leqslant \frac{1}{\lambda}, \quad \text{for all } \lambda > 0. \tag{4.4.2}$$

Theorem 4.4.1 If A is the infinitesimal generator of a contraction semi-

group $\{S(t)\}$, then $(\lambda I - A)$ is invertible for every $\lambda > 0$ and

$$(\lambda I - A)^{-1} = R(\lambda). \tag{4.4.3}$$

In particular, for every $\lambda > 0$, $\|(\lambda I - A)^{-1}\| \leqslant \lambda^{-1}$.

Proof We will show that for every $u \in V$, $R(\lambda)u \in D(A)$ and

$$(\lambda I - A)R(\lambda)u = u. \tag{4.4.4}$$

We will also show that for every $u \in D(A)$,

$$R(\lambda)(\lambda I - A)u = u. \tag{4.4.5}$$

This will prove the theorem.

Let $h > 0$. Then for $u \in V$,

$$\left(\frac{S(h) - I}{h}\right)R(\lambda)u = \frac{1}{h}\int_0^\infty e^{-\lambda\tau}(S(\tau + h)u - S(\tau)u)\, d\tau$$

$$= \frac{1}{h}\int_h^\infty e^{-\lambda(\tau-h)}S(\tau)u\, d\tau - \frac{1}{h}\int_0^\infty e^{-\lambda\tau}S(\tau)u\, d\tau$$

$$= \left(\frac{e^{\lambda\tau} - 1}{h}\right)\int_0^\infty e^{-\lambda h}S(\tau)u\, d\tau - \frac{e^{\lambda h}}{h}\int_0^h e^{-\lambda\tau}S(\tau)u\, d\tau$$

$$\to \lambda R(\lambda)u - u$$

as $h \to 0$. Thus $R(\lambda)u \in D(A)$ and

$$AR(\lambda)u = \lambda R(\lambda)u - u$$

which proves (4.4.4).

Now let $u \in D(A)$. Then

$$R(\lambda)Au = \int_0^\infty e^{-\lambda\tau}S(\tau)Au\, d\tau = \int_0^\infty e^{-\lambda\tau}\frac{d}{d\tau}(S(\tau)u)\, d\tau$$

$$= \lambda\int_0^\infty e^{-\lambda\tau}S(\tau)u\, d\tau - u$$

$$= \lambda R(\lambda)u - u$$

which proves (4.4.5). ∎

Thus the generator A of a semigroup of contractions has to be closed, densely defined and further $(\lambda I - A)^{-1}$ exists as a bounded linear operator for every $\lambda > 0$ and its norm is bounded by λ^{-1}. We will now show that these conditions are also sufficient for A to generate a contraction semigroup.

We assume henceforth that A satisfies the conditions laid down in the preceding paragraph and $R(\lambda) = (\lambda I - A)^{-1}$.

Lemma 4.4.1 Let A be as above. Then

$$\lim_{\lambda\to\infty} \lambda R(\lambda)u = u \tag{4.4.6}$$

for every $u \in V$.

Proof If $u \in D(A)$, then

$$\|\lambda R(\lambda)u - u\| = \|AR(\lambda)u\| = \|R(\lambda)Au\| \leqslant \frac{1}{\lambda}\|Au\|$$

which tends to zero as $\lambda \to +\infty$. Now $D(A)$ is dense in V and $\|\lambda R(\lambda)\| \leqslant 1$. Thus for any $u \in V$, choose $v \in D(A)$ close enough to u. Then

$$\begin{aligned}\|\lambda R(\lambda)u - u\| &\leqslant \|\lambda R(\lambda)(u - v)\| + \|\lambda R(\lambda)v - v\| + \|v - u\| \\ &\leqslant 2\|u - v\| + \|\lambda R(\lambda)v - v\|\end{aligned}$$

which can be made as small as we please. Thus (4.4.6) stands proved. ∎

We now introduce an important class of bounded linear operators which "approximate" A in the sense of Lemma 4.4.2 below. For $\lambda > 0$ we set

$$A_\lambda = \lambda AR(\lambda) = \lambda^2 R(\lambda) - \lambda I. \tag{4.4.7}$$

Note that for any $u \in V$, $R(\lambda)u \in D(A)$ and so A_λ is defined for all $u \in V$. Further the last equality shows that A_λ is a bounded linear operator on V. The operators $\{A_\lambda\}$ for $\lambda > 0$ are called the **Yosida approximations** of A and will play an important role in the sequel.

Lemma 4.4.2 Let $u \in D(A)$. Then

$$\lim_{\lambda\to\infty} A_\lambda u = Au. \tag{4.4.8}$$

Proof $\lim\limits_{\lambda\to\infty} A_\lambda u = \lim\limits_{\lambda\to\infty} \lambda AR(\lambda)u = \lim\limits_{\lambda\to\infty} \lambda R(\lambda)Au = Au$ by Lemma 4.4.1. ∎

Since A_λ is a bounded linear operator, it generates the C_0-semigroup $\exp(tA_\lambda)$. Now

$$\begin{aligned}\|\exp(tA_\lambda)\| &= e^{-t\lambda}\|e^{t\lambda^2 R(\lambda)}\| \leqslant \exp(-t\lambda)\exp(t\lambda^2\|R(\lambda)\|) \\ &\leqslant e^{-t\lambda}e^{t\lambda} = 1.\end{aligned}$$

Thus $\{\exp(tA_\lambda)\}$ is a contraction semigroup for every $\lambda > 0$.

We can now prove the following result:

Theorem 4.4.2 Let A be a closed, densely defined operator such that for every $\lambda > 0$, $(\lambda I - A)^{-1}$ exists as a bounded linear operator with

$$\|(\lambda I - A)^{-1}\| \leqslant \lambda^{-1}.$$

Then A is the infinitesimal generator of a contraction semigroup.

Proof Step 1 Clearly A_λ, A_μ commute. So do $\exp(tA_\lambda)$, $\exp(tA_\mu)$. Hence if $u \in V$,

$$\|\exp(tA_\lambda)u - \exp(tA_\mu)u\| = \left\|\int_0^1 \frac{d}{ds}(\exp(tsA_\lambda)\exp(t(1-s)A_\mu)u)\,ds\right\|$$

$$\leqslant \int_0^1 t\|\exp(tsA_\lambda)\exp(t(1-s)A_\mu)(A_\lambda u - A_\mu u)\|\,ds$$

$$\leqslant t\|A_\lambda u - A_\mu u\|.$$

Step 2 Let $u \in D(A)$. Then by Step 1 above

$$\|\exp(tA_\lambda)u - \exp(tA_\mu)u\| \leqslant t\|A_\lambda u - Au\| + t\|A_\mu u - Au\| \quad (4.4.9)$$

which tends to zero as $\lambda, \mu \to \infty$ using Lemma 4.4.2. Thus for every $u \in D(A)$, $\{\exp(tA_\lambda)u\}$ is Cauchy and so

$$\lim_{\lambda\to\infty} \exp(tA_\lambda)u$$

exists. Further it follows from (4.4.9) that this convergence is uniform on bounded intervals (with respect to t). Further $D(A)$ is dense and

$$\|\exp(tA_\lambda)\| \leqslant 1$$

and so $\{\exp(tA_\lambda)u\}$ converges for any $u \in V$ as $\lambda \to \infty$ and again this convergence is uniform on bounded intervals, with respect to t. We therefore define

$$S(t)u = \lim_{\lambda\to\infty} \exp(tA_\lambda)u, \quad u \in V. \quad (4.4.10)$$

Clearly $\|S(t)u\| \leqslant \|u\|$ and $S(t)$ is linear. Also the semigroup property holds. Further $S(0)u = u$ for all u. Finally, from the uniform convergence in t on bounded intervals we deduce that $S(t)u \to u$ as $t \downarrow 0$. Thus $\{S(t)\}$ is a C_0-semigroup of contractions.

Step 3 We show that A is the generator of $\{S(t)\}$. Let B be the infinitesimal generator of $\{S(t)\}$. Let $u \in D(A)$. Then

$$S(t)u - u = \lim_{\lambda\to\infty}(\exp(tA_\lambda)u - u) = \lim_{\lambda\to\infty}\int_0^t \exp(\tau A_\lambda)A_\lambda u\,d\tau$$

$$= \lim_{\lambda\to\infty}\int_0^t \exp(\tau A_\lambda)(A_\lambda u - Au)\,d\tau + \lim_{\lambda\to\infty}\int_0^t \exp(\tau A_\lambda)Au\,d\tau.$$

The first term is zero as $\|\exp(tA_\lambda)\| \leqslant 1$ and $A_\lambda u \to Au$ as $\lambda \to \infty$ for $u \in D(A)$. The second term is $\int_0^t S(\tau)Au\,d\tau$ by the definition of S and the Dominated Convergence Theorem. Thus if $u \in D(A)$

$$\frac{S(t)u - u}{t} = \frac{1}{t}\int_0^t S(\tau)Au\,d\tau \to Au \quad \text{as } t \downarrow 0$$

by Lemma 4.3.1. Hence $u \in D(B)$ and $Bu = Au$. Thus $D(A) \subset D(B)$ and $B|_{D(A)} = A$. Let $u \in D(B)$. Then by hypothesis, there exists $v \in D(A)$ such that

$$(I - A)v = (I - B)u.$$

But then $Av = Bv$ and so $(I - B)(v - u) = 0$. Since $(1 - B)^{-1}$ must exist by Theorem 4.4.1, we have that $v - u = 0$ or $v = u$. Thus $u \in D(A)$ and so $D(B) = D(A)$. Thus $B = A$. ∎

Remark 4.4.1 In the above proof we never used the fact that A was closed. In fact the closedness of A follows from the invertibility of $I - A$. For, let $u_n \in D(A)$ such that $u_n \to u$ and $Au_n \to v$. Then if $R = (I - A)^{-1}$, we have $RAu_n \to Rv$. But

$$RAu_n = Ru_n - u_n$$

and so

$$Rv = Ru - u$$

Thus $u = R(u - v) \in D(A)$ and $Rv = u + RAu - u = RAu$. Since R is one-one, it follows that $v = Au$. ∎

We can combine Theorems 4.4.1 and 4.4.2 and state the following result.

Theorem 4.4.3 **(Hille-Yosida)** A linear unbounded operator A on a Banach space V is the infinitesimal generator of a contraction semigroup if, and only if,

(i) A is closed, (ii) A is densely defined, and (iii) for every $\lambda > 0$, $(\lambda I - A)^{-1}$ is a bounded linear operator and

$$\|(\lambda I - A)^{-1}\| \leqslant \frac{1}{\lambda}. \quad ∎ \tag{4.4.11}$$

The importance of the above theorem lies in the fact that the study of the solvability of the initial value problem (4.3.6) is reduced to the study of the solvability and a priori estimates of the solutions of the problems

$$\lambda u - Au = v, \quad u \in D(A), v \in V, \lambda > 0, \tag{4.4.12}$$

which are often easier. In the case of V being a Hilbert space it is even easier as we shall see later.

Remark 4.4.2 If we work in a complex Banach space V and A is the infinitesimal generator of a contraction semigroup, then for $\lambda \in \mathbb{C}$ with $\mathrm{Re}\,\lambda > 0$,

$$R(\lambda)u = \int_0^\infty e^{-\lambda\tau} S(\tau)u \, d\tau$$

is defined for $u \in V$ and it can be easily shown that $R(\lambda) = (\lambda I - A)^{-1}$ and that

$$\|R(\lambda)\| \leqslant (\mathrm{Re}\,\lambda)^{-1}. \quad ∎$$

The following example shows that we cannot extend these results to cover the case $\mathrm{Re}\,\lambda = 0$.

Example 4.4.1 Let V be the space of bounded uniformly continuous functions on $(0, \infty)$. Let

$$S(t)f(s) = f(s + t), \quad t \geqslant 0, s > 0.$$

Then $\{S(t)\}$ is a contraction semi-group and

$$D(A) = \{f \in V \mid f' \in V\};\ Af = f', f \in D(A).$$

If Re $\lambda \leqslant 0$, the equation

$$\lambda\phi - A\phi = 0,\ \phi \in D(A)$$

has the non-trivial solution $\phi(s) = e^{\lambda s} \in V$. Thus $\lambda I - A$ is not invertible for Re $(\lambda) \leqslant 0$. ∎

Let $\{S(t)\}$ now be a C_0-semigroup such that

$$\|S(t)\| \leqslant e^{\omega t},\ \omega \geqslant 0 \tag{4.4.13}$$

for all $t \geqslant 0$. Then $S_1(t) = e^{-\omega t}S(t)$ will be a contraction semigroup. If A is the generator of $S(t)$, then $A - \omega I$ is the generator of S_1 and if A is the generator of S_1, $A + \omega I$ is that of S. Thus we can deduce the following result:

Theorem 4.4.4 A is the infinitesimal generator of a C_0-semigroup satisfying (4.4.13) if and only if it is closed, densely defined and for every $\lambda > \omega$, the inverse $(\lambda I - A)^{-1}$ exists and

$$\|(\lambda I - A)^{-1}\| \leqslant (\lambda - \omega)^{-1}. \ ∎ \tag{4.4.14}$$

To characterize the infinitesimal generators of general C_0-semigroups, $\{S(t)\}$ satisfying

$$\|S(t)\| \leqslant Me^{\omega t},\ M \geqslant 1,\ \omega \geqslant 0, \tag{4.4.15}$$

we usually renorm the Banach space so that $\{S(t)\}$ becomes a contraction semigroup in the new (equivalent) norm. We just state the result and omit a proof. For details, see Pazy [1].

Theorem 4.4.5 A is the infinitesimal generator of a C_0-semigroup $\{S(t)\}$ satisfying (4.4.15) if and only if it is closed, densely defined and for every $\lambda > \omega$ the inverse $(\lambda I - A)^{-1}$ exists and

$$\|(\lambda I - A)^{-n}\| \leqslant M(\lambda - \omega)^{-n}$$

for all $n = 0, 1, 2, \ldots$. ∎

In the proof of the Hille-Yosida Theorem we expressed the semigroup generated by A in terms of $\exp(tA_\lambda)$ as $\lambda \to \infty$. We now derive another definition for $S(t)$ in terms of A itself. We prove this only for the contraction case. But the same result will be available for general C_0-semigroups (cf. Pazy [1]).

Lemma 4.4.3 Let $T : V \to V$ be a bounded linear operator such that $\|T\| \leqslant 1$. Then for any positive integer n and any $u \in V$,

$$\|e^{n(T-I)}u - T^n u\| \leqslant \sqrt{n}\|u - Tu\|. \tag{4.4.16}$$

Proof *Step 1* Let k, n be non-negative integers. If $k \geqslant n$,

$$\|T^k u - T^n u\| = \|\sum_{j=n}^{k-1} (T^{j+1}u - T^j u)\|$$

$$\leqslant \|u - Tu\| \sum_{j=1}^{k-1} \|T^j\|$$

$$\leqslant (k-n)\|u - Tu\|.$$

Thus for any two non-negative integers k, n we have

$$\|T^k u - T^n u\| \leqslant |k-n| \, \|u - Tu\|.$$

Step 2 We now have for $t > 0$,

$$\|e^{t(T-I)}u - T^n u\| = \left\| e^{-t} \sum_{k=0}^{\infty} \frac{t^k}{k!} (T^k u - T^n u) \right\|$$

$$\leqslant e^{-t} \left(\sum_{k=0}^{\infty} \frac{t^k}{k!} |k-n| \right) \|u - Tu\|.$$

Let us consider the series

$$\sum_{k=0}^{\infty} \frac{t^k}{k!} |k-n| \leqslant \left(\sum_{k=0}^{\infty} \frac{t^k}{k!} \right)^{1/2} \left(\sum_{k=0}^{\infty} \frac{t^k}{k!} (k-n)^2 \right)^{1/2}$$

by the Cauchy-Schwarz inequality. But

$$\sum_{k=0}^{\infty} \frac{t^k}{k!} (k-n)^2 = n^2 \sum_{k=0}^{\infty} \frac{t^k}{k!} - (2n-1)t \sum_{k=0}^{\infty} \frac{t^k}{k!} + t^2 \sum_{k=0}^{\infty} \frac{t^k}{k!}$$

$$= (t^2 - (2n-1)t + n^2)e^t.$$

Thus if $t = n$, we get

$$\|e^{n(T-I)}u - T^n u\| \leqslant \|u - Tu\| e^{-t} e^{t/2} \sqrt{n} e^{t/2}$$

$$= \sqrt{n} \|u - Tu\|. \quad \blacksquare$$

Theorem 4.4.6 Let $\{S(t)\}$ be a contraction semigroup with infinitesimal generator A. Then for any $u \in V$,

$$S(t)u = \lim_{n\to\infty} \left(I - \frac{t}{n} A \right)^{-n} u = \lim_{n\to\infty} \left(\frac{n}{t} R\left(\frac{n}{t}\right) \right)^n u. \tag{4.4.17}$$

Proof First of all notice that by the Hille-Yosida theorem $\left(I - \frac{t}{n} A \right)^{-1}$ exists for all $t > 0$. Also by definition

$$\frac{n}{t} R\left(\frac{n}{t}\right) = \frac{n}{t} \left(\frac{n}{t} I - A \right)^{-1} = \left(I - \frac{t}{n} A \right)^{-1}.$$

Now we already know that (cf. Proof of Theorem 4.4.2)

$$S(t)u = \lim_{\lambda\to\infty} (\exp(tA_\lambda))u.$$

Setting $\lambda = \frac{n}{t}$ we can write

$$S(t) = \lim_{n\to\infty} (\exp (tA_{n/t}))u.$$

But

$$tA_{n/t} = t\left[\frac{n^2}{t^2} R\left(\frac{n}{t}\right) - \frac{n}{t} I\right] \quad \text{(cf. (4.4.7))}$$
$$= n\left(\frac{n}{t} R\left(\frac{n}{t}\right) - I\right).$$

Now $\left\|\frac{n}{t} R\left(\frac{n}{t}\right)\right\| \leqslant 1$ by Theorem 4.4.1 and so we can use $T = \frac{n}{t} R\left(\frac{n}{t}\right)$ in Lemma 4.4.3. Thus

$$\left\|\left[\exp\left\{n\left(\frac{n}{t} R\left(\frac{n}{t}\right) - I\right)\right\}\right]u - \left(\frac{n}{t} R\left(\frac{n}{t}\right)\right)^n u\right\| \leqslant \sqrt{n}\left\|\frac{n}{t} R\left(\frac{n}{t}\right)u - u\right\|.$$

If $u \in D(A)$, then from the proof of Lemma 4.4.1,

$$\left\|\frac{n}{t} R\left(\frac{n}{t}\right)u - u\right\| \leqslant \frac{t}{n}\|Au\|.$$

Combining all these we get

$$\left\|(\exp (tA_{n/t}))u - \left(\frac{n}{t} R\left(\frac{n}{t}\right)\right)^n u\right\| \leqslant \frac{t}{\sqrt{n}} \|Au\|, \quad u \in D(A).$$

Thus for fixed t and for $u \in D(A)$ we deduce that

$$S(t)u = \lim_{n\to\infty} \left(\frac{n}{t} R\left(\frac{n}{t}\right)\right)^n u. \tag{4.4.18}$$

But $\exp (tA_{n/t})$ and $\left(\frac{n}{t} R\left(\frac{n}{t}\right)\right)^n$ are both bounded in norm by unity and $D(A)$ is dense in V. Thus (4.4.18) is true for all $u \in V$ and the result is proved. ∎

In view of the first expression for $S(t)u$ in (4.4.17), this result is called an **exponential formula** for $S(t)$.

Remark 4.4.3 The exponential formula (4.4.17) has the following interpretation from the numerical analyst's point of view. Consider the initial value problem (4.3.6). An "implicit scheme" of discretization is to divide the interval $[0, t]$ into n equal parts, and set $\{u_n\}$ the approximate solution evaluated at times $\frac{jt}{n}$, $j = 0, 1, \ldots, n$. The formula for u_n is given by

$$\frac{u_n\left(\frac{jt}{n}\right) - u_n\left(\frac{(j-1)t}{n}\right)}{(t/n)} = Au_n\left(\frac{jt}{n}\right), \; u_n(0) = u_0 \in D(A) \tag{4.4.19}$$

or, recursively,

$$u_n\left(\frac{jt}{n}\right) = \left(1 - \frac{t}{n} A\right)^{-j} u_0. \tag{4.4.20}$$

Thus

$$u_n(t) = \left(1 - \frac{t}{n} A\right)^{-n} u_0$$

and the exponential formula (4.4.19) says that $u_n(t) \to u(t)$. If $u_0 \notin D(A)$ then $u(t) = S(t)u_0$ is only a generalized solution but the convergence of the implicit scheme is still true. ∎

We will close this section with a few regularity results for contraction semigroups.

If A is the infinitesimal generator of a C_0-semigroup we define

$$D(A^2) = \{u \in D(A) \mid Au \in D(A)\} \tag{4.4.21}$$

and more generally, if $k \geqslant 2$ is any integer,

$$D(A^k) = \{u \in D(A^k) \mid Au \in D(A^{k-1})\}. \tag{4.4.22}$$

Lemma 4.4.4 Let A be the infinitesimal generator of a contraction semigroup $\{S(t)\}$. Then $D(A^2)$ is dense in $D(A)$ (for the graph norm).

Proof Let $u \in D(A)$. Define

$$u_\lambda = \lambda R(\lambda)u, \quad \lambda > 0 \tag{4.4.23}$$

Then $u_\lambda \in D(A)$ and by Lemma 4.4.1, $u_\lambda \to u$ in V as $\lambda \to \infty$. Also since $R(\lambda)$ has range in $D(A)$, $u_\lambda \in D(A)$. Now by definition of $R(\lambda)$,

$$u = \frac{1}{\lambda}(\lambda I - A)u_\lambda$$

or

$$Au_\lambda = \lambda u_\lambda - \lambda u \in D(A).$$

Thus $u_\lambda \in D(A^2)$. Again

$$Au_\lambda = \lambda^2 R(\lambda)u - \lambda u = A_\lambda u.$$

which converges to Au by Lemma 4.4.2, since $u \in D(A)$. Thus $u_\lambda \to u$ in $D(A)$ as $\lambda \to \infty$. ∎

Theorem 4.4.7 Let $u_0 \in D(A^2)$, where A is the infinitesimal generator of a contraction semigroup $\{S(t)\}$. Then the solution

$$u(t) = S(t)u_0$$

of the initial value problem (4.3.6) satisfies

$$u \in C^2([0,\ \ ); V) \cap C^1([0, \infty); D(A)) \cap C([0, \infty); D(A^2)).$$

More generally, if $u_0 \in D(A^k)$, $k \geqslant 2$, then u is such that

$$u \in \bigcap_{j=0}^{k} C^{k-j}([0, \infty); D(A^j)).$$

Proof Let $u_0 \in D(A^2)$. Then $Au_0 \in D(A)$. Hence the function

$$v(t) = S(t)Au_0$$

is differentiable such that

$$\frac{dv}{dt}(t) = Av(t),\ v(0) = Au_0. \tag{4.4.24}$$

Also $v(t) \in D(A)$, for $t \geqslant 0$. But

$$v(t) = S(t)Au_0 = AS(t)u_0 = \frac{du}{dt}(t) = Au(t)$$

Hence $u(t) \in D(A^2)$ and as $v(t) \in C^1([0, \infty); V) \cap C([0; \infty); D(A))$ the result follows.

Let $k > 2$ and assume the result true for $k - 1$. Let $u_0 \in D(A^k)$. Then $Au_0 \in D(A^{k-1})$. By induction hypothesis, again as $v(t) = \dfrac{du}{dt}(t)$ satisfies (4.4.24),

$$v \in \bigcap_{j=0}^{k-1} C^{k-1-j}([0; \infty); D(A^j)),$$

i.e.

$$u \in \bigcap_{j=0}^{k-1} C^{k-j}([0; \infty); D(A^j)).$$

But $v = Au \in C([0; \infty); D(A^{k-1}))$ and so $u \in C([0, \infty); D(A^k))$. Hence the result for all k. ∎

Corollary Let $u_0 \in D(A)$, A being the infinitesimal generator of a C_0-semigroup of contractions $\{S(t)\}$. Then if $u(t) = S(t)u_0$, we have

$$\left\|\frac{du}{dt}(t)\right\| \leqslant \|Au_0\|. \tag{4.4.25}$$

Proof If $u_0 \in D(A^2)$ then $v(t)$ satisfies (4.4.24) and as $S(t)$ is a contraction,

$$\left\|\frac{du}{dt}(t)\right\| = \|v(t)\| \leqslant \|Au_0\|.$$

If $u_0 \in D(A)$, by Lemma 4.4.4, choose $u_0^n \in D(A^2)$ such that $u_0^n \to u_0$, $Au_0^n \to Au_0$. Now if $u^n(t) = S(t)u_0^n$, we have

$$\|u^n(t) - u^m(t)\| \leqslant \|u_0^n - u_0^m\|,$$

$$\left\|\frac{du^n}{dt}(t) - \frac{du^m}{dt}(t)\right\| \leqslant \|Au_0^m - Au_0^m\|.$$

Thus both u^n and $\dfrac{du^n}{dt}$ converge uniformly to u and $\dfrac{du}{dt}$ respectively and as $\left\|\dfrac{du^n}{dt}(t)\right\| \leqslant \|Au_0^n\|$, it follows that,

$$\left\|\frac{du}{dt}(t)\right\| \leqslant \|Au_0\|. \quad ∎$$

Theorem 4.4.8 Let $u_0 \in D(A^2)$. If $u_\lambda(t)$ is the solution of the initial value problem,

$$\frac{du_\lambda}{dt}(t) = A_\lambda u_\lambda(t),\ u_\lambda(0) = u_0, \tag{4.4.26}$$

where A_λ is the Yosida-approximation of A, then

$$\frac{du_\lambda}{dt}(t) \to \frac{du}{dt}(t), \text{ as } \lambda \to \infty$$

uniformly on bounded intervals (with respect to t).

Proof Let $v_\lambda(t) = \dfrac{du_\lambda}{dt}(t)$. Then v_λ satisfies

$$\frac{dv_\lambda}{dt}(t) = A_\lambda v_\lambda(t),\ v_\lambda(0) = A_\lambda u_0. \tag{4.4.27}$$

But recall that $v_\lambda(t) = (\exp(tA_\lambda))(A_\lambda u_0)$ since A_λ is bounded and so (cf. Step 1 of the proof of Theorem 4.4.2),

$$\|v_\lambda(t) - v_\mu(t)\| \leqslant t\|A_\lambda^2 u_0 - A_\mu^2 u_0\|.$$

Now

$$A_\lambda u_0 = \lambda AR(\lambda)u_0 = \lambda R(\lambda)Au_0 \quad (\text{since } u_0 \in D(A^2) \subset D(A)).$$

Thus $A_\lambda^2 u_0 = (\lambda R(\lambda))^2 A^2 u_0$. Set $\bar{u}_\lambda = \lambda R(\lambda)A^2 u_0$. Then

$$\begin{aligned}\|A_\lambda^2 u_0 - A^2 u_0\| &\leqslant \|\lambda R(\lambda)\bar{u}_\lambda - \bar{u}_\lambda\| + \|\lambda R(\lambda)A^2 u_0 - A^2 u_0\| \\ &\leqslant 2\|\lambda R(\lambda)A^2 u_0 - A^2 u_0\| \to 0, \text{ as } \lambda \to \infty.\end{aligned}$$

Thus $A_\lambda^2 u_0 \to A^2 u_0$ and so as $\lambda, \mu \to \infty$,

$$\|A_\lambda^2 u_0 - A_\mu^2 u_0\| \to 0.$$

Thus $\{v_\lambda\}$ is uniformly Cauchy on bounded intervals (of t) and so is convergent. But $v_\lambda = A_\lambda u_\lambda = A(\lambda R(\lambda))u_\lambda$. Now

$$\begin{aligned}\|\lambda R(\lambda)u_\lambda - u\| &= \|\lambda R(\lambda)u_\lambda - \lambda R(\lambda)u\| + \|\lambda R(\lambda)u - u\| \\ &\leqslant \|u_\lambda - u\| + \|\lambda R(\lambda)u - u\|\end{aligned}$$

which tends to zero as $\lambda \to \infty$, since $u_\lambda(t) = (\exp(tA_\lambda))u_0 \to S(t)u_0 = u(t)$. Thus $\lambda R(\lambda)u_\lambda \to u$ and $A(\lambda R(\lambda))u_\lambda = v_\lambda$ is convergent. Since A is closed $v_\lambda \to Au = \dfrac{du}{dt}$. Thus

$$\frac{du_\lambda}{dt} \to \frac{du}{dt}$$

and the convergence is uniform on bounded intervals of t. ∎

4.5 CONTRACTION SEMIGROUPS ON HILBERT SPACES

In this section we will assume that V is a Hilbert space whose inner-product is denoted by $(\cdot\,,\cdot)$. Let A be the infinitesimal generator of a

contraction semigroup $\{S(t)\}$. Let $u \in D(A)$. Then as

$$\|S(t)u\| \leqslant \|u\|.$$

We get

$$(S(t)u - u, u) \leqslant 0.$$

Dividing this inequality by t and letting $t \downarrow 0$ we get that

$$(Au, u) \leqslant 0 \quad \text{for all } u \in D(A). \tag{4.5.1}$$

We say that A is a **dissipative operator** ($-A$ is monotone). Also by the Hille-Yosida Theorem $\mathcal{R}(I - A) = V$ (since $I - A$ is invertible). Thus we say that A is **maximal dissipative** ($-A$ is maximal monotone, cf. Remark 4.1.4). The word 'maximal' figures in the definition because if A is dissipative and $\mathcal{R}(I - A) = V$, it is maximal in the sense that there exists no linear operator B with the same properties and such that B is an extension of A. Indeed if $D(B) \supset D(A)$ and $B|_{D(A)} = A$, with B having the same properties, then let $u \in D(B)$. Let $v \in D(A)$ such that

$$(I - A)v = (I - B)u.$$

Then $Av = Bv$ and so $(I - B)(v - u) = 0$. Multiplying this by $(v - u)$ we get

$$\|v - u\|^2 - (B(v - u), (v - u)) = 0$$

or

$$0 \leqslant \|v - u\|^2 = (B(v - u), v - u) \leqslant 0$$

whence $v = u$ or, $u \in D(A)$. Thus $B = A$.

We now show that the only generators of contraction semigroups in a Hilbert space are maximal dissipative operators.

Theorem 4.5.1 Let A be maximal dissipative. Then A is closed, densely defined and for every $\lambda > 0$, $(\lambda I - A)$ is invertible and

$$\|(\lambda I - A)^{-1}\| \leqslant \lambda^{-1}. \tag{4.5.2}$$

Proof Step 1 Let $v \in V$ such that $(v, u) = 0$ for all $u \in D(A)$. Let $w \in D(A)$ such that $w - Aw = v$. Then taking the inner-product with w, we get

$$0 = (v, w) = \|w\|^2 - (Aw, w) \geqslant \|w\|^2.$$

Thus $w = 0$ which implies that $v = 0$. Hence $D(A)$ is dense in V.

Step 2 Let $v \in V$ and $u - Au = v$. Such a $u \in D(A)$ is unique. (For if $u - Au = 0$, then $\|u\|^2 - (Au, u) = 0$ or $\|u\| \leqslant 0$ which gives $u = 0$.) Further

$$\|u\|^2 \leqslant \|u\|^2 - (Au, u) = (v, u) \leqslant \|v\| \, \|u\|.$$

Thus $\|u\| \leqslant \|v\|$. Thus the map $v \to u$ is a bounded linear operator with norm $\leqslant 1$. Clearly $u = (I - A)^{-1}v$. Thus $\|(I - A)^{-1}\| \leqslant 1$.

Step 3 Let $u_n \in D(A)$ such that $u_n \to u$ and $Au_n \to v$. Thus $u_n - Au_n \to u - v$ and $u_n = (I - A)^{-1}(u_n - Au_n)$. Thus $u = (I - A)^{-1}(u - v)$ and so $u \in D(A)$. Further $u - Au = u - v$ or $v = Au$. Thus A is closed.

Step 4 Assume finally that we have $\mathcal{R}(\lambda_0 I - A) = V$. Consider, for $v \in V$, equation $\lambda u - Au = v$. We rewrite it as

$$\lambda_0 u - Au = (\lambda_0 - \lambda)u + v$$

or

$$u = (\lambda_0 I - A)^{-1}[v + (\lambda_0 - \lambda)u] \tag{4.5.3}$$

since by an argument identical to that in Step 2 we can show that $(\lambda_0 I - A)^{-1}$ exists and its norm is $\leqslant \lambda_0^{-1}$. The equation (4.5.3) is now a fixed point equation for the map

$$u \to (\lambda_0 I - A)^{-1}[v + (\lambda_0 - \lambda)u] = F(u), \text{ say.}$$

If $u_1, u_2 \in V$, then

$$\|F(u_1) - F(u_2)\| \leqslant \frac{1}{\lambda_0} |\lambda_0 - \lambda| \, \|u_1 - u_2\|$$

and this is a contraction (and so F will have a unique fixed point) if $|\lambda_0 - \lambda| < \lambda_0$ or $0 < \lambda < 2\lambda_0$. Thus $\mathcal{R}(\lambda I - A) = V$ for all $0 < \lambda < 2\lambda_0$. We can now proceed as in Step 2 to show that for $0 < \lambda < 2\lambda_0$, $(\lambda I - A)^{-1}$ exists with norm $\leqslant \lambda^{-1}$. Now we can iterate this procedure to show that $(\lambda I - A)^{-1}$ exists for all $\lambda > 0$, with norm $\leqslant \lambda^{-1}$. ∎

By virtue of the above theorem and the Hille-Yosida Theorem we see that every maximal dissipative operator is the generator of a contraction semigroup and the regularity results of the previous section hold for the solutions of initial value problems. The advantages of the above result are the following: in order to verify that A generates a C_0-semigroup it suffices to check that (i) A is dissipative and (ii) $\mathcal{R}(\lambda_0 I - A) = V$ for *some* $\lambda_0 > 0$. The first is generally very easy. The second investigates the solution of a *single* equation of the type considered in Chapter 3. It will usually involve a combination of an existence theorem and a regularity theorem as we shall see in the next section.

We conclude this section with the study of some important special cases of maximal dissipative operators.

While we could solve the initial value problems in $[0, \infty)$ when $u_0 \in D(A)$, and not for general $u_0 \in V$, we show that if A is *self-adjoint* then we can solve the problem in the classical sense for any initial data. The price we pay for this is the lack of differentiability at $t = 0$.

Theorem 4.5.2 Let A be a self-adjoint maximal dissipative operator. Let $u_0 \in V$. Then there exists a unique u such that

$$u \in C([0, \infty); V) \cap C^1((0, \infty); V) \cap C((0, \infty); D(A))$$

such that

$$\left.\begin{aligned} \frac{du(t)}{dt} &= Au(t), \quad t > 0, \\ u(0) &= u_0. \end{aligned}\right\} \tag{4.5.4}$$

Further

$$\left.\begin{aligned} \|u(t)\| &\leqslant \|u_0\|, \quad t \geqslant 0 \\ \left\|\frac{du(t)}{dt}\right\| &= \|Au(t)\| \leqslant \frac{1}{t}\|u_0\|, \quad t > 0. \end{aligned}\right\} \tag{4.5.5}$$

Proof Step 1 (Uniqueness) Let u_1, u_2 be two solutions of (4.5.4). Set $\phi(t) = \|u_1(t) - u_2(t)\|^2$ which is continuous and such that $\phi(0) = 0$. Taking the scalar product of (4.5.4) with $u_1(t) - u_2(t)$ for $u = u_1$ and $u = u_2$ and using the dissipativity of A, we get

$$\frac{d}{dt}\phi(t) \leqslant 0,$$

whence it follows that $\phi(t) \equiv 0$. Thus $u_1(t) = u_2(t)$ for all $t \geqslant 0$.

Step 2 Let $u_0 \in D(A^2)$ so that the solution $u(t)$ exists in the classical sence. If $\{S(t)\}$ is the semigroup generated by A, then $u(t) = S(t)u_0$. Let $u_\lambda(t)$ satisfy

$$\frac{du_\lambda(t)}{dt} = A_\lambda u_\lambda(t), \; t \geqslant 0; \; u_\lambda(0) = u_0. \tag{4.5.6}$$

Then by Theorem 4.4.8, $\frac{du_\lambda}{dt} \to \frac{du(t)}{dt}$ uniformly on bounded intervals of t, as $\lambda \to \infty$. Further since A is self-adjoint, so is A_λ. For, let $R(\lambda)u_i = v_i$, $i = 1, 2$. Then $v_i - \lambda Av_i = u_i$, $i = 1, 2$. Thus

$$\begin{aligned} (R(\lambda)u_1, u_2) &= (v_1, u_2) = (v_1, v_2) - \lambda(v_1, Av_2) \\ &= (v_1, v_2) - \lambda(Av_1, v_2) = (u_1, v_2) \\ &= (u_1, R(\lambda)u_2). \end{aligned}$$

Thus $R(\lambda)$ is a symmetric bounded operator and hence self-adjoint. So A_λ is also self-adjoint. Now we can apply Theorem 4.2.1 to A_λ to deduce that

$$\left\|\frac{du_\lambda(t)}{dt}\right\| \leqslant \frac{1}{t}\|u_0\|, \quad t > 0.$$

As $\lambda \to \infty$ we obtain

$$\left\|\frac{du(t)}{dt}\right\| \leqslant \frac{1}{t}\|u_0\|, \quad t > 0$$

when $u_0 \in D(A^2)$.

Step 3 Let $u_0 \in V$. Choose $u_0^n \in D(A^2)$ such that $u_0^n \to u_0$ as $n \to \infty$. Let $u^n(t)$ be the solution of

$$\frac{du^n(t)}{dt} = Au^n(t); \; u^n(0) = u_0^n \tag{4.5.7}$$

Then we know that

$$\|u^n(t) - u^m(t)\| \leqslant \|u_0^n - u_0^m\|, \quad t \geqslant 0$$

$$\left\|\frac{du^n(t)}{dt} - \frac{du^m(t)}{dt}\right\| \leqslant \frac{1}{t}\|u_0^n - u_0^m\|, \quad t > 0.$$

Thus $\{u^n(t)\}$ converges uniformly in $[0, \infty)$ and $\left\{\frac{du^n(t)}{dt}\right\}$ converges uniformly in $[\delta, \infty)$ for any $\delta > 0$. Set

$$u(t) = \lim_{n\to\infty} u^n(t).$$

Then $\frac{du^n(t)}{dt} \to \frac{du(t)}{dt}$ uniformly in $[\delta, \infty)$ for any $\delta > 0$ and so

$$u \in C([0, \infty); V) \cap C^1((0, \infty); V).$$

Step 4 We now verify that $u(t)$ solves (4.5.4). Clearly $u(0) = \lim_{n\to\infty} u_0^n = u_0$. Also as $\{u_n(t)\}$ and $\{Au_n(t)\}$ both converge for $t > 0$, it follows that (as A is closed) $u(t) \in D(A)$ and that

$$Au(t) = \lim_{n\to\infty} Au_n(t) = \lim_{n\to\infty} \frac{du_n(t)}{dt} = \frac{du(t)}{dt}.$$

Thus $u(t)$ solves (4.5.4) and we also have

$$u \in C((0, \infty); D(A)).$$

Step 5 Taking the scalar product of (4.5.4) with u and using the dissipativity of A we get that

$$\frac{d}{dt}(\|u(t)\|^2) \leqslant 0$$

and so $\|u(t)\| \leqslant \|u_0\|$. Also

$$\left\|\frac{du^n(t)}{dt}\right\| \leqslant \frac{1}{t}\|u_0^n\|, \quad t > 0$$

from which, as $n \to \infty$, we get the other inequality in (4.5.5). ∎

We conclude the self-adjoint case with a regularity result.

Theorem 4.5.3 Let A be a self-adjoint maximal dissipative operator on V. Let u be the solution to the initial value problem (4.5.4) where $u_0 \in V$. Then

$$u \in C^k((0, \infty); D(A^l))$$

for all integers k and l.

Proof To prove the theorem, it suffices to show that

$$u \in \bigcap_{j=0}^{k} C^{k-j}((0, \infty); D(A^j)) \tag{4.5.8}$$

for any integer $k \geqslant 1$. Indeed we have already seen this for $k = 1$ in the previous theorem. We assume this for $k - 1$ $(k \geqslant 2)$ and prove the result by induction. In particular assume that

$$u \in C((0, \infty); D(A^{k-1})) \quad (k \geqslant 2). \tag{4.5.9}$$

We will now show that this implies

$$u \in C((0, \infty); D(A^k)). \tag{4.5.10}$$

If we establish (4.5.10) then consider $v(t)$ the solution of

$$\left.\begin{aligned} \frac{dv(t)}{dt} &= Av(t), \quad t \geqslant 0 \\ v(0) &= u(\epsilon) \end{aligned}\right\} \tag{4.5.11}$$

for some $\epsilon > 0$. By uniqueness and the semigroup property, $v(t) = u(t + \epsilon)$. By Theorem 4.4.7, v satisfies (4.5.8), i.e.

$$u \in \bigcap_{j=0}^{k} C^{k-j}([\epsilon, \infty); D(A^j))$$

and as $\epsilon > 0$, is arbitrary, u satisfies (4.5.8) as well.

To prove (4.5.10) assuming (4.5.9) we work with the space $D(A^{k-1}) = H$. This space is a Hilbert space with the inner product

$$(u, v)_{D(A^{k-1})} = \sum_{j=0}^{k-1} (A^j u, A^j v)$$

by the closedness of A. On H consider $\tilde{A} : D(\tilde{A}) \subset H \to H$ defined by

$$D(\tilde{A}) = D(A^k); \; \tilde{A}u = Au, \; u \in D(\tilde{A}).$$

Clearly as A is symmetric and dissipative, so is $\tilde{A}$. $\tilde{A}$ is also maximal; for if $v \in D(A^{k-1})$ there exists $u \in D(A)$ such that $u - Au = v$. Then $Au = u - v \in D(A)$ so $u \in D(A^2)$. By iterating this argument we get finally that $Au \in D(A^{k-1})$ or $u \in D(A^k)$. Then $Au = \tilde{A}u$ and so $u - \tilde{A}u = v$, $u \in D(\tilde{A})$. Thus $\mathcal{R}(1 - \tilde{A}) = H$ and so $\tilde{A}$ is maximal. So we can apply Theorem 4.5.2 to the equation

$$\frac{dv(t)}{dt} = \tilde{A}v, \quad v(0) = v_0 \in H.$$

Set $v(0) = u(\epsilon) \in D(A^{k-1}) = H$ (by (4.5.9)) and then again

$$v(t) = u(t + \epsilon)$$

so that for all $t > \epsilon$, $u(t) \in D(A^k)$. Again ϵ is arbitary and (4.5.10) follows.

Another important case is when both A and $-A$ are maximal dissipative operators. In this case it is obvious that

$$(Au, u) = 0, \quad \text{for every } u \in D(A). \tag{4.5.12}$$

Theorem 4.5.4 Let V be a Hilbert space and $A : D(A) \subset V \to V$ such that both A and $-A$ are maximal dissipative operators. Then they together generate a group of isometries.

Proof Let $u_0 \in D(A)$ and let $u(t)$ be the solution of

$$\frac{du(t)}{dt} = Au(t), \; u(0) = u_0. \tag{4.5.13}$$

Then we get

$$\tfrac{1}{2} \frac{d}{dt} \|u(t)\|^2 = (Au(t), u(t)) = 0.$$

Thus $\|u(t)\| = \|u_0\|$ for all $t \geqslant 0$, i.e. if $\{S^+(t)\}$ is the contraction semigroup generated by A then $\|S^+(t)u\| = \|u\|$ for all $t \geqslant 0$, $u \in D(A)$. But $D(A)$ is dense in V and so this holds for all $u \in V$ and so each $S^+(t)$ is an isometry on V. Similarly if $\{S^-(t)\}$ is the semigroup generated by $-A$, then $S^-(t)$ is also an isometry for each $t \geqslant 0$.

Now consider the problem

$$\left.\begin{aligned} &\frac{dv(t)}{dt} = Av(t), \quad t > 0 \\ &v(0) = u_0 \in D(A) \end{aligned}\right\}.$$

For $t \in [0, t_0]$, set $\tilde{v}(t) = v(t_0 - t)$. Then on $[0, t_0]$

$$\left.\begin{aligned} &\frac{d\tilde{v}(t)}{dt} = -\frac{dv(t_0 - t)}{dt} = -Av(t_0 - t) = -A\tilde{v}(t) \\ &\tilde{v}(0) = v(t_0) = S^+(t_0)u_0 \end{aligned}\right\}.$$

By uniqueness of the solution to the initial value problems, we get

$$\tilde{v}(t) = S^-(t)S^+(t_0)u_0.$$

In particular

$$\tilde{v}(t_0) = S^-(t_0)S^+(t_0)u_0.$$

But

$$\tilde{v}(t_0) = v(0) = u_0.$$

Thus for every $u_0 \in D(A)$

$$S^-(t_0)S^+(t_0)u_0 = u_0.$$

As $D(A)$ is dense in V, this is true for all $u_0 \in V$. Similarly

$$S^+(t_0)S^-(t_0)u_0 = u_0.$$

for every $u_0 \in V$.

If we define

$$S(t) = \begin{cases} S^+(t), & t \geqslant 0 \\ S^-(-t), & t \leqslant 0 \end{cases}$$

Then $(S(t))^{-1} = S(-t)$ and thus $\{S(t)\}_{t \in \mathbb{R}}$ is now a *group* of isometries. ∎

We will study in the next few sections applications of these special cases to various evolution equations.

4.6 THE HEAT EQUATION

Let $\Omega \subset \mathbb{R}^n$ be an open set with boundary Γ. The heat equation refers to the equation

$$\frac{\partial u}{\partial t} - \Delta u = 0 \quad \text{in } \Omega \times [0, \infty). \tag{4.6.1}$$

with appropriate initial and boundary conditions. This equation or variations thereof occur in several physical phenomena involving diffusion. It is the simplest example of a parabolic partial differential equation. In case of the heat equation u represents the temperature in Ω and is a function of $x \in \Omega$ and the time $t > 0$. The boundary conditions will depend on the physical situation we are considering. If we maintain Γ at a fixed temperature then we have a Dirichlet boundary condition on Γ. If we assume that the body is thermally insulated so that there is no exchange of heat with the surroundings then we have a Neumann boundary condition. In case the system receives heat from an external source then (4.6.1) will have an inhomogeneous term on the right hand side.

For simplicity we will assume that Ω is sufficiently smooth and that we have Γ fixed at a constant temperature, say $u = 0$. Thus we have the following initial-boundary value problem:

$$\left.\begin{aligned} \frac{\partial u(x, t)}{\partial t} - \Delta u(x, t) = 0 \quad & \text{in } \Omega \times [0, \infty) \\ u(x, t) = 0 \quad & \text{on } \Gamma \times [0, \infty) \\ u(x, 0) = u_0(x), \quad & x \in \Omega \end{aligned}\right\} \tag{4.6.2}$$

We will now study this equation in the framework of the theory of semigroups.

Theorem 4.6.1 Let $u_0 \in L^2(\Omega)$. Then there exists a unique solution u of (4.6.2) such that

$$u \in C([0, \infty); L^2(\Omega)) \cap C^1((0, \infty); L^2(\Omega)) \cap C((0, \infty); H^2(\Omega) \cap H_0^1(\Omega)).$$

Further, for every $\epsilon > 0$

$$u \in C^\infty([\epsilon, \infty) \times \bar{\Omega}).$$

Proof Let $V = L^2(\Omega)$. Define $A : D(A) \subset V \to V$ by

$$D(A) = H^2(\Omega) \cap H_0^1(\Omega); \; Au = \Delta u, \quad u \in D(A), \tag{4.6.3}$$

Clearly as

$$\int_\Omega \Delta u \cdot u = -\int_\Omega \nabla u \cdot \nabla u = -|u|_{1,\Omega}^2 \leqslant 0, \quad u \in H^2(\Omega) \cap H_0^1(\Omega),$$

we have that A is dissipative. It is maximal dissipative since there exists a unique $u \in H_0^1(\Omega)$ such that

$$(I - A)u = u - \Delta u = f$$

for every $f \in L^2(\Omega)$. Also by the regularity theorems (cf. Section 3.3) $u \in H^2(\Omega)$. Thus $\mathcal{R}(I - A) = V$. Finally we saw in Example 4.1.4 that it is self-adjoint. Hence we can apply Theorems 4.5.2 and 4.5.3 to deduce the conclusions desired. In particular note that by the regularity of Ω we have that

$$D(A^k) = \{u \in H^{2k}(\Omega) \mid u = \Delta u = \ldots \Delta^{k-1}u = 0 \text{ on } \Gamma\}.$$

Thus if $u \in D(A^k)$ for all k, it follows from the Sobolev imbedding theorem that $u \in C^\infty$. This follows then for any $t > \epsilon$, $\epsilon > 0$ by Theorem 4.5.3. ∎

Remark 4.6.1 Note that the (essential) boundary condition was incorporated into the definition of the domain. This would not be necessary for the Neumann problem. ∎

An important point to note is that however badly behaved $u_0 \in L^2(\Omega)$ may be, $u(x, t)$ is very smooth for every $t > 0$. This is known as the **strong regularizing effect** of the heat operator. In particular this shows that the heat equation is **irreversible in time,** i.e. we cannot always solve the equation:

$$\left.\begin{aligned} \frac{\partial u}{\partial t} - \Delta u = 0 \quad &\text{on } \Omega \times (0, T) \\ u(x, t) = 0 \quad &\text{on } \Gamma \times (0, T) \\ u(x, T) = u_T(x) \quad & \end{aligned}\right\} \tag{4.6.4}$$

A priori it necessary that $u_T \in D(A^j)$ for all j, i.e. $u_T \in C^\infty(\Omega)$ and $\Delta^j u_T = 0$ on Γ for all j. Even these hypotheses do not guarantee a solution to (4.6.4).

Remark 4.6.2 This irreversibility in time could have been already observed for $\Omega = \mathbb{R}^n$ from the explicit formula derived in Chapter 1 for u (cf. (1.9.16)). If u_0 has compact support then

$$u(x, t) = \frac{1}{(4\pi t)^{n/2}} \int_{\mathbb{R}^n} \exp\,(-|x - y|^2/4t) u_0(y)\, dy. \tag{4.6.5}$$

The Fourier transform (with respect to x) is

$$\tilde{u}(\xi, t) = \exp\,(-4\pi^2 |\xi|^2 t)\hat{u}_0(\xi). \tag{4.6.6}$$

When $t < 0$, (4.6.6) does not define a tempered distribution and so we cannot invert the Fourier transform. ∎

Remark 4.6.3 The explicit formula for the case $\Omega = \mathbb{R}^n$ has several uses. For instance if u_0 has compact support, it shows that $u(x, t)$ is non-zero on all of $\mathbb{R}^n$ when $t > 0$. This was already mentioned in Section 1.9 and is known as the infinite speed of propagation of signals. At first sight this does not

seem reasonable. However notice that for u_0 with compact support $u(x, t) \to 0$ exponentially as $|x| \to +\infty$. Thus the effect at large distances though non-zero, is negligible. ∎

Remark 4.6.4 Since A is maximal dissipative, it follows that $|u(\cdot, t)|_{0,\Omega} \leqslant |u_0|_{0,\Omega}$ for all $t \geqslant 0$. However in the case of $\mathbb{R}^n$ it is easy to see that this is true in all the L^p-norms assuming $u_0 \in L^p(\mathbb{R}^n)$ and has compact support. For, if $K_t(x) = (4\pi t)^{-n/2} \exp(-|x|^2/4t)$ then $u(x, t) = K_t * u_0$. By Young's inequality (cf. (1.5.4))

$$|u(\cdot, t)|_{0,p,\mathbb{R}^n} \leqslant |u_0|_{0,p,\mathbb{R}^n} |K_t|_{0,1,\mathbb{R}^n} = |u|_{0,p,\mathbb{R}^n}$$

since it is easy to see that $|K_t|_{0,1,\mathbb{R}^n} = 1$. Also it follows from (4.6.5) that

$$|u(\cdot, t)|_{0,\infty,\mathbb{R}^n} \leqslant \frac{1}{(4\pi t)^{n/2}} |u_0|_{0,1,\mathbb{R}^n}. \tag{4.6.7}$$

These assertions can be proved in any bounded open set Ω using a comparison argument based on a maximum principle which we will presently prove.

Let us now go back to the case when $u_0 \in L^2(\Omega)$. Assume that $u_0 \geqslant 0$ a.e. Then we claim that $u(\cdot, t) \geqslant 0$ for all $t \geqslant 0$. For $u(\cdot, t) = S(t)u_0$, $\{S(t)\}$ being the semigroup generated by A, A as in Theorem 4.6.1. Now

$$S(t)u_0 = \lim_{n \to \infty} \left(I - \frac{tA}{n}\right)^{-n} u_0 \tag{4.6.8}$$

by virtue of the exponential formula (Theorem 4.4.6). Now if $u_0 \geqslant 0$, by the maximum principle for the Laplace operator, if

$$-\lambda \Delta v + v = u_0, \; v \in H_0^1(\Omega) \cap C(\bar{\Omega}) \quad (\lambda > 0)$$

then $v \geqslant 0$. Thus $(I - \lambda A)^{-1} u_0 \geqslant 0$ for any $\lambda > 0$. Hence it follows from (4.6.8) that $S(t)u_0 \geqslant 0$ or $u(\cdot, t) \geqslant 0$. We now prove a more general version of this result.

Theorem 4.6.2 (Maximum Principle) Let $u \in C(\bar{\Omega} \times [0, T])$ for $T > 0$ such that u is C^1 with respect to t and C^2 with respect to x on $\Omega \times (0, T)$. Assume further that

$$\frac{\partial u}{\partial t} - \Delta u \leqslant 0 \quad \text{on } \Omega \times (0, T). \tag{4.6.9}$$

Then

$$\max_{\bar{\Omega} \times [0,T]} u = \max_{\partial_T' \Omega} u \tag{4.6.10}$$

where $\partial_T' \Omega = (\bar{\Omega} \times \{0\}) \cup (\Gamma \times [0, T])$.

Proof Let $\epsilon > 0$ and set

$$v(x, t) = u(x, t) - \epsilon t. \tag{4.6.11}$$

Then

$$\frac{\partial v}{\partial t} - \Delta v = \frac{\partial u}{\partial t} - \Delta u - \epsilon < 0.$$

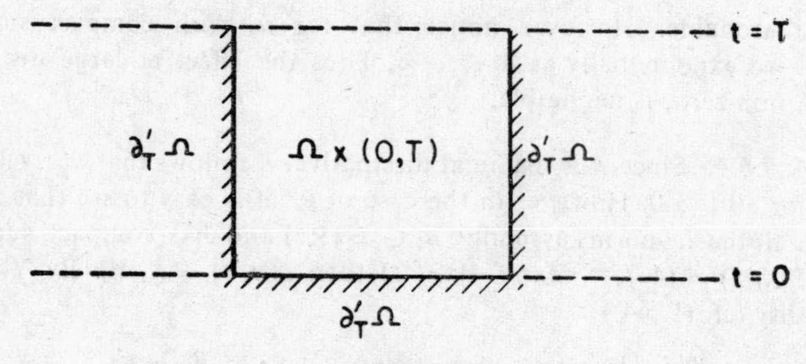

Fig. 18

Assume if possible that the maximum of v is attained at (x_0, t_0), $x_0 \in \Omega$, $0 < t_0 \leqslant T$. Then $\Delta v(x_0, t_0) \leqslant 0$ and $\frac{\partial v}{\partial t}(x_0, t_0) \geqslant 0$ (and $= 0$, if $t_0 < T$). Thus $\frac{\partial v}{\partial t} - \Delta v \geqslant 0$, a contradiction. Thus

$$\max_{\bar{\Omega}\times[0,T]} u = \max_{\bar{\Omega}\times[0,T]} (v + \epsilon t) \leqslant \max_{\bar{\Omega}\times[0,T]} v + \epsilon T$$
$$= \max_{\partial'_T \Omega} v + \epsilon T \leqslant \max_{\partial'_T \Omega} u + \epsilon T$$

since $v \leqslant u$. But as ϵ is arbitrary, we let it tend to zero to obtain

$$\max_{\partial'_T \Omega} u \leqslant \max_{\bar{\Omega}\times[0,T]} u \leqslant \max_{\partial'_T \Omega} u$$

which proves (4.6.10). ∎

Corollary If $u \leqslant 0$ on $\Gamma \times [0, \infty)$ and $u_0 \leqslant 0$ on Ω then $u(x, t) \leqslant 0$ for the $t \geqslant 0$, $x \in \Omega$. Similarly if $u \geqslant 0$ on $\Gamma \times [0, \infty)$ and $u_0 \geqslant 0$ then $u(x, t) \geqslant 0$ for all $x \in \Omega$, $t \geqslant 0$ for the solution of (4.6.2). ∎

The above maximum principle helps us to prove results on Ω knowing them for $\mathbb{R}^n$ via a comparison argument. Let $u_0 \in L^2(\Omega)$ and define $v_0 \in L^2(\mathbb{R}^n) \cap L^1(\mathbb{R}^n)$ by

$$v_0(x) = \begin{cases} |u_0(x)|, & x \in \Omega \\ 0 \quad , & x \in \mathbb{R}^n \setminus \Omega \end{cases}$$

Let v be the solution to the heat equation in $\mathbb{R}^n$ with initial data v_0. Then $v \geqslant 0$ in $\mathbb{R}^n$. Now if u is the solution to the heat equation in Ω with homogeneous Dirichlet boundary conditions and initial data u_0 (cf. (4.6.2)), then consider $w = v - u$ on Ω. We have

$$\left.\begin{aligned} &\frac{\partial w}{\partial t} - \Delta w = 0 && \text{on } \Omega \\ &w \geqslant 0 && \text{on } \Gamma \times [0, \infty) \\ &w_0 = v_0 - u_0 \geqslant 0 && \text{on } \Omega \end{aligned}\right\}$$

Thus $w \geqslant 0$ on Ω or $u \leqslant v$. Similarly $-u \leqslant v$. Hence

$$|u| \leqslant v \quad \text{on } \Omega.$$

Then we can deduce from (4.6.7) that

$$\begin{aligned} |u|_{0,\infty,\Omega} &\leqslant |v|_{0,\infty,\Omega} \leqslant |v|_{0,\infty,\mathbb{R}^n} \leqslant (4\pi t)^{-n/2}|v_0|_{0,1,\mathbb{R}^n} \\ &= (4\pi t)^{-n/2}|u_0|_{0,1,\Omega}. \end{aligned}$$

Similarly

$$|u|_{0,p,\Omega} \leqslant |v|_{0,p,\Omega} \leqslant |v|_{0,p,\mathbb{R}^n} \leqslant |v_0|_{0,p,\mathbb{R}^n} = |u_0|_{0,p,\Omega}$$

if $u_0 \in L^p(\Omega)$.

4.7 THE WAVE EQUATION

The wave equation is the simplest example of a hyperbolic equation of second order. If $x \in \mathbb{R}^n$ represents the space variable and t the time variable it can model waves in pipes or vibrating strings when $n = 1$, waves on the surface of water when $n = 2$ and waves in optics or acoustics when $n = 3$. The initial value problem for the wave equation is written as:

$$\left.\begin{aligned} \frac{\partial^2 u}{\partial t^2}(x, t) - \Delta u(x, t) &= 0 \\ u(x, 0) &= f(x) \\ \frac{\partial u}{\partial t}(x, 0) &= g(x) \end{aligned}\right\} \qquad (4.7.1)$$

where Δ is the Laplace operator in the *space variables* alone and f and g are the initial displacement and initial velocity respectively. In the case when $x \in \Omega$ a subset of $\mathbb{R}^n$ we may also prescribe boundary conditions.

Theorem 4.7.1 Let Ω be a bounded open set of class C^∞. Let $f \in H^2(\Omega) \cap H_0^1(\Omega)$ and $g \in H_0^1(\Omega)$. Then there exists a unique solution u of the problem:

$$\left.\begin{aligned} \frac{\partial^2 u}{\partial t^2} - \Delta u &= 0 && \text{in } \Omega \times [0, \infty) \\ u &= 0 && \text{on } \Gamma \times [0, \infty) \\ u(x, 0) &= f(x) && \text{in } \Omega \\ \frac{\partial u}{\partial t}(x, 0) &= g(x) && \text{in } \Omega \end{aligned}\right\} \qquad (4.7.2)$$

such that

$$u \in C([0, \infty); H^2(\Omega) \cap H_0^1(\Omega)) \cap C^1([0, \infty); H_0^1(\Omega)) \cap C^2([0, \infty); L^2(\Omega)).$$

Further

$$\left|\frac{\partial u}{\partial t}(\cdot, t)\right|_{0,\Omega}^2 + |u|_{1,\Omega}^2 = |g|_{0,\Omega}^2 + |f|_{1,\Omega}^2 \qquad (4.7.3)$$

for all $t \geqslant 0$.

If in addition $f, g \in H^k(\Omega)$ for all integers $k \geqslant 0$ and on Λ

$$f = \Delta f = \ldots = \Delta^j f = \ldots = 0 \quad \text{for all } j \geqslant 0 \tag{4.7.4}$$

$$g = \Delta g = \ldots = \Delta^j g = \ldots = 0 \quad \text{for all } j \geqslant 0 \tag{4.7.5}$$

then

$$u \in C^\infty(\bar{\Omega} \times [0, \infty)).$$

Proof Since Ω is bounded, by Poincaré's inequality (cf. Theorem 2.3.4) we can equip the space $H_0^1(\Omega)$ with the inner-product

$$(u, v) = \int_\Omega \nabla u \cdot \nabla v.$$

Let $V = H_0^1(\Omega) \times L^2(\Omega)$ equipped with the inner-product

$$(\mathbf{u}, \mathbf{v}) = \int_\Omega \nabla u_1 \cdot \nabla v_1 + \int_\Omega u_2 v_2$$

where $\mathbf{u} = (u_1, u_2)$, $\mathbf{v} = (v_1, v_2)$ are in V.

If we set $v = \dfrac{\partial u}{\partial t}$ then we get the first order system for the pair (u, v):

$$\left.\begin{aligned} \frac{\partial u}{\partial t} - v &= 0 \\ \frac{\partial v}{\partial t} - \Delta u &= 0. \end{aligned}\right\}$$

We define the unbounded operator $A : D(A) \subset V \to V$ as follows:

$$D(A) = (H^2(\Omega) \cap H_0^1(\Omega)) \times H_0^1(\Omega).$$

$$A\mathbf{u} = (v, \Delta u), \ \mathbf{u} = (u, v) \in D(A).$$

Clearly if $\mathbf{u} \in D(A)$, then

$$(A\mathbf{u}, \mathbf{u}) = \int_\Omega \nabla v \cdot \nabla u + \int \Delta u \cdot v = 0 \tag{4.7.6}$$

by Green's formula. Let $\mathbf{h} = (h_1, h_2) \in V$. Consider the equation $(1 - A)\mathbf{u} = \mathbf{h}$, i.e.

$$u - v = h_1; \quad v - \Delta u = h_2;$$

$$u \in H^2(\Omega) \cap H_0^1(\Omega), \ v \in H_0^1(\Omega).$$

Adding the two equations we get

$$u - \Delta u = h_1 + h_2 \tag{4.7.7}$$

and as $h_1 + h_2 \in L^2(\Omega)$, there exists a unique $u \in H^2(\Omega) \cap H_0^1(\Omega)$ satisfying (4.7.7) by the existence and regularity results of Chapter 3. Then $v = u - h_1 \in H_0^1(\Omega)$ exists. Thus we have shown that $\mathcal{R}(I - A) = V$ and that A is dissipative. Thus A is maximal dissipative.

In the same way $-A$ is also maximal dissipative. Thus A and $-A$ generate a group of isometries (cf. Theorem 4.5.4). Thus for $\mathbf{u}_0 = (f, g) \in D(A)$

there exists a unique solution $\mathbf{u} = (u, v) \in V$ to the equation

$$\frac{d}{dt}\mathbf{u} = A\mathbf{u},\ \mathbf{u}(0) = \mathbf{u}_0.$$

and u will then satisfy (4.7.2). The regularity of u then follows as usual. The conservation relation (4.7.3) is a consequence of the fact that the semigroup generated by A is a group of isometries.

Finally we notice that

$$D(A^k) = \left\{ (u, v) \,\middle|\, \begin{array}{l} u \in H^{k+1}(\Omega),\ v \in H^k(\Omega) \text{ with} \\ \Delta^j u = 0 \text{ on } \Gamma,\ 0 \leqslant j \leqslant \left[\frac{k}{2}\right],\ \Delta^j v = 0,\ 0 \leqslant j \leqslant \left[\frac{k+1}{2}\right] - 1 \end{array} \right\}$$

Thus from (4.7.4) and (4.7.5) $\mathbf{u}_0 = (f, g) \in D(A^k)$ for all k and the regularity follows from Theorem 4.4.7 and the Sobolev inclusion theorems. ∎

Remark 4.7.1 The condition $g = 0$ on Γ is necessary since $u = 0$ on Γ for all t implies $\frac{\partial u}{\partial t} = 0$ on Γ. ∎

Remark 4.7.2 The relation (4.7.3) states that the energy is conserved in time. ∎

Remark 4.7.3 Since we have a group instead of semigroup, it immediately follows that unlike the heat equation, the wave equation is reversible in time. We can solve the problem

$$\left.\begin{aligned} \frac{d\mathbf{u}}{dt} &= A\mathbf{u} \quad \text{in } (0, T) \\ \mathbf{u}(T) &= \mathbf{u}_T \end{aligned}\right\} \tag{4.7.8}$$

for given data $\mathbf{u}_T \in D(A)$. ∎

We have solved here an initial-boundary value problem when Ω was a bounded open set. For the case of a general domain Ω with Γ bounded, see the exercises at the end of this chapter.

We conclude this section with some explicit solutions to the wave equation when $\Omega = \mathbb{R}^n$ and use them to highlight some further important properties of solutions to the wave equation and contrast them with those of the heat equation.

Henceforth we will consider the problem:

$$\left.\begin{aligned} \frac{\partial^2 u}{\partial t^2} - \Delta u &= 0 && \text{in } \mathbb{R}^n \times (0, \infty) \\ u(x, 0) &= f(x) && \text{in } \mathbb{R}^n \\ \frac{\partial u}{\partial t}(x, 0) &= g(x) && \text{in } \mathbb{R}^n \end{aligned}\right\} \tag{4.7.9}$$

When $n = 1$, the equation becomes

$$\frac{\partial^2 u}{\partial t^2} - \frac{\partial^2 u}{\partial x^2} = 0$$

If we set $x + t = \xi$ and $x - t = \eta$, this reduces to

$$\frac{\partial^2 u}{\partial \xi \, \partial \eta} = 0$$

which has the general solution $F(\xi) + G(\eta)$. Thus $u(x, t)$ may be written in the form

$$u(x, t) = F(x + t) + G(x - t).$$

Using the initial conditions we can find F and G to get (cf. John [1]) the well-known **D'Alembert's solution:**

$$u(x, t) = \tfrac{1}{2}[f(x + t) + f(x - t)] + \tfrac{1}{2} \int_{x-t}^{x+t} g(\xi) \, d\xi. \tag{4.7.10}$$

The above formula shows that the wave equation—unlike the heat equation—has no smoothing effect on the solution. If $g \equiv 0$, then

$$u(x, t) = \tfrac{1}{2}(f(x + t) + f(x - t))$$

which is just as smooth as f. In fact if f is C^∞ on $\mathbb{R} \setminus \{x_0\}$ then u is C^∞ on $\mathbb{R}_x \times \mathbb{R}_t$ except on the lines $x + t = x_0$ and $x - t = x_0$. These lines are the **characteristic lines** through $(x_0, 0)$. Thus the singularities propagate along the characteristics.

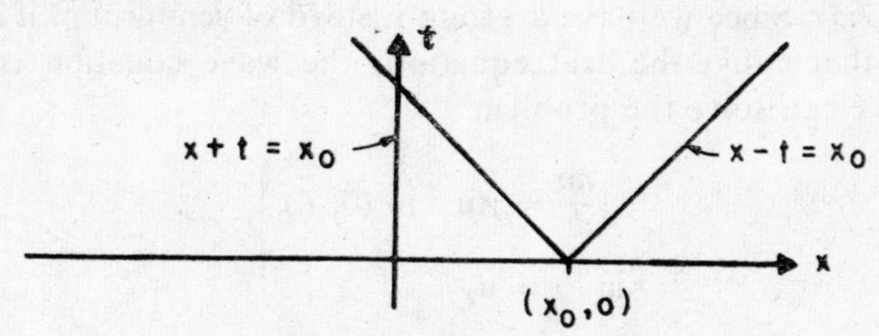

Fig. 19

Another important feature of the wave equation is that it has a finite speed of propagation of signals again unlike the heat equation. To see this, formula (4.7.10) tells us that $u(x, t)$ depends only on the values of f and g in the interval $[x - t, x + t]$. Thus if f and g have compact support in $\mathbb{R}$, so will u. The interval $[x-t, x+t]$ is called the **domain of dependence** of the solution at (x, t).

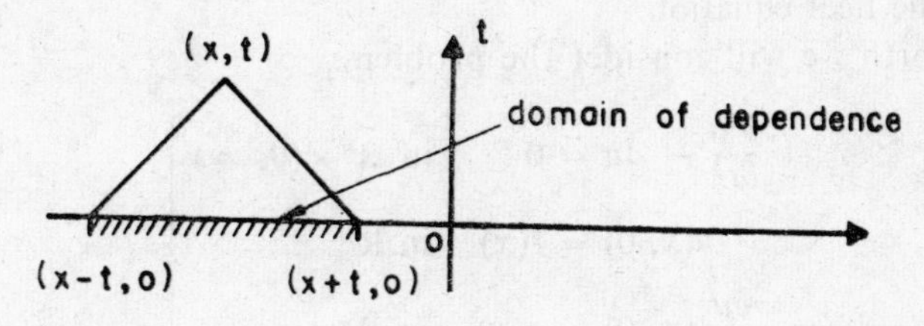

Fig. 20

In short the heat equation smooths out irregular data, has an infinite speed of propagation and is irreversible in time while the wave equation has no smoothing effect, has a finite speed of propagation and is reversible in time.

When $n = 3$ the explicit solution can be obtained by the method of spherical means (cf. John [1]). We omit the details and just record the formula:

$$u(x, t) = \frac{1}{4\pi t} \int_{|y-x|=t} g(y)\, dS(y) + \frac{\partial}{\partial t}\left(\frac{1}{4\pi t} \int_{|y-x|=t} f(y)\, dS(y)\right)$$
$$= \frac{1}{4\pi t^2} \int_{|y-x|=t} \left(tg(y) + f(y) + \sum_{i=1}^{3} \frac{\partial f}{\partial y_i}(y)(y_i - x_i)\right) dS(y). \tag{4.7.11}$$

When $n = 2$, we can consider $u(x, t)$ the solution of the wave equation as a particular case of the situation when $n = 3$ with data independent of x_3. Then $u(x_1, x_2, t)$ will be obtained from (4.7.11) with $x_3 = 0$. The surface element on the sphere $|y - x| = t$ when $x_3 = 0$ becomes

$$dS(y) = \sqrt{1 + \left(\frac{\partial y_3}{\partial y_1}\right)^2 + \left(\frac{\partial y_3}{\partial y_2}\right)^2}\, dy_1\, dy_2 = \frac{t}{|y_3|}\, dy_1\, dy_2$$

since the sphere is given by

$$\sqrt{(y_1 - x_1)^2 + (y_2 - x_2)^2 + y_3^2} = t.$$

Also the hemispheres $\{(y_1, y_2, y_3) \mid y_3 > 0\}$ and $\{(y_1, y_2, y_3) \mid y_3 < 0\}$ make the same contributions to the integrals. Thus

$$u(x_1, x_2, t) = \frac{1}{2\pi} \int_{r<t} \frac{g(y_1, y_2)}{\sqrt{t^2 - r^2}}\, dy_1\, dy_2 + \frac{\partial}{\partial t}\left(\frac{1}{2\pi} \int_{r<t} \frac{f(y_1, y_2)}{\sqrt{t^2 - r^2}}\, dy_1\, dy_2\right) \tag{4.7.12}$$

where

$$r = \sqrt{(x_1 - y_1)^2 + (x_2 - y_2)^2}.$$

This method of obtaining the 2-dimensional solution from the 3-dimensional one is known as *Hadamard's method of descent* (cf. John [1]).

Some very interesting conclusions can be drawn from formulae (4.7.11) and (4.7.12). These are explained in the comments at the end of this chapter.

4.8 THE SCHRODINGER EQUATION

The Schrodinger equation occurs in physics. If $i = \sqrt{-1}$, we look for a complex-valued function $u(x, t) : \Omega \times [0, \infty) \to \mathbb{C}$, where Ω is an open set in $\mathbb{R}^n$, such that

$$\left.\begin{aligned} i\frac{\partial u}{\partial t} - \Delta u &= 0 \quad \text{in } \Omega \times [0, \infty) \\ u &= 0 \quad \text{on } \Gamma \times (0, \infty) \\ u(x, 0) &= u_0(x) \quad \text{on } \Omega \end{aligned}\right\} \tag{4.8.1}$$

Writing $u = u_1 + iu_2$ in terms of its real and imaginary parts, we get on separating the real and imaginary parts,

$$\left.\begin{aligned} \frac{\partial u_1}{\partial t} - \Delta u_2 = 0 \quad & \text{on } \Omega\times(0; \infty) \\ \frac{\partial u_2}{\partial t} + \Delta u_1 = 0 \quad & \text{on } \Omega\times(0, \infty) \\ u_1 = u_2 = 0 \quad & \text{on } \Gamma\times(0, \infty) \\ u_1(x, 0) = u_{1,0}(x),\ u_2(x, 0) = u_{2,0}(x) \quad & \text{on } \Omega \end{aligned}\right\} \tag{4.8.2}$$

Theorem 4.8.1 Let Ω be a bounded open set of class C^2. If $u_{1,0}$ and $u_{2,0}$ the real and imaginary parts of u_0 are in $H^2(\Omega) \cap H_0^1(\Omega)$ then there exists a unique solution u of (4.8.1) such that

$$\int_\Omega |u_1(x, t)|^2\, dx + \int_\Omega |u_2(x, t)|^2\, dx = \text{constant} \tag{4.8.3}$$

for $t \geqslant 0$.

Proof: Let $V = (L^2(\Omega))^2$. We define $A : D(A) \subset V \to V$ by

$$D(A) = (H^2(\Omega) \cap H_0^1(\Omega))^2,\ Au = (\Delta u_2, -\Delta u_1),\ u = (u_1, u_2) \in D(A).$$

Then (4.8.2) can be written as

$$\frac{du}{dt}(t) = Au(t).$$

Clearly if $u \in D(A)$ then

$$(Au, u) = \int_\Omega \Delta u_2 \cdot u_1 - \int_\Omega \Delta u_1 \cdot u_2 = 0$$

by Green's formula. Also we saw in Section 3.4 (cf. (3.4.11)) that $\mathcal{R}(I - A) = V$, by the Galerkin method. The same method will also show that $\mathcal{R}(I + A) = V$. Thus both A and $-A$ are maximal dissipative and so generate together a group of isometries. The result now follows from the general result of Theorem 4.5.4. ▮

In appearance the Schrodinger equation looks similar to the heat equation. In particular if $\Omega = \mathbb{R}^n$ we have an explicit formula for the solution of

$$\left.\begin{aligned} i\frac{\partial u}{\partial t} - \Delta u = 0 \quad & \text{in } \mathbb{R}^n\times(0, \infty) \\ u(x, 0) = u_0(x) \quad & \text{in } \mathbb{R}^n \end{aligned}\right\} \tag{4.8.4}$$

given by

$$u(x, t) = (-4\pi it)^{-n/2} \int_{\mathbb{R}^n} \exp\,(-i|x - y|^2/4t) u_0(y)\, dy. \tag{4.8.5}$$

Of course while for $p = 2$ it is true (by (4.8.3)) that

$$|u(\cdot, t)|_{0,\Omega} = |u_0|_{0,\Omega}$$

it is not true (unlike the heat equation) that

$$|u(\cdot, t)|_{0,p,\Omega} \leqslant |u_0|_{0,p,\Omega}.$$

Even in $\mathbb{R}^n$ this is false since the kernel

$$(-4\pi\, it)^{-n/2} \exp(-i|x|^2/4t)$$

is not integrable and so we cannot use Young's inequality as we did in the case of the heat equation. However as in the case of the heat equation we deduce from (4.8.5) that if $u_0 \in L^1(\mathbb{R}^n)$, then

$$|u(\cdot, t)|_{0,\infty,\mathbb{R}^n} \leqslant Ct^{-n/2}|u_0|_{0,1,\mathbb{R}^n}. \tag{4.8.6}$$

In the heat equation we proved a similar inequality for bounded domains as well using a comparison argument which is no longer available to us in the case of the Schrodinger equation. Indeed such an inequality can never hold for a bounded domain Ω. For if we have $C_t > 0$ such that

$$|u(\cdot, t)|_{0,\infty,\Omega} \leqslant C_t|u_0|_{0,1,\Omega} \tag{4.8.7}$$

This will lead to

$$|u(\cdot, t)|_{0,\infty,\Omega} \leqslant C_t'|u_0|_{0,\Omega} = C_t'|u(\cdot, t)|_{0,\Omega}, \tag{4.8.8}$$

since Ω is bounded, and since (4.8.3) is true. If $\{S(t)\}$ is the corresponding semigroup we have

$$|S(t)u_0|_{0,\infty,\Omega} \leqslant C'|S(t)u_0|_{0,1,\Omega}$$

for all $u_0 \in D(A)$ and hence for all $u_0 \in V$ by density. But then if $u \in V$ we can always find u_0 such that $S(t)u_0 = u$, since $\{S(t)\}$ is a group of isometries. Thus

$$|u|_{0,\infty,\Omega} \leqslant C_t'|u|_{0,\Omega}$$

for all $u \in V$ which implies that $(L^2(\Omega))^2 \subset (L^\infty(\Omega))^2$ which is absurd.

4.9 THE INHOMOGENEOUS EQUATION

Let A be the infinitesimal generator of a C_0-semigroup $\{S(t)\}$ on a Banach space V. Let $F : [0, T] \to V$ be a given mapping. We now investigate the existence of solutions to the initial value problem

$$\left.\begin{aligned} \frac{du(t)}{dt} &= Au(t) + f(t), \quad 0 < t < T \\ u(0) &= u_0. \end{aligned}\right\} \tag{4.9.1}$$

Definition 4.9.1 A function $u : [0, T] \to V$ is said to be a **classical solution** of (4.9.1) if u is continuous on $[0, T]$, continuously differentiable on $(0, T)$, $u(t) \in D(A)$ for $0 < t < T$ and (4.9.1) is satisfied on $(0, T)$. ∎

If u is a classical solution of (4.9.1) then consider the function defined

by

$$w(s) = S(t-s)u(s), \quad 0 \leqslant s \leqslant T. \tag{4.9.2}$$

Then

$$\frac{dw}{ds}(s) = -AS(t-s)u(s) + S(t-s)Au(s) + S(t-s)f(s)$$
$$= S(t-s)f(s)$$

since S and A commute on $D(A)$. Integrating this from 0 to t, we get

$$w(t) - w(0) = \int_0^t S(t-s)f(s)\,ds$$

or,

$$u(t) = S(t)u_0 + \int_0^t S(t-s)f(s)\,ds. \tag{4.9.3}$$

Thus a classical solution has to be of the form (4.9.3) and hence is always **unique.**

If the function $f: [0, T] \to V$ is integrable then (4.9.3) still makes sense. Then u defined by that formula is called a **generalized solution** (or, **mild solution)** of (4.9.1). Thus a generalized solution always exists but it need not be a classical solution as the following example shows.

Example 4.9.1 (Pazy [1]) Let $x \in V$ such that $S(t)x \notin D(A)$ for any $t \geqslant 0$. Let $f(t) = S(t)x$. Then f is even continuous. However the generalized solution of (4.9.1) with $u_0 = 0 \in D(A)$ is given by

$$u(t) = tS(t)x$$

which is not differentiable and hence not a classical solution. ∎

We now examine situations when generalized solutions are classical.

Theorem 4.9.1 Let $f: [0, T] \to V$ be continuous and let

$$v(t) = \int_0^t S(t-s)f(s)\,ds, \quad 0 \leqslant t \leqslant T. \tag{4.9.4}$$

If (4.9.1) has a unique classical solution for every $u_0 \in D(A)$ then the following hold:

(i) $v(t)$ is continuously differentiable on $(0, T)$.
(ii) $v(t) \in D(A)$ for all $t \in (0, T)$ and $Av(t)$ is continuous on $(0, T)$.

Proof Let u be a classical solution of (4.9.1) so that

$$u(t) = S(t)u_0 + v(t)$$

by (4.9.3). If $u_0 \in D(A)$ then so does $S(t)u_0$ and as $u(t) \in D(A)$ by definition, it follows that $v(t) \in D(A)$. Also

$$Av(t) = Au(t) - AS(t)u_0 = u'(t) - f(t) - \frac{d}{dt}(S(t)u_0)$$

which is continuous. Thus (ii) is verified. Since $S(t)u_0$ and $u(t)$ are continuously differentiable, so is $v(t)$ and (i) is also true. ∎

The conditions (i) and (ii) are equivalent to each other. If (i) is true, then for $h > 0$,

$$\begin{aligned}\frac{S(h)-I}{h}v(t) &= \frac{1}{h}\left[\int_0^t S(t+h-s)f(s)\,ds - \int_0^t S(t-s)f(s)\,ds\right]\\ &= \frac{1}{h}\left[\int_0^{t+h} S(t+h-s)f(s)\,ds - \int_0^t S(t-s)f(s)\,ds\right]\\ &\quad - \frac{1}{h}\int_t^{t+h} S(t+h-s)f(s)\,ds\\ &= \frac{v(t+h)-v(t)}{h} - \frac{1}{h}\int_t^{t+h} S(t+h-s)f(s)\,ds\\ &\to v'(t) - f(t)\end{aligned}$$

Thus $v(t) \in D(A)$ and $Av(t) = v'(t) - f(t)$ which is continuous. Thus (ii) is true. Conversely, if (ii) is true the above calculation shows that

$$D^+v(t) = Av(t) + f(t)$$

and $D^+v(t)$ is continuous. Hence $D^-v(t)$ also exists and is equal to $D^+v(t)$. Thus $v(t)$ is continuously differentiable and so (i) is true.

Theorem 4.9.2 If condition (i) (or, equivalently (ii)) is verified then the generalized solution is classical for $u_0 \in D(A)$.

Proof In this case we have $v(t) \in D(A)$ and so if $u_0 \in D(A)$, it follows that $u(t) \in D(A)$. Also since both $v(t)$ and $S(t)u_0$ are differentiable, so is $u(t)$. Further

$$\begin{aligned}Au(t) &= AS(t)u_0 + Av(t)\\ &= \frac{d}{dt}(S(t)u_0) + \frac{dv}{dt}(t) - f(t)\\ &= \frac{d}{dt}u(t) - f(t)\end{aligned}$$

and so (4.9.1) is satisfied. ∎

Corollary Let $f \in C^1([0, T]; V)$. Then (4.9.1) has a unique classical solution for every $u_0 \in D(A)$.

Proof Let $v(t)$ be given by (4.9.4). It is easy to see that

$$v'(t) = S(t)f(0) + \int_0^t S(t-s)f'(s)\,ds$$

which is continuous on $(0, T)$. Thus (4.9.1) has a unique classical solution by Theorem 4.9.2. ∎

For further results in this direction, see Pazy [1].

Remark 4.9.1 The formula (4.9.3) defining the generalized solution is known as the **Variation of Parameters Formula** keeping in mind the formula for the solution of linear ordinary differential equations with constant coefficients. ∎

COMMENTS

1. *General Remarks.* The systematic theory of semigroups rapidly developed starting with the independent discovery of the generation theorem for contraction semigroups by Hille and Yosida. Now it is a major topic of research in functional analysis with extensions in several directions. The material presented in this chapter gives a rudimentary introduction to the subject from the point of view of applications to partial differential equations. For greater detail in this regard, the interested reader is referred to the book by Pazy [1].

2. *Exponential Formulae.* The exponential map $A \to e^A$ is well known in matrix theory and, as seen earlier, can be used to solve linear systems of ordinary differential equations with constant coefficients. The notion of a semigroup generalizes that of the exponential map for those unbounded operators which satisfy the hypotheses of the Hille-Yosida Theorem (Theorem 4.4.5). In fact the proof of Yosida expresses the semigroup as the limit of $\exp(tA_\lambda)$ as $\lambda \to \infty$. The proof of Hille was based on the exponential formula (4.4.17) which he proved directly. This formula again displays the semigroup generated by A as an 'exponential function.' Another type of exponential formula can be described as follows; if $u \in D(A)$ then $A_h u = \dfrac{S(h)u - u}{h}$, converges to Ax as $h \downarrow 0$. It can be shown that (cf. Pazy [1]) for any $u \in V$,

$$S(t)u = \lim_{h \downarrow 0} \exp(tA_h)u$$

and that the limit is uniform in t on any bounded interval.

The above exponential formula has an interesting consequence (cf. Pazy [1]). If V is the space of bounded uniformly continuous functions on $\mathbb{R}$ with the semigroup defined as in Example 4.3.2, then for $f \in V$

$$(A_h f)(s) = \frac{f(s+h) - f(s)}{h} = (\Delta_h f)(s),$$

viz. the forward difference quotient at s with step size h. It can be seen that

$$(A_h f)^k(s) = \frac{1}{h^k} \sum_{m=0}^{k} (-1)^{k-m} \binom{k}{m} f(s+mh) = (\Delta_h^k f)(s)$$

Thus by the above exponential formula, we get

$$f(s+t) = \lim_{h \downarrow 0} \sum_{k=0}^{\infty} \frac{t^k}{k!} (\Delta_h^k f)(s)$$

which is a generalization of the Taylor's series to uniformly continuous functions.

3. *The Laplace Transform.* In Section 4.4 we defined the *resolvent operator* $R(\lambda)$ by the formula (4.4.1) which is none other than the Laplace transform of the continuous function $t \to S(t)u$. It is possible to recover $S(t)$ from $R(\lambda)$ by inverting the Laplace transform (cf. Pazy [1]).

4. *Maximal Dissipative Operators.* We saw in this chapter that the generators of contraction semigroups in a Hilbert space are precisely the maximal dissipative operators. The results for the Hilbert space case are due to Phillips [1] and were generalized to the case of a Banach space by Lumer and Phillips [1] (cf. also Pazy [1]). The case of self-adjoint maximal dissipative operators is the simplest example of generators of **Analytic Semigroups** (cf. Pazy [1]). These semigroups have smoothing properties. For instance if $\{S(t)\}$ is an analytic semigroup generated by A then for every $t > 0$ and every integer $k \geqslant 0$, we have

$$S(t) : V \to D(A^k).$$

This was seen in the self adjoint case. The results of this were applied to the heat equation and we saw that for any L^2-data, the solution become smooth for any $t > 0$. An interesting consequence of this is the *Weierstrass Approximation Theorem* (cf. Folland [1]). If f is continuous with compact support in $\mathbb{R}^n$ let

$$u(x, t) = (4\pi t)^{-n/2} \int_{\mathbb{R}^n} \exp(-|x-y|^2/4t) f(y) \, dy.$$

Then $u(x, t)$ is the solution of the heat equation in $\mathbb{R}^n$ with initial data, f. Thus $u(x, t) \to f(x)$ as $t \downarrow 0$, uniformly in $\mathbb{R}^n$. Now if $x \in \mathbb{C}^n$, $u(x, t)$ is an entire function and so on any compact set can be uniformly approximated by the partial sums of its Taylor series which are polynomials. Thus f can in turn be uniformly approximated by polynomials in any compact set K of $\mathbb{R}^n$.

The other important maximal dissipative class is when A and $-A$ are both maximal dissipative, which leads to a group of isometries. A special case (in complex spaces) is *Stone's Theorem* which states that A generates a group of unitary operators $\{S(t)\}$ if and only if iA is self-adjoint (cf. Pazy [1]).

5. *The Wave Equation.* An interesting feature of the solution of the wave equation when $n = 3$ is that the solution at (x, t) depends only on

the data on the sphere

$$S(x, t) = \{y \mid |y - x| = t\}$$

according to the formula (4.7.11). The set $S(x, t)$ has 3-dimensional measure zero. Now assume that the initial data f and g have support in $B(0; \rho)$ for some fixed $\rho > 0$. Then $S(x, t)$ will intersect $B(0; \rho)$ if and only if $t - \rho < |x| < t + \rho$, i.e., the point x lies in a spherical shell centre origin and bounded by balls of radii $t - \rho$ and $t + \rho$. Now given any $x \in \mathbb{R}^n$ if t is large enough $(t > |x| + \rho)$ the sphere $S(x, t)$ will not intersect $B(0; \rho)$ and consequently $u(x, t) = 0$. Thus after sufficiently large time the signal at any point completely dies out. This is known as **Huygens' Principle.** Thus if a musical instrument is emitting notes a listener at distance d from the instrument will hear, at time t, only the note emitted at time $t - d$ and no other, thus avoiding a cacaphony caused by accumulation of different notes!

The Huygens' principle is valid only in special cases and is not a general property of hyperbolic equations. In fact it is not true even for the wave equation when $n = 2$. Here the domain of dependence is a solid disc $r \leqslant t$ in $\mathbb{R}^2$. Hence the support of u will continue to expand keeping the energy constant. Thus $u(x, t)$ will become smaller and smaller as $t \to \infty$ but will never become zero. Thus if a stone is dropped in water the ripples will move out from the centre of disturbance. As time increases the disturbance will be barely noticeable but will continue to exist and will not die out.

6. *Bibliography.* For a nice introduction to the theory of semigroups with applications to partial differential equations, see Pazy [1] or Friedman [1]. A classical reference for semigroups is the book by Hille and Phillips [1] or Dunford and Schwartz [1]. See also Brezis [1] and John [1] and the references cited therein for additional information on parabolic and hyperbolic equations.

EXERCISES 4

4.1 Let A be an $n \times n$ real matrix. Let $\|A\|_1$, $\|A\|_\infty$ be the norms induced on A when $\mathbb{R}^n$ is provided with the norms

$$\|x\|_1 = \sum_{i=1}^{n} |x_i|$$

and

$$\|x\|_\infty = \max_{1 \leqslant i \leqslant n} |x_i|$$

respectively. Show that

$$\|A\|_1 = \max_{1 \leqslant j \leqslant n} \sum_{i=1}^{n} |a_{ij}|.$$

Use it to compute $\|A\|_\infty$.

4.2 Let V, W be Banach spaces and $A : D(A) \subset V \to W$ a densely defined linear operator. If $G(A)$ is dense in $V \times W$, show that $D(A^*) = \{0\}$.

4.3 Let V be a Hilbert space and $A : D(A) \subset V \to V$ a densely defined linear operator. Let $J : V \times V \to V \times V$ be defined by

$$J(u, v) = (-v, u).$$

Show that

$$J(G(A^*)) = G(A)^{\perp}.$$

4.4 Let V be a Hilbert space and $A : D(A) \subset V \to V$ a closed and densely defined operator. Show that A^* is also densely defined.

4.5 Let V be a Banach space and A, B bounded linear operators on V. If $AB = BA$, show that

$$e^{A+B} = e^A e^B.$$

4.6 Let A be a 2×2 matrix with distinct eigenvalues. Find constants α and β such that

$$e^A = \alpha I + \beta A.$$

4.7 (a) Let V be a Banach space and $A : D(A) \subset V \to V$ the infinitesimal generator of a C_0-semigroup $\{S(t)\}$ on V. If $u \in D(A^2)$, show that

$$S(t)u - u = tAu + \int_0^t (t - \tau)S(\tau)A^2u \, d\tau.$$

(b) Deduce that for $u \in D(A^2)$,

$$\|Au\|^2 \leqslant 4M^2\|A^2u\| \, \|u\|,$$

if $\|S(t)\| \leqslant M$ for all t.

4.8 (a) Let $V = L^2(\mathbb{R})$. Let $\{S(t)\}$ be the semigroup defined by

$$(S(t)f)(s) = f(s + t).$$

Find its infinitesimal generator A.

(b) Use the previous exercise to show that if $u \in H^2(\mathbb{R})$,

$$|u|^2_{1, \mathbb{R}} \leqslant 4|u|_{2, \mathbb{R}}|u|_{0, \mathbb{R}}.$$

4.9 (a) Let A and B be linear operators on a Hibert space V such that $D(A) \subset D(B)$ and such that $A + tB$ is dissipative for all $t \in [0, 1]$. Assume that there exists $0 \leqslant \alpha < 1$ and $\beta \geqslant 0$ such that

$$\|Bu\| \leqslant \alpha\|Au\| + \beta\|u\|, \quad \text{for every } u \in D(A).$$

If there exists $t_0 \in [0, 1]$ such that $A + t_0B$ is maximal dissipative, show that $A + tB$ is maximal dissipative for all $t \in [0, 1]$.

(b) Deduce that if A is maximal dissipative and B dissipative with $D(B) \supset D(A)$ and if there exist $0 \leqslant \alpha < 1$ and $\beta \geqslant 0$ such that

$$\|Bu\| \leqslant \alpha\|Au\| + \beta\|u\|, \quad \text{for every } u \in D(A)$$

then $A + B$ is maximal dissipative.

(c) Show by example that the above result is not true in general if $\alpha = 1$.

4.10 Apply the preceding exercise to show the existence of a unique solution to the problem:

$$\left.\begin{aligned} &\frac{\partial u}{\partial t} - \frac{\partial^2 u}{\partial x^2} + c\frac{\partial u}{\partial x} = 0 \quad \text{in } (0, 1)\times(0, \infty) \\ &u(0, t) = u(1, t) = 0, \qquad t > 0 \\ &u(x, 0) = u_0(x) \qquad\quad \text{in } (0, 1) \end{aligned}\right\}$$

where $u_0 \in H^2((0, 1)) \cap H_0^1((0, 1))$ and $c \in \mathbb{R}$, $c \neq 0$.

4.11 Let $\Omega \subset \mathbb{R}^n$ be a smooth bounded open set and $u_0 \in L^2(\Omega)$. Let u be the solution of the homogeneous Dirichlet boundary value problem for the heat equation with initial data u_0. Show that for all $T > 0$,

$$\tfrac{1}{2}|u(T)|_{0,\Omega}^2 + \int_0^T |\nabla u(t)|_{0,\Omega}^2\, dt = \tfrac{1}{2}|u_0|_{0,\Omega}^2.$$

4.12 Let $u \in L^2(\Omega)$, $u \geqslant 0$ a.e., Ω a bounded smooth open set in $\mathbb{R}^n$. Show that there exists a sequence $\{u_m\}$ in $H^2(\Omega) \cap H_0^1(\Omega)$ such that $u_m \geqslant 0$ a.e. for each m and such that $u_m \to u$ in $L^2(\Omega)$.

4.13 Let A be a maximal dissipative operator on a Hilbert space V. Show that there exists a unique solution to the problem

$$\left.\begin{aligned} &\frac{du}{dt}(t) + \lambda u(t) = Au(t), \quad t > 0 \\ &u(0) = u_0 \end{aligned}\right\}$$

where $\lambda \in \mathbb{R}$ and $u_0 \in D(A)$.

4.14 Let Ω be a smooth open set in $\mathbb{R}^n$ with bounded boundary Γ. Show that there exists a unique solution u to the problem

$$\left.\begin{aligned} &\frac{\partial^2 u}{\partial t^2} - \Delta u = 0 \qquad \text{in } \Omega\times(0, \infty) \\ &u = 0 \qquad \text{on } \Gamma\times(0, \infty) \\ &\left.\begin{aligned} u(x, 0) &= f(x) \\ u_t(x, 0) &= g(x) \end{aligned}\right\} \text{in } \Omega \end{aligned}\right\}$$

when $f \in H^2(\Omega) \cap H_0^1(\Omega)$ and $g \in H_0^1(\Omega)$.

4.15 (The Klein-Gordon Equation). Let $m \neq 0$, $m \in \mathbb{R}$. Let $\Omega \subset \mathbb{R}^n$ be a smooth bounded open set with boundary Γ. Show that there exists a unique u such that

$$\left.\begin{aligned} &\frac{\partial^2 u}{\partial t^2} - \Delta u + m^2 u = 0 \quad \text{in } \Omega\times(0, \infty) \\ &u = 0 \quad \text{on } \Gamma\times(0, \infty) \\ &u(x, 0) = f \quad \text{in } \Omega \\ &u_t(x, 0) = g \quad \text{in } \Omega \end{aligned}\right\}$$

when $f \in H^2(\Omega) \cap H_0^1(\Omega)$ and $g \in H_0^1(\Omega)$. Show also that

$$u \in C([0, \infty); H^2(\Omega) \cap H^1(\Omega)) \cap C^1([0, \infty); H_0^1(\Omega)) \cap C^2([0, \infty); L^2(\Omega)).$$

4.16 Let V be a Banach space and A the infinitesimal generator of a C_0-semigroup. Let $f \in C([0, T]; D(A))$ where $D(A)$ is equipped with the graph norm. Show then that the generalized solution of

$$\left.\begin{aligned} \frac{du}{dt} - Au = f, \quad t > 0 \\ u(0) = u_0 \in D(A) \end{aligned}\right\}$$

is a classical solution.

FIVE

Some Techniques from Nonlinear Analysis

5.1 INTRODUCTION

In this chapter we will illustrate a few simple techniques to solve some nonlinear problems. The problems we have in mind will be boundary value problems for semilinear elliptic equations. We may consider, as a typical example,

$$\left.\begin{aligned} -\Delta u &= f(u) + g \quad \text{in } \Omega \\ u &= 0 \qquad\qquad\; \text{on } \Gamma \end{aligned}\right\} \tag{5.1.1}$$

where $\Omega \subset \mathbf{R}^n$ is a bounded open set with boundary Γ, g a given function defined on Ω and $f : \mathbf{R} \to \mathbf{R}$ a given nonlinear function. We will define a suitable weak formulation of such problems and investigate the existence and multiplicity results for the solutions of these problems.

The techniques we have chosen to illustrate can be broadly classified as follows:

(i) fixed point methods,
(ii) approximation methods, and
(iii) variational methods,

Given a problem of the form (5.1.1), we will seek a suitable weak formulation and cast it into the form

$$F(u) = 0 \tag{5.1.2}$$

where u is in a suitable function space (usually the Sobolev space $H_0^1(\Omega)$). The equation (5.1.2) may be equivalently written as

$$u = G(u) \tag{5.1.3}$$

and thus we are led to consider fixed point theorems to solve (5.1.3). We will consider a few standard fixed point theorems and illustrate their use.

Another approach to solve (5.1.2) will be to construct approximations to the solution and study their convergence. One such method is to construct successive approximations by an iterative procedure. We will illustrate this in the method of monotone iterations. Another is to approximate the space by a finite dimensional subspace, solve the possibly simpler finite dimensional analogue to (5.1.2) and then use a limiting argument. This is

the spirit of the Galerkin method already illustrated in the linear case in Chapter 3. We will study some nonlinear applications of this method.

Our third approach to solve (5.1.2) is to look for a functional $J : V \to \mathbb{R}$, where V is our function space, such that its Fréchet derivative J' is equal to F, i.e., for instance if V is a Hilbert space,

$$(J'(u), v) = (F(u), v), \quad \text{for every } v \in V, \tag{5.1.4}$$

for any $u \in V$. Then solving (5.1.2) is equivalent to looking for $u \in V$ such that $J'(u) = 0$, i.e., looking for critical point of J. The obvious critical points are local maxima or minima (the latter will be more probable in applications, since J will represent an 'energy'). We will thus study situations when J admits à local minimum. Of course these are not the only critical points available and we will also see examples where critical points can be obtained by other methods.

The methods studied in this chapter are by no means exhaustive. There is a vast literature in nonlinear analysis both in the form of books and an exponentially growing collection of published articles in leading journals. We will try to indicate some of these in course of this chapter as well as in the comments.

5.2 SOME FIXED POINT THEOREMS

As mentioned in the previous section, our first approach to solve nonlinear problems will be to use fixed point theorems. In this section we will prove some important fixed point theorems and illustrate their use in semilinear elliptic boundary value problems.

Our first result is very well known and so we merely state it without proof.

Theorem 5.2.1 **(Contraction Mapping Theorem)** Let (X, d) be a complete metric space and $f : X \to X$ a contraction mapping, i.e., there exists a constant C such that $0 < C < 1$ and

$$d(f(x), f(y)) \leqslant Cd(x, y) \tag{5.2.1}$$

for every $x, y \in X$. Then f admits a unique fixed point $\tilde{x}$ which can be obtained as follows:

$$\tilde{x} = \lim_{n \to \infty} x_n \tag{5.2.2}$$

where

$$x_n = f(x_{n-1}), \; x_0 \in X, \text{ arbitrary.} \quad \blacksquare \tag{5.2.3}$$

In fact, we have already used the above result twice in the preceding chapters (cf. Theorems 3.1.3 and 4.5.1.) It has the merit of being constructive. The unique fixed point $\tilde{x}$ can be obtained as the limit of the successive approximations x_n, *starting from any arbitrary point* x_0. Amongst the well-known applications to this theorem are the *Picard's theorem* for ordinary

differential equations (cf. for instance, Simmons [1] or [2]) and the *implicit function theorem* (cf. Dieudonne [1]).

Remark 5.2.1 A word of caution on the terminology. In the previous chapter we used the term contraction to linear operators $\{S(t)\}$ of a semi-group when $\|S(t)\| \leqslant 1$. However in the present context they will not be called contraction mappings. For a linear map to be a contraction mapping its norm must be strictly less than unity. ∎

We will now prove another famous and oft used fixed point theorem, viz. the *Brouwer fixed point theorem.* The proof of this theorem follows from the 'no retraction' theorem for the unit ball in $\mathbb{R}^n$. The usual proofs of this latter result are topological using either the techniques of algebraic topology or those of degree theory. In the recent years several 'elementary' proofs using just calculus have appeared. The proof we will be presenting is due to Kannai [1].

Lemma 5.2.1 Let $B = B(0; 1)$ be the unit ball in $\mathbb{R}^n$ and $f: B \to \mathbb{R}^n$ a map of class C^2. Let $M(x)$ denote the matrix

$$M(x) = \begin{bmatrix} \frac{\partial f_2(x)}{\partial x_1} & \cdots & \frac{\partial f_n(x)}{\partial x_1} \\ \vdots & & \vdots \\ \frac{\partial f_2(x)}{\partial x_n} & \cdots & \frac{\partial f_n(x)}{\partial x_n} \end{bmatrix}, \; f = (f_1, \ldots, f_n). \tag{5.2.4}$$

Let $E_i(x)$ be the determinant got from the above matrix on suppressing its ith row. Then

$$\sum_{i=1}^{n} (-1)^i \frac{\partial E_i}{\partial x_i}(x) = 0, \text{ for all } x \in B.$$

Proof If $i \neq j$, define $C_{ij}(x)$ to be the determinant got from $M(x)$ by (i) suppressing its ith row and (ii) replacing the row

$$\left(\frac{\partial f_2}{\partial x_j}(x), \ldots, \frac{\partial f_n}{\partial x_j}(x)\right) \tag{5.2.5}$$

by

$$\left(\frac{\partial^2 f_2}{\partial x_i\, \partial x_j}(x), \ldots, \frac{\partial^2 f_n}{\partial x_i\, \partial x_j}(x)\right).$$

Then

$$\frac{\partial E_i}{\partial x_i} = \sum_{j \neq i} C_{ij}$$

Now $\frac{\partial^2 f_k}{\partial x_i\, \partial x_j} = \frac{\partial^2 f_k}{\partial x_j\, \partial x_i}$ and hence $C_{ji} = (-1)^{j-i-1} C_{ij}$. Hence

$$\begin{aligned} \sum_{i=1}^{n} (-1)^i \frac{\partial E_i}{\partial x_i} &= \sum_{i=1}^{n} (-1)^i \Big(\sum_{j<i} C_{ij} + \sum_{j>i} C_{ij}\Big) \\ &= \sum_{i=1}^{n} \Big\{\sum_{j<i} (-1)^i C_{ij} + \sum_{j>i} (-1)^i (-1)^{j-i-1} C_{ji}\Big\} \\ &= 0. \end{aligned}$$ ∎

We use the above result in the proof of the no retraction theorem.

Theorem 5.2.2 (no differentiable retraction) There does not exist any C^2 map $f: B \to \partial B = S = \{x \in \mathbb{R}^n \mid |x| = 1\}$ such that $f(x) = x$ for all $x \in S$.

Proof Assume the contrary. Since $|f(x)|^2 \equiv 1$ we have that the matrix $f'(x)$ of partial derivatives is singular, i.e. if $J(x)$ is the Jacobian of f at x then $J(x) \equiv 0$. We can write

$$J(x) = \sum_{i=1}^{n} (-1)^{i+1} \frac{\partial f_1}{\partial x_i}(x) E_i(x),\ x \in B,$$

where E_i is as defined previously. Thus on integrating overall of B we get,

$$\begin{aligned} 0 = \int_B J(x)\,dx &= \sum_{i=1}^{n} \int_B (-1)^{i+1} \frac{\partial f_1}{\partial x_i}(x) E_i(x)\,dx \\ &= \sum_{i=1}^{n} \int_B (-1)^{i+1} \frac{\partial}{\partial x_i}(f_1(x)E_i(x)) \\ &\quad + \sum_{i=1}^{n} \int_B (-1)^{i} \frac{\partial E_i}{\partial x_i}(x) f_1(x)\,dx \end{aligned}$$

and by Lemma 5.2.1, the latter term is zero. Thus we get

$$\int_B \sum_{i=1}^{n} (-1)^{i+1} \frac{\partial}{\partial x_i}(f_1(x)E_i(x))\,dx = 0$$

or, by Green's formula,

$$\int_S \sum_{i=1}^{n} (-1)^{i+1} f_1(x) E_i(x) x_i\,dS = 0. \tag{5.2.6}$$

But $f_i(x) = x_i$ for $x \in S$. Hence the gradient field $\nabla(f_i(x) - x_i)$ must be normal to S (since all the tangential derivatives will vanish). But the unit normal to S at x is x itself and so there exist scalars $\lambda_i(x)$ such that

$$\nabla(f_i(x) - x_i) = \lambda_i x,\ x \in S.$$

Then $M(x)$ has the following form:

$$M(x) = \begin{bmatrix} \lambda_2 x_1 & \cdots & \lambda_n x_1 \\ 1 + \lambda_2 x_2 & \cdots & \lambda_n x_2 \\ \vdots & & \vdots \\ \lambda_2 x_n & \cdots & 1 + \lambda_n x_n \end{bmatrix}.$$

Hence $(-1)^{i+1} E_i(x) x_i$ is equal to the determinant

$$\begin{vmatrix} x_1 & \lambda_2 x_1 & \cdots & \lambda_n x_1 \\ x_2 & 1 + \lambda_2 x_2 & \cdots & \lambda_n x_2 \\ \vdots & \vdots & & \vdots \\ x_n & \lambda_2 x_n & \cdots & 1 + \lambda_n x_n \end{vmatrix}$$

$$= \begin{vmatrix} x_1 & 0 & \dots & 0 \\ x_2 & 1 & \dots & 0 \\ \vdots & \vdots & & \vdots \\ x_n & 0 & \dots & 1 \end{vmatrix} = x_1.$$

Thus from (5.2.6) it follows that

$$0 = \int_S f_1(x)x_1 \, dS = \int_S x_1^2 \, dS > 0,$$

a contradiction. Hence the theorem is proved. ∎

Theorem 5.2.3 **(Brouwer)** Let $f : B \to B$ be a continuous map. Then f has a fixed point.

Proof If f has no fixed points then $f(x) - x \neq 0$ for all $x \in B$. Since B is compact we have a constant $C > 0$ such that $|f(x) - x| \geqslant C > 0$ for all $x \in B$. We can then find $g : B \to B$, g of class C^2 such that $|g(x) - f(x)| < C/2$ for all $x \in B$ so that $|g(x) - x| \geqslant C/2 > 0$ for all $x \in B$. We now define $h : B \to S$ as follows: $h(x)$ is the intersection of the line from $g(x)$ to x with S (cf. Fig. 21).

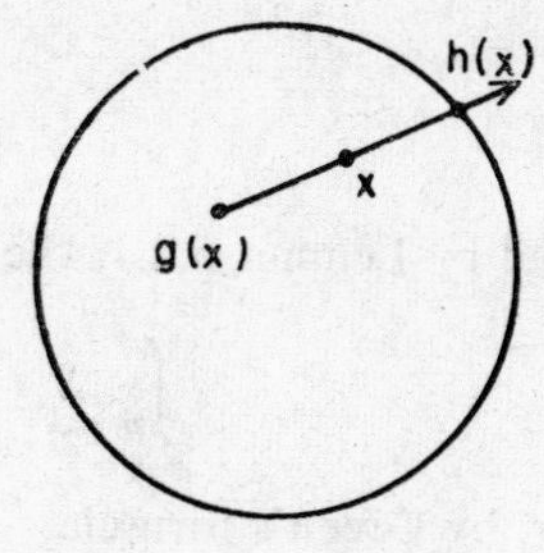

FIG. 21

Since $g(x) \neq x$, this line and hence $h(x)$ is always well defined. The map $x \to h(x)$ is of class C^2. Indeed

$$h(x) = \lambda x + (1 - \lambda)g(x), \ \lambda \geqslant 1, \ |h(x)| = 1.$$

Now

$$|h(x)|^2 = \lambda^2|x|^2 + (1 - \lambda)^2|g(x)|^2 + 2\lambda(1 - \lambda)(x, g(x)) = 1,$$

i.e.

$$\lambda^2|x - g(x)|^2 + 2\lambda(x - g(x), g(x)) + |g(x)|^2 - 1 = 0$$

which has exactly one root $\lambda \geqslant 1$ and as $x - g(x) \neq 0$ the map $x \to \lambda(x)$ is C^2 and so is the map $x \to h(x)$. But if $x \in S$, obviously $h(x) = x$ and so $h : B \to S$ is a C^2 retraction, a contradiction to the preceding theorem. ∎

Corollary If $B = B(0; R)$, $R > 0$, and $f : B \to B$ is a continuous map, then f has a fixed point. ∎

The above Corollary can be generalized as follows:

Theorem 5.2.4 **(Brouwer)** Let $K \subset \mathbb{R}^n$ be a compact and convex set, $f : K \to K$ a continuous map. Then f has a fixed point.

Proof Let $P_K : \mathbb{R}^n \to K$ be the projection onto K (cf. Theorem 3.1.1.).

Since K is compact, there exists $R > 0$ such that $K \subset B(0; R)$. Define $\tilde{f}: B(0; R) \to B(0; R)$ by

$$\tilde{f}(x) = f(P_K(x)).$$

The range of $\tilde{f}$ is contained in K. By the preceding corollary $\tilde{f}$ has a fixed point $x_0 \in B(0; R)$ and

$$x_0 = \tilde{f}(x_0) \in K.$$

But then $P_K x_0 = x_0$ and so $x_0 = f(x_0)$. ∎

The following consequence of Brouwer's theorem will be useful later.

Theorem 5.2.5 Let $f : \mathbb{R}^n \to \mathbb{R}^n$ be continuous such that $(f(x), x) \geqslant 0$ for all x such that $|x| = R > 0$. Then there exists x_0 such that $|x_0| \leqslant R$ and $f(x_0) = 0$.

Proof If not, $f(x) \neq 0$ for all $x \in B(0; R)$. Define

$$g(x) = -\frac{R}{|f(x)|} f(x)$$

which maps $B(0; R)$ into itself and is continuous. Hence it has a fixed point x_0. Since $x_0 = g(x_0)$, we have $|x_0| = |g(x_0)| = R > 0$. But then

$$0 < R^2 = |x_0|^2 = (x_0, x_0) = (x_0, g(x_0)) = -\frac{R}{|f(x_0)|} (x_0, f(x_0)) \leqslant 0$$

by hypothesis, which is a contradiction. ∎

Remark 5.2.2 The above result is also obviously true if $(f(x), x) \leqslant 0$ for all $|x| = R > 0$. ∎

Remark 5.2.3 The above result not only gives the existence of a solution, but also provides an estimate for the solution. ∎

We now investigate the possibility of extending Brouwer's theorem to infinite dimensions. The following example shows, however, that in the form of Theorem 5.2.3, this is not possible.

Example 5.2.1 Let $V = l_2$ the space of square summable sequences. Let B be the unit ball in V and S its boundary. Define for $x = \{x_n\}$ in B,

$$f(x) = (\sqrt{1 - \|x\|^2}, x_1, x_2, \ldots)$$

where

$$\|x\|^2 = \sum_{n=1}^{\infty} |x_n|^2.$$

Then $f : B \to S \subset B$ and is continuous. But $x = f(x)$ implies that $\|x\| = 1$ and so

$$x = (x_1, x_2, \ldots, x_n \ldots) = (0, x_1, x_2, \ldots) = f(x)$$

and so successively $x_1 = 0$, $x_2 = x_1 = 0$ and so on. Thus $x = 0$ and $\|x\| = 1$, a contradiction. Thus f has no fixed point in B. ∎

The reason why f fails to have a fixed point is that B is not compact. The correct generalization is in the form of Theorem 5.2.4 and before we prove it we need the following preliminary result.

Lemma 5.2.2 Let V be a Banach space and $K \subset V$ a compact set. For every $\epsilon > 0$ there exists a finite dimensional subspace $V_\epsilon \subset V$ and a continuous map $g_\epsilon : K \to V_\epsilon$ such that

$$\|g_\epsilon(v) - v\| \leqslant \epsilon, \quad \text{for every } v \in K. \tag{5.2.7}$$

If, in addition, K is convex, then $g_\epsilon(v) \in K$ for every $v \in K$.

Proof Since K is compact, there exist $x_1, \ldots, x_n \in K$ $(n = n(\epsilon))$ such that

$$K \subset \bigcup_{i=1}^{n} B(x_i; \epsilon). \tag{5.2.8}$$

Let

$$V_\epsilon = \text{span}\,\{x_1, \ldots, x_n\} \subset V.$$

Then $\dim V_\epsilon \leqslant n < +\infty$. Define

$$b_i(x) = \begin{cases} \epsilon - \|x - x_i\|, & \text{if } x \in B(x_i; \epsilon) \\ 0, & \text{otherwise.} \end{cases}$$

By (5.2.8) for every $x \in K$, there exist i such that $b_i(x) \neq 0$. Thus $\sum_{i=1}^{n} b_i(x) \neq 0$ for every $x \in K$. Also b_i is clearly continuous. Now define

$$g_\epsilon(x) = \frac{\sum_{i=1}^{n} b_i(x) x_i}{\sum_{i=1}^{n} b_i(x)} = \sum_{i=1}^{n} \beta_i(x) x_i \in V_\epsilon$$

where

$$\beta_i = b_i(x) \Big/ \left(\sum_{j=1}^{n} b_j(x) \right).$$

Since $0 \leqslant \beta_i \leqslant 1$, $\sum_{i=1}^{n} \beta_i = 1$, we have that $g_\epsilon(x)$ is a convex combination of the x_i and hence belongs to K if K is convex. We now only need to prove (5.2.7). But if $b_i(x) \neq 0$, we have $\|x - x_i\| < \epsilon$. Hence

$$\|g_\epsilon(x) - x\| \leqslant \sum_{\{i \mid \beta_i \neq 0\}} \beta_i \|x_i - x\| \leqslant \left(\sum_{i=1}^{n} \beta_i \right) \epsilon = \epsilon. \quad ∎$$

Theorem 5.2.6 **(Schauder)** Let V be a Banach space and let $K \subset V$ be a convex, compact set. If $f : K \to K$ is a continuous map, then f has a fixed point.

Proof For $\epsilon > 0$, let V_ϵ, g_ϵ be as in Lemma 5.2.2. Then as K is convex, $g_\epsilon(K) \subset K$. Let K_ϵ be the convex hull of $x_1, \ldots, x_n$. Then $K_\epsilon \subset K$ as well. Now define $f_\epsilon : K_\epsilon \to K_\epsilon$ by

$$f_\epsilon(x) = g_\epsilon(f(x)).$$

Since K_ϵ is a compact convex set in the finite dimensional space V_ϵ, it follows from Brouwer's theorem that it has a fixed point $x_\epsilon \in K_\epsilon \subset K$. Thus

$$x_\epsilon = g_\epsilon(f(x_\epsilon)).$$

Since K is compact, we can extract a convergent subsequence. Let $x_\epsilon \to x \in K$ as $\epsilon \to 0$. Now

$$\|x - f(x)\| \leqslant \|x - x_\epsilon\| + \|x_\epsilon - f(x_\epsilon)\| + \|f(x_\epsilon) - f(x)\|.$$

The first term on the right tends to zero as $\epsilon \to 0$. So does the last term since f is continuous. Further

$$\|x_\epsilon - f(x_\epsilon)\| = \|g(f(x_\epsilon)) - f(x_\epsilon)\| \leqslant \epsilon$$

since $f(x_\epsilon) \in K$, by (5.2.7). Hence $\|x - f(x)\|$ can be made arbitrarily small and so $x = f(x)$. ∎

In many applications we may not have K compact. Instead we may have f to be a compact map, i.e. it maps bounded sets to relatively compact sets. In such situations the following corollary will be useful.

Corollary Let K be a closed, bounded and convex set in a Banach space V and $f : K \to K$ a compact map. Then f has a fixed point.

Proof Since $\overline{f(K)}$ is compact, so is $\hat{K}$, its closed convex hull. Since K is convex and $f(K) \subset K$, it follows that $\hat{K} \subset K$. Now $f|_{\hat{K}}$ maps $\hat{K}$ into itself and so has a fixed point. ∎

We give one last fixed point theorem which is a slight but very useful variation of Schauder's theorem.

Theorem 5.2.7 **(Schaeffer)** Let V be a Banach space and $f : V \to V$ a compact map. Assume that there exists $C > 0$ such that whenever

$$u = \sigma f(u), \; 0 < \sigma < 1, \tag{5.2.9}$$

then $\|u\| < C$. Then there exist a fixed point u of f such that $\|u\| \leqslant C$.

Proof The ball $B(0; C)$ admits the following continuous projection from V onto itself:

$$Pv = \begin{cases} v, & \text{if } \|v\| \leqslant C \\ C \dfrac{v}{\|v\|}, & \text{if } \|v\| > C. \end{cases} \tag{5.2.10}$$

Define $\tilde{f}: B(0; C) \to B(0; C)$ by

$$\tilde{f}(v) = P(f(v)).$$

Then $\tilde{f}$ is compact and, by the preceding corollary, has a fixed point u. Thus

$$u = P(f(u)). \tag{5.2.11}$$

If $\|f(u)\| > C$, then

$$u = \frac{C}{\|f(u)\|} f(u) = \sigma f(u) \tag{6.2.12}$$

where $\sigma = C/\|f(u)\| < 1$. Thus $\|u\| < C$, but we know from (5.2.12) that $\|u\| = C$, a contradiction. Hence $\|f(u)\| \leqslant C$ and so $P(f(u)) = f(u)$ and thus it follows from (5.2.11) that u is a fixed point of f. ∎

Remark 5.2.4 In verifying the existence of a $C > 0$ as in the hypothesis of the above theorem, we do not care if (5.2.10) has a solution or not! We assume such a u exists and deduce the value of C. Such an estimate is known as an *a priori* estimate. ∎

We conclude this solution with some applications of these fixed point theorems to semilinear elliptic boundary value problems.

Let $\Omega \subset \mathbb{R}^n$ be a bounded open set with boundary Γ. Let $f: \mathbb{R} \to \mathbb{R}$ be a continuous function such that if $u \in H_0^1(\Omega)$ then $f(u) \in H^{-1}(\Omega)$. Let us consider the boundary value problem

$$\left.\begin{aligned} -\Delta u &= f(u) \quad \text{in } \Omega \\ u &= 0 \quad \text{on } \Gamma \end{aligned}\right\}. \tag{5.2.13}$$

If we set, for $g \in H^{-1}(\Omega)$, $Gg = w \in H_0^1(\Omega)$ such that

$$\left.\begin{aligned} -\Delta w &= g \quad \text{in } \Omega \\ w &= 0 \quad \text{on } \Gamma \end{aligned}\right\} \tag{5.2.14}$$

then $G: H^{-1}(\Omega) \to H_0^1(\Omega)$ is a continuous linear operator (cf. Section 3.2.1). The operator G is also continuous from $L^2(\Omega)$ into $H_0^1(\Omega)$. If we define

$$F(u) = G(f(u)) \tag{5.2.15}$$

then (5.2.13) is equivalent to finding $u \in H_0^1(\Omega)$ such that

$$u = F(u).$$

Example 5.2.2 Let $f: \mathbb{R} \to \mathbb{R}$ be Lipschitz continuous with Lipschitz constant L, i.e.

$$|f(x) - f(y)| \leqslant L|x - y|, \quad x, y \in \mathbb{R}. \tag{5.2.16}$$

Assume that $L < \lambda_1$, where λ_1 is the first eigenvalue of the Dirichlet problem for the Laplace operator (cf. Section 3.6), i.e. $\lambda_1 > 0$ is the smallest real number such that there exists $w_1 \in H_0^1(\Omega)$, $w_1 \neq 0$, and $-\Delta w_1 = \lambda_1 w_1$.

If $u \in L^2(\Omega)$ then $f(u) \in L^2(\Omega)$; for,

$$\begin{aligned}|f(u)| &\leqslant |f(u) - f(0)| + |f(0)| \\ &\leqslant L|u| + |f(0)|\end{aligned}$$

and as Ω is bounded, the constant function $f(0) \in L^2(\Omega)$ and $u \in L^2(\Omega)$. Thus $|f(u)| \in L^2(\Omega)$. It is continuous as well, since

$$|f(u) - f(v)| \leqslant L|u - v|$$

which implies that

$$|f(u) - f(v)|_{0,\Omega} \leqslant L|u - v|_{0,\Omega}. \tag{5.2.17}$$

If F were defined by (5.2.15) then for $u, v \in L^2(\Omega)$ we have

$$\begin{aligned}|F(u) - F(v)|_{0,\Omega} &= |G(f(u)) - G(f(v))|_{0,\Omega} \\ &\leqslant L\|G\|\,|u - v|_{0,\Omega}\end{aligned}$$

where $\|G\|$ refers to the norm of G considered as a mapping of $L^2(\Omega)$ into itself. Indeed, if $G(g) = w \in H_0^1(\Omega)$ for $g \in L^2(\Omega)$,

$$\begin{aligned}\lambda_1|w|^2_{0,\Omega} \leqslant |w|^2_{1,\Omega} &= \int_\Omega \nabla w \cdot \nabla w = \int_\Omega g \cdot w \\ &\leqslant |g|_{0,\Omega}|w|_{0,\Omega}\end{aligned}$$

whence,

$$|w|_{0,\Omega} \leqslant \frac{1}{\lambda_1}\,|g|_{0,\Omega}.$$

Thus $\|G\| \leqslant 1/\lambda_1$ and so,

$$|F(u) - F(v)|_{0,\Omega} \leqslant \frac{L}{\lambda_1}\,|u - v|_{0,\Omega}.$$

Thus when $L < \lambda_1$, F is a contraction mapping and so has a unique fixed point. Thus when f is Lipschitz continuous with $L < \lambda_1$, (5.2.13) has a unique solution in $H_0^1(\Omega)$. ∎

Example 5.2.3 Let $f: \mathbb{R} \to \mathbb{R}$ be uniformly bounded and such that $u \to f(u)$ is a continuous mapping on $L^2(\Omega)$ (e.g., f is a bounded, Lipschitz continuous function). Then (5.2.13) has at least one solution.

Again we consider F as in the previous example. We notice that F is compact; for if $u \in L^2(\Omega)$, then $u \to f(u) \to F(u) \in H_0^1(\Omega)$ is continuous into $H_0^1(\Omega)$ and $H_0^1(\Omega) \to L^2(\Omega)$ is a compact inclusion by the Rellich-Kondrasov Theorem. Thus $F: L^2(\Omega) \to L^2(\Omega)$ is compact since it is the composition of $F: L^2(\Omega) \to H_0^1(\Omega)$ and the compact inclusion mentioned above. Further let

$$u = \sigma F(u), \quad 0 < \sigma < 1.$$

Then u satisfies

$$\left.\begin{aligned}-\Delta u &= \sigma f(u) \quad \text{in } \Omega \\ u &= 0 \quad \text{on } \Gamma\end{aligned}\right\} \tag{5.2.18}$$

and so

$$\lambda_1|u|_{0,\Omega}^2 \leqslant |u|_{1,\Omega}^2 = \int_\Omega f(u)\cdot u \leqslant M\int_\Omega |u|$$
$$\leqslant M\,(\text{meas }(\Omega))^{1/2}|u|_{0,\Omega}, \quad \text{where } |f(x)| \leqslant M, \quad x \in \mathbb{R}.$$

Thus

$$|u|_{0,\Omega} < C = \frac{M}{\lambda_1}(\text{meas }(\Omega))^{1/2} + \epsilon$$

where $\epsilon > 0$. Hence F has a fixed point, by Schaeffer's theorem and the solution satisfies $|u|_{0,\Omega} \leqslant C_0$ where

$$C_0 = \frac{M}{\lambda_1}(\text{meas }(\Omega))^{1/2}. \quad ∎$$

Remark 5.2.5 We could also consider F as a map of $H_0^1(\Omega)$ into itself. Again $u \in H_0^1(\Omega) \to u \in L^2(\Omega) \to f(u) \in L^2(\Omega) \to F(u) \in H_0^1(\Omega)$ and as the first map is compact, $F: H_0^1(\Omega) \to H_0^1(\Omega)$ is compact. Again we can apply Schaeffer's theorem to get the existence of a solution u to (5.2.13) satisfying

$$|u|_{1,\Omega} \leqslant \frac{M}{\sqrt{\lambda_1}}(\text{meas }(\Omega))^{1/2}. \quad ∎$$

5.3 THE GALERKIN METHOD

An example of the Galerkin method for linear problems was given in Section 3.4. We will now illustrate its use via a nonlinear example.

Let $\Omega \subset \mathbb{R}^n$, $n \leqslant 4$ be a bounded open set. Then if $u \in H^1(\Omega)$, by the Sobolev inclusion theorem (Theorem 2.4.1) $u \in L^4(\Omega)$. Thus we have, for $v \in H_0^1(\Omega)$,

$$\left|\int_\Omega u^3 v\right| \leqslant |u|_{0,4,\Omega}^3 |v|_{0,4,\Omega} \leqslant C\|u\|_{1,\Omega}^3\|v\|_{1,\Omega}$$

and hence $u^3 \in H^{-1}(\Omega)$.

We can therefore discuss the existence of weak solutions in $H_0^1(\Omega)$ of the problem

$$\left.\begin{aligned} -\Delta u - \lambda u + u^3 &= f \quad \text{in } \Omega \\ u &= 0 \quad \text{on } \Gamma \end{aligned}\right\} \tag{5.3.1}$$

where $f \in L^2(\Omega)$ (or even $H^{-1}(\Omega)$) is a given function and $\lambda \in \mathbb{R}$. The weak formulation of (5.3.1) will be to find $u \in H_0^1(\Omega)$ such that

$$\int_\Omega \nabla u \cdot \nabla v - \lambda \int_\Omega uv + \int_\Omega u^3 v = \int_\Omega fv. \tag{5.3.2}$$

for every $v \in H_0^1(\Omega)$.

Theorem 5.3.1 Let $\Omega \subset \mathbb{R}^n$ a bounded open set, $n \leqslant 4$. Then there exists at least one solution to the equation (5.3.2) when $\lambda < \lambda_1$, where λ_1 is the first eigenvalue for the Dirichlet problem for the Laplace operator.

Proof Since $H_0^1(\Omega)$ is separable, let $\{w_1, \ldots, w_m, \ldots\}$ be an orthonormal basis for this space. Define

$$W_m = \text{span}\,\{w_1, \ldots, w_m\}.$$

Step 1 We look for $u_m \in W_m$ such that

$$\int_\Omega \nabla u_m \cdot \nabla v - \int_\Omega u_m v + \int_\Omega u_m^3 v = \int_\Omega fv, \; v \in W_m. \tag{5.3.3}$$

Notice that it suffices that (5.3.3) is verified for $v = w_i$, $1 \leqslant i \leqslant m$. Let $\xi = (\xi_i) \in \mathbb{R}^m$. To each such ξ, we associate a unique element $v \in W_m$ by

$$v = \sum_{i=1}^{m} \xi_i w_i. \tag{5.3.4}$$

This establishes a linear bijection between $\mathbb{R}^m$ and W_m and further, since the $\{w_i\}$ are orthonormal in $H_0^1(\Omega)$,

$$|v|_{1,\Omega}^2 = \sum_{i=1}^{n} |\xi_i|^2 = |\xi|^2. \tag{5.3.5}$$

(Here we assume that $H_0^1(\Omega)$ is equipped with the norm $|\cdot|_{1,\Omega}$.) We now define $F : \mathbb{R}^m \to \mathbb{R}^m$ by

$$(F(\xi))_i = \int_\Omega \nabla v \cdot \nabla w_i - \lambda \int_\Omega v w_i + \int_\Omega v^3 w_i - \int_\Omega f w_i$$

where v is as given in (5.3.4). Thus (5.3.3) will have a solution if there exists a ξ such that $F(\xi) = 0$. Now

$$\begin{aligned}(F(\xi), \xi) &= \sum_{i=1}^{m} (F(\xi))_i \xi_i \\ &= |v|_{1,\Omega}^2 - \lambda |v|_{0,\Omega}^2 + \int_\Omega v^4 - \int_\Omega fv \\ &\geqslant |v|_{1,\Omega}^2 - \lambda |v|_{0,\Omega}^2 - |f|_{0,\Omega} |v|_{0,\Omega}\end{aligned}$$

since $\int_\Omega v^4 \geqslant 0$ for any v. If we recall the variational characterization of the first eigenvalue (cf. Theorem 3.6.2), we have that

$$|v|_{0,\Omega}^2 \leqslant \frac{1}{\lambda_1} |v|_{1,\Omega}^2.$$

Thus

$$\begin{aligned}(F(\xi), \xi) &\geqslant \left(1 - \frac{\lambda}{\lambda_i}\right)|v|_{1,\Omega}^2 - \frac{|f|_{0,\Omega}}{\sqrt{\lambda_1}} |v|_{1,\Omega} \\ &\left(1 - \frac{\lambda}{\lambda_1}\right)|\xi|^2 - \frac{|f|_{0,\Omega}}{\sqrt{\lambda_1}} |\xi|\end{aligned}$$

Since $\lambda < \lambda_1$, we have $\left(1 - \frac{\lambda}{\lambda_1}\right) > 0$. Hence if $|\xi| = R$ is chosen large enough ($R > |f|_{0,\Omega}\lambda_1^{-1/2}(1 - \lambda/\lambda_1)^{-1}$), we have

$$(F(\xi), \xi) \geqslant 0 \quad \text{for} \quad |\xi| = R.$$

Hence by a consequence of Brouwer's theorem (Theorem 5.2.5) there exists a ξ^m, such that

$$|\xi^m| \leqslant R,\ F(\xi^m) = 0.$$

Set $u_m = \sum_{i=1}^{m} \xi_i^m w_i$ and $u_m \in W_m$ will be a solution of (5.3.3). Further by (5.3.3)

$$|u_m|_{1,\Omega} \leqslant R \tag{5.3.6}$$

and as seen before R is independent of m; it only depends on λ and f.

Step 2 Since $\{u_m\}$ is uniformly bounded in $H_0^1(\Omega)$, we can extract a weakly convergent subsequence which we will continue to denote by $\{u_m\}$. Let

$$u_m \to u \text{ in } H_0^1(\Omega) \text{ weakly.}$$

Step 3 Let $v \in \mathcal{D}(\Omega)$. Then there exists $v_m \in W_m$ such that $v_m \to v$ strongly in $H_0^1(\Omega)$. In fact we can choose

$$v_m = \sum_{i=1}^{m} \left(\int_\Omega \nabla v \cdot \nabla w_i \right) w_i.$$

Now by (5.3.3),

$$\int_\Omega \nabla u_m \cdot \nabla v_m - \lambda \int_\Omega u_m v_m + \int_\Omega u_m^3 v_m - \int_\Omega f v_m = 0.$$

It is easy to pass to the limit in all but the nonlinear terms; $u_m \to u$ weakly in $H_0^1(\Omega)$ and so strongly in $L^2(\Omega)$, $v_m \to v$ strongly in $H_0^1(\Omega)$ and $L^2(\Omega)$. Thus,

$$\lim_{m\to\infty} \int_\Omega \nabla u_m \cdot \nabla v_m = \int_\Omega \nabla u \cdot \nabla v$$

$$\lim_{m\to\infty} \int_\Omega u_m v_m = \int_\Omega uv$$

$$\lim_{m\to\infty} \int_\Omega f v_m = \int_\Omega f v.$$

Step 4 We now pass to the limit in the nonlinear term. Now

$$\int_\Omega u_m^3 v_m - \int_\Omega u^3 v = \int_\Omega u_m^3 (v_m - v) + \int_\Omega (u_m^3 - u^3) v.$$

But

$$\left| \int_\Omega u_m^3 (v_m - v) \right| \leqslant |u_m|_{0,4,\Omega}^3 |v_m - v|_{0,4,\Omega}$$

$$\leqslant C |u_m|_{1,\Omega}^3 |v_m - v|_{1,\Omega}$$

$$\leqslant C R^3 |v_m - v|_{1,\Omega}$$

which tends to zero as $m \to \infty$.

Further

$$\left|\int_\Omega \left(u_m^3 - u^3\right)v\right| = \left|\int_\Omega (u_m - u)(u_m^2 + u_m u + u^2)v\right|$$
$$\leqslant |v|_{0,\infty,\Omega}|u_m - u|_{0,\Omega}|u_m^2 + u_m u + u^2|_{0,\Omega}$$
$$\leqslant |v|_{0,\infty,\Omega}|u_m - u|_{0,\Omega}(|u_m|_{0,4,\Omega}^2 + |u|_{0,4,\Omega}^2)$$
$$\leqslant C|u_m - u|_{0,\Omega}$$

which also tends to zero as $m \to \infty$. Thus

$$\lim_{m\to\infty} \int_\Omega u_m^3 v_m = \int_\Omega u^3 v.$$

Thus for every $v \in \mathcal{D}(\Omega)$

$$\int_\Omega \nabla u \cdot \nabla v - \lambda \int_\Omega uv + \int_\Omega u^3 v - \int_\Omega fv = 0.$$

and this is true for all $v \in H_0^1(\Omega)$ as well since $\mathcal{D}(\Omega)$ is dense in $H_0^1(\Omega)$ and the above expression is linear in v. Hence u is a solution of (5.3.2). ∎

Remark 5.3.1 If $n \leqslant 3$ then $H_0^1(\Omega) \to L^4(\Omega)$ is a compact inclusion and we can straight away pass to the limit in $\int_\Omega u_m^3 v_m$ since now $u_m \to u$ strongly $L^4(\Omega)$. It is for the limiting case $n = 4$ that we need the argument of step 4. In general most methods which work for the cases where the Sobolev inclusion is compact will fail for the limiting case. The study of semilinear problems where there is a lack of compactness is a topic of great current research interest. ∎

Remark 5.3.2 The above method will not work if the term u^3 appeared on the right hand side of the equation. ∎

Remark 5.3.3 While we have only proved the existence of a solution for the case $\lambda < \lambda_1$, the existence can also be proved when $\lambda \geqslant \lambda_1$. We will do this later using a variational method. The Galerkin method can also be made to work but a lot more of effort will be necessary to get the estimate of the type (5.3.6). This is described in Kesavan [2]. These remarks also hold when u^3 is replaced by $|u|^{p-1}u$ where p is chosen such that the Sobolev inclusion $H_0^1(\Omega) \to L^{p+1}(\Omega)$ holds. ∎

5.4 MONOTONE ITERATIONS

We now present another approximation method for the solution of semilinear problems. This method, known as the *method of monotone iterations,* uses the maximum principle (cf. Section 3.5) for elliptic problems in a very crucial way.

Let $\Omega \subset \mathbb{R}^n$ a smooth bounded open set for $n \leqslant 6$ so that if $u \in H^1(\Omega)$ then $u \in L^3(\Omega)$ and so $u^2 \in H^{-1}(\Omega)$. Let $f \in L^2(\Omega)$ be given, $f \not\equiv 0$.

Consider problem

$$\left.\begin{aligned} -\Delta u &= u^2 + f \text{ in } \Omega \\ u &= 0 \qquad \text{on } \Gamma \end{aligned}\right\} \tag{5.4.1}$$

The weak formulation of the above problem would be to find $u \in H_0^1(\Omega)$ such that

$$\int_\Omega \nabla u \cdot \nabla v = \int_\Omega u^2 v + \int_\Omega fv, \quad \text{for every } v \in H_0^1(\Omega). \tag{5.4.2}$$

Theorem 5.4.1 Let $f < 0$ and smooth enough. Then (5.4.1) has a solution $u \in H_0^1(\Omega)$ such that $u \leqslant 0$ in Ω.

Proof: *Step 1* Let $v_0 \in H_0^1(\Omega)$ such that $-\Delta v_0 = f$ in Ω. Then by regularity results, v_0 is also smooth and by the strong maximum principle (cf. Corollary to Theorem 3.5.2) $v_0 < 0$ in Ω. Define

$$K = 2|\min_{x \in \Omega} v_0(x)| > 0.$$

Step 2 Given $u \in H_0^1(\Omega)$, define $F(u) \in H_0^1(\Omega)$ by

$$w = F(u),$$
$$-\Delta w + Kw = Ku + u^2 + f \text{ in } \Omega. \tag{5.4.3}$$

Note that if u is smooth then so is w. Note also that a fixed point of F is a solution of (5.4.1).

Step 3 Let $v_0 \leqslant u \leqslant 0$ in $\bar{\Omega}$, $u \in H_0^1(\Omega)$. Then $Ku + u^2 = u(u + K)$ which is non-positive. Hence by the weak maximum principle (Theorem 3.5.1) if $w = F(u)$, $w \leqslant 0$, and further

$$\begin{aligned} -\Delta(w - v_0) + K(w - v_0) &= Ku + u^2 + f - f - Kv_0 \\ &= (Ku + u^2) - Kv_0 \geqslant 0 \end{aligned}$$

which implies by the weak maximum principle that $w - v_0 \geqslant 0$. Thus if $v_0 \leqslant u \leqslant 0$, we have

$$v_0 \leqslant F(u) \leqslant 0.$$

Similarly if $v_0 \leqslant u_1 \leqslant u_2 \leqslant 0$, then $F(u_1) \leqslant F(u_2)$; for if $w_i = F(u_i)$, $i = 1, 2$

$$\begin{aligned} -\Delta(w_2 - w_1) + K(w_2 - w_1) &= K(u_2 - u_1) + (u_2^2 - u_1^2) \\ &= (u_2 - u_1)(K + u_2 + u_1) \geqslant 0. \end{aligned}$$

Thus $w_2 \geqslant w_1$ by the weak maximum principle again.

Step 4 Set $u_0 = 0$ and define $u_{k+1} = F(u_k)$. Then by step 3 above we have

$$v_0 \leqslant \ldots \leqslant u_{k+1} \leqslant u_k \leqslant \ldots \leqslant u_0 = 0.$$

It then follows that $|u_k| \leqslant |v_0|$ for every k and so $\{u_k\}$ is uniformly bounded

in $L^2(\Omega)$. In fact

$$|u_k|_{0,\Omega} \leqslant |v_0|_{0,\Omega}. \tag{5.4.4}$$

Further

$$-\Delta u_{k+1} + Ku_{k+1} = Ku_k + u_k^2 + f$$

and so on taking the scalar product with u_{k+1},

$$\begin{aligned}|u_{k+1}|_{1,\Omega}^2 + K|u_{k+1}|_{0,\Omega}^2 &= K\int_\Omega u_k u_{k+1} + \int_\Omega u_k^2 u_{k+1} + \int_\Omega f u_{k+1} \\ &\leqslant (K|u_k|_{0,\Omega} + |f|_{0,\Omega})|u_{k+1}|_{0,\Omega} \\ &\leqslant C|u_{k+1}|_{1,\Omega}\end{aligned}$$

since $\int_\Omega u_k^2 u_{k+1} \leqslant 0$ and we have (5.4.4). Thus

$$|u_{k+1}|_{1,\Omega} \leqslant C$$

for all k. Thus given any subsequence of $\{u_k\}$ we can extract a weakly convergent subsequence (in $H_0^1(\Omega)$) which will converge strongly in $L^2(\Omega)$ and a.e. But as the $\{u_k\}$ is a decreasing sequence, the a.e. limit will always be $u(x) = \inf_k u_k(x)$. Thus the limit is independent of the convergent subsequence and thus we deduce that the *entire* sequence $\{u_k\}$ converges weakly in $H_0^1(\Omega)$ to a function u. This convergence is strong in $L^2(\Omega)$ by the Rellich-Kondrasov Theorem.

Step 5 Let $v \in \mathscr{D}(\Omega)$. Then we have for *all* k

$$\int_\Omega \nabla u_{k+1} \cdot \nabla v + K\int_\Omega u_{k+1} v = K\int_\Omega u_k v + \int_\Omega u_k^2 v + \int_\Omega fv.$$

Since $u_k \to u$ weakly in $H_0^1(\Omega)$ and hence strongly in $L^2(\Omega)$, we can pass to the limit in all the linear terms. Also

$$\begin{aligned}\left|\int_\Omega u_k^2 v - \int_\Omega u^2 v\right| &= \left|\int_\Omega (u_k - u)(u_k + u)v\right| \\ &\leqslant \int_\Omega |u_k - u|\,|u_k + u|\,|v| \\ &\leqslant |v|_{0,\infty,\Omega}\,|u_k - u|_{0,\Omega}|u_k + u|_{0,\Omega} \\ &\leqslant C|u_k - u|_{0,\Omega}\end{aligned}$$

which tends to zero as $k \to \infty$. Hence for every $v \in \mathscr{D}(\Omega)$

$$\int_\Omega \nabla u \cdot \nabla v + K\int_\Omega uv = K\int_\Omega uv + \int_\Omega u^2 v + \int_\Omega fv$$

and as $\mathscr{D}(\Omega)$ is dense in $H_0^1(\Omega)$, this is true for all $v \in H^1(\Omega)$. Cancelling the common term on both sides, it follows that $u \in H_0^1(\Omega)$ satisfies (5.4.2) and is thus a weak solution of (5.4.1). Finally as all the u_k are $\leqslant 0$ and as $u_k \to u$ a.e., it follows that $u \leqslant 0$ a.e. ∎

Remark 5.4.1 Again, as in the previous section, if $n \leqslant 5$, since the injection $H_0^1(\Omega) \to L^3(\Omega)$ will be compact, it is easier to pass to the limit in step 5. ∎

Remark 5.4.2 If we start from v_0 and set $v_{k+1} = F(v_k)$, we will get an increasing sequence and

$$v = \sup_k v_k$$

will also be a solution to the problem and further

$$v_0 \leqslant v \leqslant u \leqslant 0.$$

But now

$$-\Delta(u - v) = u^2 - v^2 \leqslant 0$$

since $v \leqslant u \leqslant 0$ and by the maximum principle we get $u - v \leqslant 0$ or $u \leqslant v$. Thus

$$\sup_k v_k = v = u = \inf_k u_k.$$

In fact the solution u produced above is the only non-positive solution of (5.4.1). For if $\tilde{u} \leqslant 0$ is any solution of (5.4.1), then

$$-\Delta\tilde{u} = \tilde{u}^2 + f = \tilde{u}^2 - \Delta v_0.$$

Thus

$$-\Delta(\tilde{u} - v_0) = \tilde{u}^2 \geqslant 0$$

and so $v_0 \leqslant \tilde{u} \leqslant 0$. But then by step 3 of the proof above

$$v_1 = F(v_0) \leqslant F(\tilde{u}) = \tilde{u}.$$

Iterating this we get that $\tilde{u} \geqslant v_k$ for all k and so $\tilde{u} \geqslant v = u$. Then again

$$-\Delta(u - \tilde{u}) = u^2 - \tilde{u}^2 \geqslant 0$$

since $0 \geqslant \tilde{u} \geqslant u$ and so $u \geqslant \tilde{u}$, i.e. $u = \tilde{u}$. ∎

By using a variational method we will see later that another solution exists for this problem.

Example 5.4.1 If $f > 0$, it is possible that (5.4.1) has no solution as we now demonstrate. Let $w_1 > 0$ be the normalized eigenfunction corresponding to the eigenvalue λ_1 of the Dirichlet problem for the Laplace operator i.e. $w_1 \in H_0^1(\Omega)$, $-\Delta w_1 = \lambda_1 w_1$, $\int_\Omega w_1^2 = 1$, and $w_1 > 0$. (For the existence of w_1 cf. Theorem 3.6.3). Let $f = tw_1$, $t > 0$ to be chosen. Let, if possible, $u \in H_0^1(\Omega)$ such that $-\Delta u = u^2 + f$. Taking the scalar product with w_1, we get

$$\lambda_1 \int_\Omega w_1 u = \int_\Omega \nabla u \cdot \nabla w_1 = \int_\Omega u^2 w_1 + \int_\Omega f w_1. \qquad (5.4.5)$$

$$= \int_\Omega u^2 w_1 + t\left(\text{since} \int w_1^2 = 1\right).$$

Since $w_1 > 0$ and smooth, $\sqrt{w_1} \in L^2(\Omega)$ and we apply the Cauchy-Schwarz inequality to the term

$$\int_\Omega w_1 u = \int_\Omega \sqrt{w_1}(\sqrt{w_1}u).$$

Thus this integral is bounded above by

$$\left(\int_\Omega w_1\right)^{1/2}\left(\int_\Omega w_1 u^2\right)^{1/2}.$$

Substituting in (5.4.5), we get

$$t + \int_\Omega u^2 w_1 \leqslant \lambda_1 c\left(\int_\Omega u^2 w_1\right)^{1/2}, \; c = \left(\int_\Omega w_1\right)^{1/2}$$

This is possible only if

$$\lambda_1^2 c^2 \geqslant 4t$$

and in that case

$$\frac{\lambda_1 c - \sqrt{\lambda_1^2 c^2 - 4t}}{2} \leqslant \left(\int_\Omega u^2 w_1\right)^{1/2} \leqslant \frac{\lambda_1 c + \sqrt{\lambda_1^2 c^2 - 4t}}{2}.$$

Thus if $t > \lambda_1^2 c^2/4$, (5.4.1) cannot have any solution. ∎

For more details on the method of monotone iterations, see the survey article of Amann [1].

5.5 VARIATIONAL METHODS

Our next approach to solving equations of the form

$$F(u) = 0 \tag{5.5.1}$$

in a *Hilbert space* V will be variational; we look for a functional

$$J : V \to \mathbb{R}$$

such that its Fréchet derivative $J' = F$, i.e.

$$(J'(u), v) = (F(u), v), \text{ for every } u, v \in V. \tag{5.5.2}$$

Then solutions of (5.5.1) will be critical points of J. We will illustrate this idea by reconsidering the examples studied in Sections 5.3 and 5.4.

Let $\Omega \subset \mathbb{R}^n$ be a bounded open set with boundary Γ. We assume that $n \leqslant 3$ so that the inclusions $H_0^1(\Omega) \subset L^p(\Omega)$ are compact for $1 \leqslant p \leqslant 4$. We first consider the problem:

$$\left.\begin{aligned} -\Delta u - \lambda u + u^3 &= f \quad \text{in } \Omega \\ u &= 0 \quad \text{on } \Gamma \end{aligned}\right\} \tag{5.5.3}$$

where $f \in L^2(\Omega)$ is given. We now define $J : H_0^1(\Omega) \to \mathbb{R}$ given by

$$J(v) = \tfrac{1}{2}\int_\Omega |\nabla v|^2 - \frac{\lambda}{2}\int_\Omega u^2 + \tfrac{1}{4}\int_\Omega u^4 - \int_\Omega fu. \tag{5.5.4}$$

A simple calculation of the Fréchet derivative (see Appendix 3) shows that

$$(J'(u), v) = \int_{\Omega} \nabla u \cdot \nabla v - \lambda \int_{\Omega} uv + \int_{\Omega} u^3 v - \int fv, \qquad (5.5.5)$$

where $(\cdot, \cdot)$ represents the duality product between $H^{-1}(\Omega)$ and $H_0^1(\Omega)$, and so critical points of J (i.e. u such that $J'(u) = 0$) will yield weak solutions of (5.5.3).

Theorem 5.5.1 For any given $\lambda \in \mathbb{R}$, there exists at least one weak solution u of (3.5.3) such that

$$J(u) = \min_{v \in H_0^1(\Omega)} J(v) \qquad (5.5.6)$$

Proof: *Step 1* The functional J is *coercive*, i.e. $J(v) \to +\infty$ if $|v|_{1,\Omega} \to +\infty$. If not, assume that there exists a sequence $\{v_m\}$ such that $|v_m|_{1,\Omega} \to \infty$ as $m \to \infty$ while $J(v_m) \leqslant C$ for all m. Let $w_m = v_m / |v_m|_{1,\Omega}$ so that $|w_m|_{1,\Omega} = 1$ for all m. Hence we can assume (on the extraction of a subsequence again indexed by m for convenience) that $w_m \to w$ in $H_0^1(\Omega)$ weakly. Now

$$\tfrac{1}{2}|v_m|_{1,\Omega}^2 - \frac{\lambda}{2} \int_{\Omega} v_m^2 + \tfrac{1}{4} \int_{\Omega} v_m^4 - \int_{\Omega} f v_m \leqslant C. \qquad (5.5.7)$$

Dividing throughout by $|v_m|_{1,\Omega}^4$ and passing to the limit as $m \to \infty$ we get (since $w_m \to w$ strongly in $L^4(\Omega)$)

$$\int_{\Omega} w^4 \leqslant 0.$$

Thus $w = 0$ and so $w_m \to 0$ weakly in $H_0^1(\Omega)$ and hence strongly in $L^2(\Omega)$. Now dividing (5.5.7) by $|v_m|_{1,\Omega}^2$ and ignoring the non-negative term $\int_{\Omega} v_m^4$, we get

$$\tfrac{1}{2} - \frac{\lambda}{2} \int_{\Omega} w_m^2 - \frac{1}{|v_m|_{1,\Omega}} \int_{\Omega} f w_m \leqslant \frac{C}{|v_m|_{1,\Omega}^2}$$

and as $m \to \infty$ we get

$$\tfrac{1}{2} \leqslant 0,$$

a contradiction. Hence $J(v) \to +\infty$ as $|v|_{1,\Omega} \to \infty$.

Step 2 Let $\{u_m\}$ be a minimizing sequence for J, i.e.

$$J(u_m) \to \inf_{v \in H_0^1(\Omega)} J(v).$$

By Step 1, $\{u_m\}$ must be bounded in $H_0^1(\Omega)$. Hence (by extracting a subsequence) we may assume that $u_m \to u$ weakly in $H_0^1(\Omega)$. Then, since the norm is a *weakly lower semicontinuous function* we have that

$$|u|_{1,\Omega} \leqslant \liminf_{m \to \infty} |u_m|_{1,\Omega}.$$

Also since $u_m \to u$ strongly in $L^4(\Omega)$, we have

$$\int_\Omega u_m^4 \to \int_\Omega u^4 \quad \text{as } m \to \infty.$$

Now taking the lim inf of the sequence $\{J(u_m)\}$ we get

$$\tfrac{1}{2}|u|_{1,\Omega}^2 - \frac{\lambda}{2}\int_\Omega u^2 + \tfrac{1}{4}\int_\Omega u^4 - \int_\Omega fu \leqslant \liminf_{m\to\infty} J(u_m) = \inf_{v \in H_0^1(\Omega)} J(v)$$

or

$$\inf_{v\in H_0^1(\Omega)} J(v) \leqslant J(u) \leqslant \inf_{v\in H_0^1(\Omega)} J(v)$$

or

$$J(u) = \min_{v\in H_0^1(\Omega)} J(v)$$

which means that $J'(u) = 0$, i.e. u is a solution of (5.5.3). ∎

Remark 5.5.1 We used the fact that the norm is a weakly lower semicontinuous function. If X is a topological space and $f: X \to \mathbb{R}$ a function, it is said to be lower semicontinuous if for every $\alpha \in \mathbb{R}$, the set

$$\{x \in X \mid f(x) \leqslant \alpha\}$$

is closed in X. It can then be shown that if $x_m \to x$ in X, then

$$f(x) \leqslant \liminf_{m\to\infty} f(x_m).$$

In the case of the norm of a normed linear space, the set $\{x \mid \|x\| \leqslant \alpha\}$ is a closed convex set and hence weakly closed by the Hahn-Banach Theorem. Hence the norm is weakly lower semicontinuous (i.e. lower semicontinuous for the weak topology). ∎

Remark 5.5.2 In the above theorem we have proved the existence of a solution for all $\lambda \in \mathbb{R}$ unlike Theorem 5.3.1 where it was proved only for $\lambda < \lambda_1$. ∎

The proof of Theorem 5.5.1 holds in a more general setting. If J is a coercive functional on a Hilbert space V, i.e. $J(v) \to +\infty$ when $\|v\| \to +\infty$ and is weakly lower semicontinuous, then J attains a global minimum, i.e. there exists $u \in V$ such that

$$J(u) = \min_{v\in V} J(v).$$

In particular if J is coercive, then it is bounded below. However many natural functionals J which we may encounter may not be bounded at all neither above nor below. Thus we need to look for results where other types of critical points may be identified. We describe and illustrate the use of one such result below.

We first introduce a new type of compactness condition known as the **Palais-Smale Condition.**

Definition 5.5.1 Let V be a Banach space and $J : V \to \mathbb{R}$ a C^1 functional. It is said to satisfy the Palais-Smale Condition (PS) if the following holds: whenever $\{u_m\}$ is a sequence in V such that $\{J(u_m)\}$ is bounded and $J'(u_m) \to 0$ strongly in V^* (the dual space), then $\{u_m\}$ has a strongly convergent subsequence. ∎

The (PS) condition is fairly strong. It is not satisfied by even "nice" functions very often.

Example 5.5.1 Let $V = \mathbb{R}$ and $J(x) = c$ a constant. Then J does not satisfy (PS). For consider the sequence $\{m\}$, of integers. $J(m) = c$, is bounded and $J'(m) = 0$. But $\{m\}$ has no convergent subsequence. ∎

It can be shown that if J satisfies (PS) and is bounded below, then J attains a minimum. However we will now state a very useful theorem wherein the (PS) condition will play an important role.

Theorem 5.5.2 **(Mountain Pass Theorem of Ambrosetti and Rabinowitz)** Let $J : V \to \mathbb{R}$ be a C^1 functional satisfying (PS). Let $u_0, u_1 \in V$, $c_0 \in \mathbb{R}$ and $R > 0$ such that

(i) $\|u_1 - u_0\| > R$

(ii) $J(u_0), J(u_1) < c_0 \leqslant J(v)$, for all v such that $\|v - u_0\| = R$.

Then J has a critical value $c \geqslant c_0$ defined by

$$c = \inf_{\gamma \in \mathcal{P}} \max_{t \in [0, 1]} J(\gamma(t)) \tag{5.5.8}$$

where $\mathcal{P}$ is the collection of all continuous paths $\gamma : [0, 1] \to V$ such that $\gamma(0) = u_0$, $\gamma(1) = u_1$. ∎

In particular, we have $u \in V$ such that $J'(u) = 0$ and $J(u) = c \geqslant c_0$. It is clear that $c \geqslant c_0$ since every path γ connecting u_0 to u_1 will have to cross the sphere $\{v \mid \|v - u_0\| = R\}$ as u_1 lies outside the ball centre u_0 and radius R. On this sphere the value of J is at least c_0. Hence the maximum value of $J(\gamma(t))$ for any such γ is at least c_0. Hence $c \geqslant c_0$.

The proof of Theorem 5.5.2 is based on the following lemma which we state without proof. The proof is quite technical and lengthy. It can be found in Ambrosetti and Rabinowitz [1] where it is presented as a variation of a result proved in Clark [1].

Lemma 5.5.1 (*Deformation Lemma*) Let $J : V \to \mathbb{R}$ be a C^1 functional satisfying (PS). Let $c, s \in \mathbb{R}$. Define

$$K_c = \{v \in V \mid J(v) = c,\ J'(v) = 0\}.$$

$$A_s = \{v \in V \mid J(v) \leqslant s\}.$$

Assume $K_c = \emptyset$. Then there exists $\bar{\epsilon} > 0$ and a continuous homotopy

$\eta : [0, 1]\times V \to V$ such that for all $0 < \epsilon \leqslant \bar{\epsilon}$,

(i) $\eta(0, v) = v$, for all $v \in V$
(ii) $\eta(t, v) = v$, for all $t \in [0, 1]$, $v \notin J^{-1}([c - \epsilon, c + \epsilon])$
(iii) $\eta(1, A_{c+\epsilon}) \subset A_{c-\epsilon}$. ∎

We have stated the deformation lemma in a form much less general than it is known. But we will need it only in the above form. Using the above lemma we can prove Theorem 5.5.2.

Proof of Theorem 5.5.2 Let c be as defined in (5.5.8). If it is not a critical value, then $K_c = \emptyset$. Let η and $\bar{\epsilon}$ be as in the deformation lemma. Now by condition (ii) of the theorem, we can choose ϵ small enough so that $0 < \epsilon < \bar{\epsilon}$ and $J(v_0), J(u_1) \notin [c - \epsilon, c + \epsilon]$ (since $c \geqslant c_0$). Let $\gamma \in \mathcal{P}$. Define the path $\zeta : [0, 1] \to V$ by

$$\zeta(t) = \eta(1, \gamma(t)).$$

Now as $\gamma(0) = u_0$ and $\gamma(1) = u_1$, it follows, by the choice of ϵ, that

$$\zeta(0) = \eta(1, u_0) = u_0 \text{ and } \zeta(1) = \eta(1, u_1) = u_1,$$

using condition (ii) of the lemma. Thus $\zeta \in \mathcal{P}$. Now, we can choose $\gamma \in \mathcal{P}$ such that

$$\max_{t\in[0,1]} J(\gamma(t)) < c + \epsilon.$$

Hence as $\gamma(t) \in A_{c+\epsilon}$, by (iii) of the lemma, $\zeta(t) \in A_{c-\epsilon}$. Thus

$$\max_{t\in[0,1]} J(\zeta(t)) \leqslant c - \epsilon$$

which contradicts the definition of c. Hence $K_c \neq \emptyset$. ∎

An important aspect of the Mountain Pass Theorem is that the solution u one obtains to the equation $J'(u) = 0$ is distinct from u_0 and u_1. Hence if u_0 is already a solution obtained by some other method, then u will be a second solution. It is in this way that we shall now use this theorem. The example we are going to give is described in a survey article by Nirenberg [4].

We consider the example described in Section 5.4. If $\Omega \subset \mathbb{R}^n$, $n \leqslant 3$, is a bounded open set with boundary Γ and $f \in L^2(\Omega)$ a smooth function, then we look at the problem:

$$\left.\begin{aligned} -\Delta u &= u^2 + f \quad \text{in } \Omega \\ u &= 0 \quad\quad\quad \text{on } \Gamma \end{aligned}\right\} \tag{5.5.9}$$

To this we associate the functional

$$J(v) = \tfrac{1}{2}\int_\Omega |\nabla v^2| - \tfrac{1}{3}\int_\Omega v^3 - \int_\Omega fv, \quad \in H_0^1(\Omega). \tag{5.5.10}$$

Then for $u, v \in H_0^1(\Omega)$,

$$(J'(u), v) = \int_\Omega \nabla u \cdot \nabla v - \int_\Omega u^2 v - \int_\Omega fv,$$

where $(\cdot, \cdot)$ denotes the duality product between $H^{-1}(\Omega)$ and $H_0^1(\Omega)$, so that $J'(u) = 0$ gives a weak solution to the problem (5.5.9).

Theorem 5.5.3 Let $f < 0$ be smooth enough. Then the problem (5.5.9) has at least two solutions.

Proof: *Step 1* We already saw (cf. Theorem 5.4.1) that (5.5.9) has a non-positive solution. We call this solution u_0. Thus $u_0 \leqslant 0$. Now if $w \in B(0; 1)$ in $H_0^1(\Omega)$, consider $v = u_0 + \epsilon w$ for $\epsilon > 0$ so that $\|v - u_0\| = \epsilon$. Now

$$J(u_0 + \epsilon w) - J(u_0) = \epsilon \int_\Omega \nabla u_0 \cdot \nabla w + \frac{\epsilon^2}{2}\int_\Omega |\nabla w|^2 - \epsilon \int_\Omega u_0^2 w$$

$$- \epsilon \int_\Omega u_0 w^2 - \frac{\epsilon^3}{3}\int_\Omega w^3 - \epsilon \int fw$$

$$= \frac{\epsilon^2}{2}\int_\Omega |\nabla w|^2 - \epsilon \int_\Omega u_0 w^2 - \frac{\epsilon^3}{3}\int_\Omega w^3,$$

since u_0 is a weak solution of (5.5.9). Also as $u_0 \leqslant 0$, $\int_\Omega u_0 w^2 \leqslant 0$. Further $|w|_{1,\Omega} = 1$ and so $\left|\int_\Omega w^3\right| \leqslant c|w|_{1,\Omega}^3 \leqslant c$. Combining all these we get

$$J(u_0 + \epsilon w) - J(u_0) \geqslant \frac{\epsilon^2}{2} - \frac{c}{3}\epsilon^3 = \epsilon^2\left(\tfrac{1}{2} - \frac{c\epsilon}{3}\right).$$

Thus by choosing $\epsilon > 0$ small enough ($\epsilon < 3/4c$) we get $J(v) - J(u_0) \geqslant (\epsilon^2/4) > 0$ for all v such that $\|v - u_0\| = \epsilon$. Now consider the function $w = \alpha w_1$, where w_1 is the first eigenfunction of the Dirichlet eigenvalue problem for the Laplace operator, i.e. $-\Delta w_1 = \lambda_1 w_1$, $w_1 > 0$, $|w_1|_{0,\Omega} = 1$. Then

$$J(w) = \frac{\alpha^2}{2}\lambda_1 - \frac{\alpha^3}{3}\int_\Omega w_1^3 - \alpha \int_\Omega fw_1.$$

If $\alpha \to +\infty$ then $J(w) \to -\infty$. (Thus J is not bounded below.) By choosing α large enough we can ensure that $J(w) < J(u_0)$. The corresponding w, we denote by u_1. Thus u_0 and u_1 satisfy the hypotheses of the Theorem 5.5.2 with $c_0 = J(u_0) + \epsilon^2/4$ and $R = \epsilon$.

Step 2 J satisfies (PS).

Let $\{v_m\}$ be a sequence in $H_0^1(\Omega)$ such that $|J(v_m)| \leqslant c$ and $J'(u_m) \to 0$ strongly in $H^{-1}(\Omega)$. Now

$$J(v_m) = \tfrac{1}{2}|v_m|_{1,\Omega}^2 - \tfrac{1}{3}\int_\Omega v_m^3 - \int_\Omega fv_m. \tag{5.5.11}$$

$$(J'(v_m), w) = \int_\Omega \nabla v_m \cdot \nabla w - \int_\Omega v_m^2 w - \int_\Omega fw, \quad w \in H_0^1(\Omega). \tag{5.5.12}$$

In particular

$$(J'(v_m), v_m) = |v_m|^2_{1,\Omega} - \int_\Omega v_m^3 - \int_\Omega f v_m$$

$$= 3J(v_m) - \tfrac{1}{2}|v_m|^2_{1,\Omega} + 2\int_\Omega f v_m. \qquad (5.5.13)$$

Hence

$$\tfrac{1}{2}|v_m|^2_{1,\Omega} = 3J(v_m) - (J'(v_m), v_m) + 2\int_\Omega f v_m$$

$$\leqslant 3c + \|J'(v_m)\|_{-1,\Omega}|v_m|_{1,\Omega} + 2|f|_{0,\Omega}|v_m|_{0,\Omega}.$$

It follows from this that $\{|v_m|_{1,\Omega}\}$ is bounded. (If not, divide by $|v_m|^2_{1,\Omega}$ and let $m \to \infty$, using $\|J'(v_m)\|_{-1,\Omega} \to 0$ to get a contradiction viz., $\frac{1}{2} \leqslant 0$.)

Thus (for a subsequence) $\{v_m\}$ converges weakly to a limit v in $H_0^1(\Omega)$. By the Rellich-Kondrasov Theorem ($n \leqslant 3$) this convergence will be strong in $L^2(\Omega)$ and $L^3(\Omega)$. Passing to the limit as $m \to \infty$ in (5.5.12) we get

$$\int_\Omega \nabla v \cdot \nabla w = \int_\Omega v^2 w - \int_\Omega f w = 0, \quad w \in H_0^1(\Omega) \qquad (5.5.14)$$

and passing to the limit in (5.5.13) we get

$$\lim_{m\to\infty} |v_m|^2_{1,\Omega} = \int_\Omega v^3 + \int_\Omega f v = |v|^2_{1,\Omega} \qquad (5.5.15)$$

by setting $w = v$ in (5.5.14). Thus $v_m \to v$ weakly and $|v_m|^2_{1,\Omega} \to |v|^2_{1,\Omega}$ and so $v_m \to v$ strongly in $H_0^1(\Omega)$. This proves that J satisfies (PS).

Step 3 Thus J satisfies all the hypotheses of the Theorem 5.5.2 and so there exists a critical point u of J (which will be a weak solution of (5.5.9)) such that $u \neq u_0$. Hence (5.5.9) has at least two solutions. ∎

For generalizations of the Mountain Pass Theorem see the comments at the end of this chapter.

5.6 POHOZAEV'S IDENTITY

To conclude this chapter, we prove below an identity which is often used to prove the *non-existence* of solutions.

Theorem 5.6.1 Let Ω be a bounded open subset of $\mathbb{R}^n$ with sufficiently smooth boundary Γ. Let $u \in H_0^1(\Omega)$ such that

$$\left.\begin{aligned} -\Delta u &= g(u) \quad \text{in } \Omega \\ u &= 0 \quad \text{on } \Gamma \end{aligned}\right\} \qquad (5.6.1)$$

where $g : \mathbb{R} \to \mathbb{R}$ is a given function. Let

$$G(s) = \int_0^s g(t)\, dt, \quad s \in \mathbb{R}. \qquad (5.6.2)$$

Then

$$\left(1-\frac{n}{2}\right)\int_\Omega g(u)u + n\int_\Omega G(u) = \tfrac{1}{2}\int_\Gamma (x\cdot\nu)\left(\frac{\partial u}{\partial \nu}\right)^2, \qquad (5.6.3)$$

where ν is the unit outer normal on Γ.

Proof Notice that $G(u) = 0$ on Γ. Now, multiplying the equation in (5.6.1) by $x_i \dfrac{\partial u}{\partial x_i}$ and integrating by parts, we get

$$\int_\Omega -\Delta u \frac{\partial u}{\partial x_i} x_i = \int_\Omega g(u)\frac{\partial u}{\partial x_i} x_i = \int_\Omega \frac{\partial}{\partial x_i}(G(u))x_i = -\int_\Omega G(u).$$

Thus

$$\begin{aligned}
-\int_\Omega G(u) &= -\int_\Omega \sum_{j=1}^n \frac{\partial^2 u}{\partial x_j^2}\frac{\partial u}{\partial x_i} x_i \\
&= \int_\Omega \sum_{j=1}^n \frac{\partial u}{\partial x_j}\frac{\partial^2 u}{\partial x_i \partial x_j} x_i + \int_\Omega \sum_{j=1}^n \frac{\partial u}{\partial x_j}\frac{\partial u}{\partial x_i}\delta_{ij} - \int_\Gamma \sum_{j=1}^n \frac{\partial u}{\partial x_j}\frac{\partial u}{\partial x_i} x_i \nu_j \\
&= \tfrac{1}{2}\int_\Omega \sum_{j=1}^n \frac{\partial}{\partial x_i}\left(\left(\frac{\partial u}{\partial x_j}\right)^2\right) x_i + \int_\Omega \left(\frac{\partial u}{\partial x_i}\right)^2 - \int_\Gamma \frac{\partial u}{\partial x_i} x_i \frac{\partial u}{\partial \nu} \\
&= -\tfrac{1}{2}\int_\Omega \sum_{j=1}^n \left(\frac{\partial u}{\partial x_j}\right)^2 + \int_\Omega \left(\frac{\partial u}{\partial x_i}\right)^2 + \tfrac{1}{2}\int_\Gamma \sum_{j=1}^n \left(\frac{\partial u}{\partial x_j}\right)^2 x_i \nu_i \\
&\quad - \int_\Gamma \frac{\partial u}{\partial x_i} x_i \frac{\partial u}{\partial \nu}
\end{aligned}$$

Summing over i we get

$$-n\int_\Omega G(u) = \left(1-\frac{n}{2}\right)\int_\Omega |\nabla u|^2 + \tfrac{1}{2}\int_\Gamma |\nabla u|^2 (x\cdot\nu) - \int_\Gamma \frac{\partial u}{\partial \nu}(x\cdot\nabla u) \qquad (5.6.4)$$

Now, by (5.6.1),

$$\int_\Omega |\nabla u|^2 = \int_\Omega g(u)\cdot u.$$

Further as $u = 0$ on Γ, $|\nabla u|^2 = \left(\dfrac{\partial u}{\partial \nu}\right)^2$ and $x\cdot\nabla u = (x\cdot\nu)\dfrac{\partial u}{\partial \nu}$ on Γ. Substituting in (5.6.4) and rearranging the terms we get (5.6.3). ∎

Definition 5.6.1 A domain Ω containing the origin is said to be star-shaped about the origin if $x\cdot\nu > 0$ for all $x \in \Gamma$. ∎

For example any ball centre origin is star-shaped about the origin for, given any $x \in \Gamma$, $\nu(x) = \dfrac{x}{R}$. Thus $x\cdot\nu = R > 0$.

Example 5.6.1 Let $\Omega \subset \mathbb{R}^n$ be a smooth bounded open set. Consider the

problem

$$\left.\begin{array}{ll} -\Delta u = u^p + \lambda u & \text{in } \Omega \\ u > 0 & \text{in } \Omega \\ u = 0 & \text{on } \Gamma \end{array}\right\} \tag{5.6.5}$$

where $p = \dfrac{n+2}{n-2}$. Note that by the strong maximum principle if $u \geqslant 0$ is solution of $-\Delta u = u^p + \lambda u$, $u \in H_0^1(\Omega)$, then either $u \equiv 0$ or $\bar{u} > 0$ in Ω.

If $\lambda \geqslant \lambda_1$, the first eigenvalue for the Dirichlet eigenvalue problem for the Laplace operator then (5.6.5) has no solution $u > 0$. For, if $w_1 > 0$ is the corresponding eigenfunction,

$$\lambda_1 \int_\Omega uw_1 = \int_\Omega \nabla u \cdot \nabla w_1 = \int_\Omega u^p w_1 + \lambda \int_\Omega uw_1 > \lambda \int_\Omega uw_1$$
$$\geqslant \lambda_1 \int_\Omega uw_1,$$

since $uw_1 > 0$. This is a contradiction.

If Ω is star-shaped about the origin ($0 \in \Omega$) then (5.6.1) has no solution for $\lambda \leqslant 0$. In this case set

$$g(s) = s^p + \lambda s$$

so that

$$G(s) = \frac{s^{p+1}}{p+1} + \frac{\lambda s^2}{2}.$$

Substituting in Pohozaev's identity (5.6.3) we get

$$\frac{n}{p+1}\int_\Omega u^{p+1} + \frac{n\lambda}{2}\int_\Omega u^2 + \left(1 - \frac{n}{2}\right)\int_\Omega (u^{p+1} + \lambda u^2) = \tfrac{1}{2}\int_\Gamma (x\cdot\nu)\left(\frac{\partial u}{\partial \nu}\right)^2.$$

i.e.

$$\left(\frac{n}{p+1} + 1 - \frac{n}{2}\right)\int_\Omega u^{p+1} + \lambda \int_\Omega u^2 = \tfrac{1}{2}\int_\Gamma (x\cdot\nu)\left(\frac{\partial u}{\partial \nu}\right)^2.$$

But

$$\frac{n}{p+1} + 1 - \frac{n}{2} = 0 \text{ if } p = \frac{n+2}{n-2}.$$

Thus

$$\lambda \int_\Omega u^2 = \tfrac{1}{2}\int_\Gamma (x\cdot\nu)\left(\frac{\partial u}{\partial \nu}\right)^2. \tag{5.6.6}$$

which is a contradiction if $\lambda < 0$. If $\lambda = 0$, (5.6.6) implies that $\dfrac{\partial u}{\partial \nu} = 0$ on Γ and so

$$\int_\Omega u^p = \int_\Omega -\Delta u = 0$$

and as $u \geqslant 0$, we get $u \equiv 0$. ∎

Remark 5.6.1 The exponent $p = \dfrac{n+2}{n-2}$ is the critical Sobolev exponent

in the sense that we have the optimal Sobolev inclusion $H_0^1(\Omega) \subset L^{p+1}(\Omega)$ with the inclusion not being compact. Such problems are of special interest and their study occurs in several interesting situations. Cf. Brezis and Nirenberg [1] for an introduction to such problems. ∎

Remark 5.6.2 If Ω were not star-shaped, we can have solutions even when $\lambda \leqslant 0$ in Example 5.6.1. In fact for an annular domain it is known that a radial solution (i.e. $u(x) = u(|x|)$) exists for (5.6.5) for all $\lambda \in (-\infty, \lambda_1)$. ∎

COMMENTS

1. *General Remarks.* In this chapter we have given an extremely brief introduction to some topics in nonlinear functional analysis. Indeed the topic is so vast and the techniques often *ad hoc* that it would require a separate treatise to even try to do justice to the topic. We have limited ourselves to a few simple techniques useful in the study of semilinear elliptic equations.

2. *The Topological Degree and Fixed Points.* The fixed point theorems of Brouwer and Schauder are usually proved using the notion of the topological degree. In finite dimensions it is called the **Brouwer degree** and in infinite dimensions it is called the **Leray-Schauder degree.** The topological degree is a powerful tool in the study of existence and multiplicity of solutions to nonlinear equations. For an exposition on this subject, see Joshi and Bose [1], Nirenberg [2], Rabinowitz [1] or J. T. Schwartz [1]. While all proofs of the Brouwer Theorem cited upto now are existence proofs, a constructive proof, closely related to the Newton's method of solving nonlinear equations, was given by Kellogg, Li and Yorke [1]. Another algorithm for calculating Brouwer fixed points is due to Scarf [1].

3. *Monotone Operators.* In this chapter we only illustrated the Galerkin method using compactness arguments. Another powerful argument is via the theory of monotone operators developed by Minty and Browder. For a description of this theory, see, for instance, Joshi and Bose [1]. See also J. L. Lions [1] for several examples from partial differential equations of the methods of monotonicity and compactness used in conjunction with the Galerkin method.

4. *Critical Point Theories.* We have seen that critical points of the 'energy' functional give solutions of partial differential equations. This approach is important since most equations in physics and engineering are derived from variational principles. Some of the important critical point theories which we have not discussed here are the Morse theory (cf. Milnor [1]) and the theory of Lusternik and Schnirelman (cf., for instance, Berger [1]). The Morse theory leads to the **Morse index** and the

Morse inequalities which can be used to study the multiplicity of solutions to nonlinear equations. The Lusternik-Schnirelman theory is also useful in the study of critical points of functionals. For instance if we consider the problem

$$\left.\begin{aligned} -\Delta u - \lambda u &= u^3 + f \text{ in } \Omega \\ u &= 0 \quad \text{ on } \Gamma \end{aligned}\right\}$$

with the associated functional

$$J(v) = \tfrac{1}{2}\int_\Omega |\nabla v|^2 - \frac{\lambda}{2}\int_\Omega v^2 - \tfrac{1}{4}\int_\Omega v^4 - \int_\Omega fv,\ v \in H_0^1(\Omega),$$

the techniques of this chapter will not work. In particular the functional is not coercive. But the existence of solutions has been studied using the Lusternik-Schnirelman theory by Bahri and Berestycki [1]. Berger [1] has also used it to study the Von Karman equations.

5. *The Mountain Pass Theorem.* This theorem is called thus because of the following geometric interpretation. If u_0 is a place surrounded by a ring of mountains and u_1 a place outside and J represents the height, then a person seeks that path from u_0 to u_1 across the mountains in which he climbs the least. This will lead to a pass in the mountains or a saddle point.

Several generalizations of the original result of Ambrosetti and Rabinowitz have been proved in the recent years. See Nirenberg [3] for some of these results and for references. More recently Pucci and Serrin [1] have proved that if we only have the inequality $J(u_0), J(u_1) \leqslant c_0$ in condition (ii) of Theorem 5.5.2 but also that $c_0 \leqslant J(v)$ for all v in an annulus $r \leqslant \|v - u_0\| \leqslant R$ (i.e. the mountain range has non-zero width) then J has a critical point different from u_0 and u_1 (with even a weaker form of (*PS*)). In particular this implies that if J has two local minima, then it must have a third critical point.

(*Proof*: Let $J(u_0) \geqslant J(u_1)$; without loss of generality we can assume $u_0 = 0$, by changing the origin. Choose $0 < R < \|u_1\|$ such that $J(v) \geqslant J(0)$ in $B(0; R)$. Now choose $r = R/2$ and apply the previous result.)

6. *Critical Sobolev Exponents.* Consider the Sobolev inclusions

$$H_0^1(\Omega) \subset L^{p+1}(\Omega),\ 1 \leqslant p \leqslant \frac{n+2}{n-2}.$$

If Ω is bounded and $p = \dfrac{n+2}{n-2}$ then the inclusion, which is the limiting case, is not compact (cf. comments on Chapter 2). Thus the map $u \to u^p$ is not compact from $H_0^1(\Omega)$ into its dual $H^{-1}(\Omega)$. Thus compactness argu-

ments will not work for problems like the following:

$$\left.\begin{aligned} -\Delta u &= \lambda u + u^p && \text{in } \Omega \\ u &> 0 && \text{in } \Omega \\ u &= 0 && \text{on } \Gamma \end{aligned}\right\}$$

We saw in Section 5.6 that the above problem has no solution if $\lambda \geqslant \lambda_1$ and no solution, when Ω is star-shaped, if $\lambda \leqslant 0$, where λ_1 is the first Dirichlet eigenvalue of the Laplace operator. Brezis and Nirenberg [1] were one of the first to consider such problems with lack of compactness. They proved that if $n \geqslant 4$ then a solution always exists for the above problem if $0 < \lambda < \lambda_1$. However when $n = 3$ and Ω a ball, a solution exists if, and only if, $\lambda_1/4 < \lambda < \lambda_1$. There are several open problems connected with such questions and the reader is referred to Brezis [2]. Cerami *et al* [1] have studied solutions of the above equations, without the condition $u > 0$ in Ω, for all $\lambda \in \mathbb{R}$.

7. *Bifurcation Theory.* An important class of nonlinear problems not discussed in this chapter is the so-called nonlinear eigenvalue problems. The general form is given by the equation

$$F(u, \lambda) = 0$$

where $u \in V$, a Banach space and λ is a parameter. Usually $\lambda \in \mathbb{R}$ (or $\mathbb{R}^n$). Such problems arise in elasticity, heat transfer, population dynamics and a variety of other situations. What we are interested in is the solution set for each value of λ. As λ changes, the number of solutions may change. Usually one can parametrize the solution "paths" as $(u(s), \lambda(s))$ and at certain solutions (u_0, λ_0) more than one path may intersect and hence the term bifurcation. For instance consider the one-dimensional equation (i.e. $V = \mathbb{R}$) with the single real parameter λ given by

$$(1 - \lambda)u + u^3 = 0, u \in \mathbb{R}.$$

If $\lambda < 1$ then $u = 0$ is the only solution. If $\lambda > 1$ then we have three solutions viz., $u = 0$, $u = \sqrt{\lambda - 1}$ and $u = -\sqrt{\lambda - 1}$. We can parametrize the solution paths by λ itself. They are two in number, viz. $u \equiv 0$ and $u^2(\lambda) = \lambda - 1$, $\lambda \geqslant 1$ which intersect at $(0, 1)$.

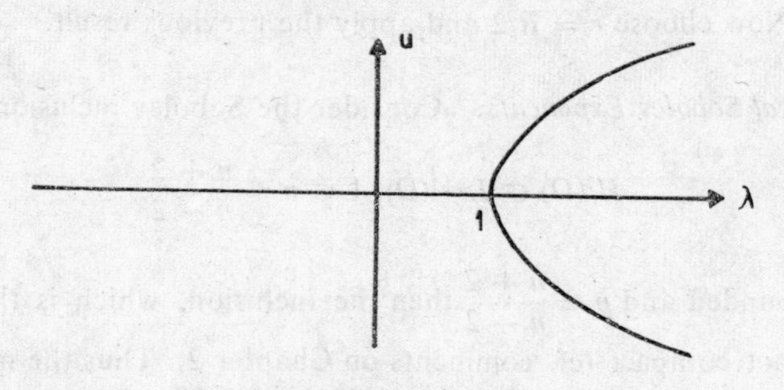

FIG. 22.

The point (0, 1) is called a **bifurcation point.** Bifurcation theory deals with the study of the local solution set near a bifurcation point and the global solution paths. For a discussion of examples and the main results of this theory see Berger [1], Nirenberg [2] and Schaeffer [1]. For computational aspects see Keller [1].

8. *Bibliography.* For general treatises on nonlinear functional analysis, see Deimling [1], Nirenberg [2] and J. T. Schwartz [1]. See also Joshi and Bose [1] for a general treatment of some topics in nonlinear analysis with applications to integral equations. For a survey of (positive) solutions of semilinear elliptic problems, see P. L. Lions [1]. For problems involving lack of compactness refer to Brezis [2]. For various aspects of bifurcation theory see Rabinowitz [2] and the references therein. The article of Nirenberg [3] gives an excellent survey of topological and variational methods for nonlinear problems.

EXERCISES 5

5.1 Show by examples that a continuous map $f : K \to K$, $K \subset \mathbb{R}^n$ need not have a fixed point if (i) K is not compact or (ii) K is not convex.

5.2 Assuming the validity of Theorem 5.2.5 prove the Brouwer fixed point theorem for a ball.

5.3 In Example 5.2.2, show that the unique solution $u \in H_0^1(\Omega)$ satisfies the estimate

$$|u|_{1,\Omega} \leqslant \left(1 - \frac{L}{\lambda_1}\right)^{-1} \left(\frac{\text{meas}\,(\Omega)}{\lambda_1}\right)^{1/2} |f(0)|.$$

5.4 (a) Let $\Omega \subset \mathbb{R}^n$ be a bounded open set. Let $f : \mathbb{R} \to \mathbb{R}$ be a continuous function such that

$$\lim_{s \to \pm\infty} \frac{f(s)}{s} = f_{\pm}$$

with f_+, f_- finite. Show that the mapping $u \to f(u)$ defined by $f(u)(x) = f(u(x))$ is continuous from $L^2(\Omega)$ into itself.

(b) Let $f_- = 0$ and $f_+ < \lambda_1$, where λ_1 is the first eigenvalue of the problem:

$$\left.\begin{aligned} -\Delta w &= \lambda w \quad \text{in } \Omega \\ w &= 0 \quad \text{on } \Gamma \end{aligned}\right\}.$$

Then show using Schaeffer's theorem that the problem

$$\left.\begin{aligned} -\Delta u &= f(u) \quad \text{in } \Omega \\ u &= 0 \quad \text{on } \Gamma \end{aligned}\right\}$$

has at least one (weak) solution in $H_0^1(\Omega)$.

5.5 Let λ_1 be as in the previous exercise and let $0 < \lambda < \lambda_1$. Consider

the problem

$$\left.\begin{aligned} -\Delta u - \lambda u + u^3 &= f \quad \text{in } \Omega \\ u &= 0 \quad \text{on } \Gamma \end{aligned}\right\}$$

where Ω is a bounded open subset of $\mathbb{R}^n$, $n \leqslant 4$. Show that there exists a constant $K > 0$ such that if $|f|_{0,\Omega} < K$ then the problem has a unique solution.

5.6 Let $n \leqslant 3$ and Ω a bounded open subset of $\mathbb{R}^n$. Let λ_1 be as before and let $w_1 > 0$ be the corresponding eigenfunction such that $\int_\Omega w_1^2 = 1$. Consider the problem

$$\left.\begin{aligned} -\Delta u &= u^2 - t w_1 \quad \text{in } \Omega \\ u &= 0 \quad \text{on } \Gamma. \end{aligned}\right\}$$

If $t > 0$ is sufficiently small, show that the solution $u \leqslant 0$ obtained in Theorem 5.4.1 can also be obtained by the Contraction Mapping Theorem.

5.7 Let $f(x) = e^x$, $x \in \mathbb{R}$. Show that f does not satisfy (PS).

5.8 Let (λ_1, w_1) be as in Exercise 5.6. Show that the problem

$$\left.\begin{aligned} -\Delta u &= ((u - w_1)^+)^2 \quad \text{in } \Omega \\ u &= 0 \quad \text{on } \Gamma \end{aligned}\right\}$$

has at least two solutions.

5.9 Let $1 < p < \dfrac{n+2}{n-2}$ so that $H^1(\mathbb{R}^n) \subset L^{p+1}(\mathbb{R}^n)$. Show that there exists a non-zero radial solution $u \in H^1(\mathbb{R}^n)$ (i.e., $u \in H^1_r(\mathbb{R}^n)$) to the problem:

$$-\Delta u + \omega u = |u|^{p-1} u \quad \text{in } \mathbb{R}^n$$

where $\omega > 0$.

5.10 Let $\Omega \subset \mathbb{R}^n$ be a bounded open set and let $w \in H^1_0(\Omega)$ be an eigenfunction to the Laplace operator, i.e., $w \neq 0$, and there exists $\lambda > 0$ such that

$$-\Delta w = \lambda w.$$

Show that $\dfrac{\partial w}{\partial \nu} \not\equiv 0$ on Γ.

APPENDIX 1

Partition of Unity

We will now give the details of the proof of the Theorem 1.2.1 on the C^∞-partition of unity. We need a few topological results in order to do this.

Lemma A1.1 Let $\Omega \subset \mathbb{R}^n$ be an open subset and let $\{U_i, i \in I\}$ be an open covering of Ω. Then there exists a locally finite open refinement $\{V_j, j \in J\}$ of this cover.

Proof Let $\{\tilde{\Omega}_k, k \geqslant 0\}$, be an increasing family of relatively compact open sets which cover Ω, i.e., $\tilde{\Omega}_k \subset \tilde{\Omega}_{k+1}$ for all $k \geqslant 0$, $\bar{\tilde{\Omega}}_k$ is compact and $\Omega = \bigcup_{k=0}^{\infty} \tilde{\Omega}_k$. Since $\bar{\tilde{\Omega}}_k$ is compact, there exists a finite subfamily of the U_i, say, $U_{1,k}, \ldots, U_{i_k,k}$ which cover $\bar{\tilde{\Omega}}_k$. Define

$$V_{i,k} = U_{i,k} \cap (\Omega \setminus \bar{\tilde{\Omega}}_{k-1}), \quad k \geqslant 1, 1 \leqslant i \leqslant i_k$$

which is an open set. Also set

$$V_{i,0} = U_{i,0} \cap \tilde{\Omega}_1, \quad 1 \leqslant i \leqslant i_0.$$

Since the $\{V_{i,0}\}$ cover $\tilde{\Omega}_0$ and the $\{V_{i,k}\}$ cover $\tilde{\Omega}_k \setminus \bar{\tilde{\Omega}}_{k-1}$, we have that the entire family $\{V_{i,k}, k \geqslant 0, 1 \leqslant i \leqslant i_k\}$ covers Ω. Each $V_{i,k}$ is contained in a U_j and hence the family is a refinement of the original one. Finally we check that this family is locally finite. Let $x \in \Omega$. Then $x \in \tilde{\Omega}_k$ for some k. But then $x \notin V_{i,l}$ for any $l \geqslant k+1$ and so we have constructed a neighbourhood of x, viz., $\tilde{\Omega}_k$, which intersects only a finite number of the $V_{i,k}$. ∎

Remark A1.2 The above lemma states that $\Omega \subset \mathbb{R}^n$ is a **paracompact space** (cf. Dugundji [1] or Dieudonne [1]). ∎

Remark A1.2 By the construction described in the previous lemma, it is obvious that the refinement $\{V_{i,k}\}$ is countable. In fact it is easy to see that any locally finite refinement has to be countable. Henceforth we will index a locally finite cover by means of the integers. ∎

Lemma A1.2 Let $\{V_m, m \geqslant 0\}$ be a locally finite open cover of Ω. Then there exists an open cover $\{W_m\}$ of Ω such that $\overline{W}_m \subset V_m$ for each $m \geqslant 0$.

Proof We set $W_0 = \emptyset$ and we define W_m inductively such that for each $m \geqslant 1$, the union

$$\bigcup_{k \leqslant m} W_k \bigcup_{k>m} V_k$$

covers Ω. Assume that the W_k have been constructed for $k < m$. Let

$$U = \bigcup_{k<m} W_k \bigcup_{k>m} V_k,$$

so that $\Omega \setminus V_m \subset U$. We will construct an open set V such that

$$\Omega \setminus V_m \subset V \subset \bar{V} \subset U.$$

This is always possible since Ω is a normal space. Now set $W_m = \Omega \setminus \bar{V}$. Clearly $\overline{W}_m \subset \Omega \setminus V \subset V_m$ and the induction is complete. If $x \in \Omega$ then by the local finiteness of the V_i, there exists an m such that $x \notin V_k$ for all $k > m$. Hence $x \in W_k$ for some $k \leqslant m$. Thus $\{W_m\}$ is a cover of Ω. ∎

As a corollary of Theorem 1.2.1 we proved that if $K \subset \mathbb{R}^n$ is a compact set and $U \supset K$, U open, then there exists $\phi \in \mathcal{D}(\mathbb{R}^n)$ such that $\phi = 1$ on K and supp $(\phi) \subset U$. We now give a direct proof of a slightly weaker result.

Lemma A1.3 Let $K \subset \mathbb{R}^n$ be compact and $U \supset K$ an open set. Then there exists a $\phi \in \mathcal{E}$ such that $\phi \geqslant 0$ and $\phi > 0$ on K, supp $(\phi) \subset U$.

Proof Let $d = \text{dist}(K, \mathbb{R}^n \setminus U)$, which is strictly positive. Define for $y \in K$,

$$\phi_y(x) = \begin{cases} \exp(-d^2/(d^2 - |x-y|^2)), & |x-y| < d \\ 0, & |x-y| \geqslant d \end{cases}$$

Then ϕ_y is a C^∞ function and set $V_y = \{x \mid \phi_y(x) > 0\}$. Then V_y is an open set contained in U and $y \in V_y$. Thus by the compactness of K, there exist a finite number of points $y_1, \ldots, y_k$ in K such that the $\{V_{y_i}, 1 \leqslant i \leqslant k\}$ cover K. The function

$$\phi(x) = \sum_{i=1}^{k} \phi_{y_i}(x)$$

has the required properties. ∎

We are now in a position to prove Theorem 1.2.1.

Proof of Theorem 1.2.1 Let $\{\Omega_i, i \in I\}$, be an open cover of Ω and $\{\widetilde{\Omega}_k, k \geqslant 0\}$, be an increasing family of relatively compact open sets as in Lemma A1.1. Let $\{V_j, j \in J\}$ be a locally finite refinement of the $\{\Omega_i\}$. Then there exists, for each integer $k \geqslant 0$, a finite set $J_k \subset J$ such that $V_j \cap \widetilde{\Omega}_k \neq \emptyset$ if and only if, $j \in J_k$, since $\overline{\widetilde{\Omega}}_k$ is compact and $\{V_j\}$ is locally

finite. Consider the open sets

$$V_{j,k} = V_j \cap \tilde{\Omega}_k, \quad j \in J_k.$$

Then the family $\{V_{j,k}, k \geqslant 0, j \in J_k\}$ is also a locally finite open cover of Ω. By Lemma A1.2, we can find another locally finite open cover $\{W_{j,k}, k \geqslant 0, j \in J_k\}$ such that $\overline{W}_{j,k} \subset V_{j,k} \subset \tilde{\Omega}_k$. Hence $\overline{W}_{j,k}$ is compact. Then by Lemma A1.3, there exists a C^∞ function $\psi_{j,k} \geqslant 0$ with $\operatorname{supp}(\psi_{j,k}) \subset V_{j,k} \subset V_j$ and $\psi_{j,k} > 0$ on $\overline{W}_{j,k}$. Now define

$$\psi = \sum_k \sum_{j \in J_k} \psi_{j,k},$$

(an empty sum meaning zero). Now if $x \in \Omega$ then there exist j and k such that $j \in J_k$, $x \in W_{j,k}$. Hence $\psi_{j,k}(x) > 0$. Thus $\psi(x) > 0$ for all $x \in \Omega$ and is a C^∞ function. Hence $1/\psi$ is also a C^∞ function and if we define

$$\psi_j = \frac{1}{\psi} \sum_{\{k \mid j \in J_k\}} \psi_{j,k}$$

then ψ_j will be a C^∞ function, $\psi_j \geqslant 0$, $\sum_{j \in J} \psi_j \equiv 1$, and $\operatorname{supp}(\psi_j) \in V_j$. Now to each $j \in J$ associate a unique $i = \tau(j)$ such that $V_j \subset \Omega_i$. Let $J_i = \tau^{-1}(i)$. Then J_i is a (possibly empty) subset of J and the $\{J_i\}$ constitute a partition of J. Now define

$$\phi_i = \sum_{j \in J_i} \psi_j$$

(the empty sum meaning zero) and the ϕ_i have the required properties. ∎

APPENDIX 2

The Topology on $\mathcal{D}(\Omega)$

We will now describe the topology on the space of C^∞ functions with compact support contained in Ω, an open subset of $\mathbf{R}^n$.

Let $\{X_k, k \geqslant 1\}$ be a family of topological spaces. Assume that there exist connecting (continuous) maps

$$f_{kl} : X_k \to X_l$$

whenever $k \leqslant l$ such that

$$f_{lm} \circ f_{kl} = f_{km}, \quad k \leqslant l \leqslant m.$$

If $x \in X$, we say that $f_{kl}(x)$ is a successor of x for any $l \geqslant k$. We now set

$$X = \bigcup_{k \geqslant 1} X_k$$

and say that $x \sim y$ in X if both x and y have a common successor. Obviously $x \sim x$ and $x \sim y$ implies $y \sim x$. Finally assume that $x \sim y$ and that $y \sim z$. Let $x \in X_k$, $y \in X_l$, $z \in X_m$ and let $u \in X_p$, $v \in X_q$ such that

$$f_{kp}(x) = f_{lp}(y) = u, \quad k \leqslant p, \, l \leqslant p,$$
$$f_{lq}(y) = f_{mq}(z) = v, \quad l \leqslant q, \, m \leqslant q.$$

Let $r \geqslant p, q$. Then $f_{pr}(u) = f_{kr}(x) = f_{lr}(y)$. Also $f_{qr}(v) = f_{mr}(z) = f_{lr}(y)$. Thus x and z have a common successor, viz. $f_{pr}(u) = f_{qr}(v) = f_{lr}(y)$. Hence $x \sim z$ and thus the relation $\sim$ is an equivalence relation. We define X_∞ to be the collection of all equivalence classes with respect to the above relation. Then we have the canonical projection maps

$$p_k : X_k \to X_\infty$$

mapping any point to its equivalence class. We then provide X_∞ with the finest topology such that each of the maps p_k is continuous, i.e. $U \subset X_\infty$ is open if, and only if, $p_k^{-1}(U)$ is open in X_k for each $k \geqslant 1$. The space X_∞ is then said to be provided with the **inductive limit topology** with respect to the family $\{X_k\}$ and the connecting maps $\{f_{kl}\}$.

Let us now consider an open set $\Omega \subset \mathbf{R}^n$. We can then find compact sets $K_m \subset \Omega$ such that $K_m \subset K_{m+1}$ for each m and such that

$$\Omega = \bigcup_{m=1}^{\infty} K_m.$$

If $K \subset \Omega$ is any compact set we consider the vector space $\mathcal{D}_K$ of C^∞ functions with support in K. Define for $\phi \in \mathcal{D}(\Omega)$,

$$\|\phi\|_N = \max_{x \in \Omega} \{|D^\alpha \phi(x)| \mid |\alpha| \leqslant N\}, \quad N \geqslant 1.$$

Define the sets

$$V_N = \{\phi \in \mathscr{D}_K \mid \|\phi\|_N < 1/N\}, \quad N \geqslant 1.$$

Then these sets V_N form a local base for a topology on $\mathscr{D}_K$. The open sets of this topology are precisely those sets which can be expressed as a union of translates of the sets V_N.

Now consider the compact sets $\{K_m\}$ described previously. Between $\mathscr{D}_{K_m}$ and $\mathscr{D}_{K_l}$, $m \leqslant l$, we have the obvious connecting map, viz. the inclusion map. With the topology described above for these spaces, these inclusion maps are continuous. Now

$$\mathscr{D}(\Omega) = \bigcup_{m=1}^{\infty} \mathscr{D}_{K_m}$$

and the equivalence classes with respect to the equivalence defined via common successors are just singletons, i.e. if $\phi \in \mathscr{D}_{K_m}$ then its successor anywhere else is itself. Thus $\phi \sim \psi$ if, and only if ϕ is an extension by zero of ψ or vice versa. Thus we can consider the space $\mathscr{D}(\Omega)$ provided with the corresponding inductive limit topology.

From the definition it is obvious that the topology $\mathscr{D}_K$ inherits as a subspace of $\mathscr{D}(\Omega)$ is the topology described previously. To see this, notice that if $K \subset \Omega$ is compact then $K \subset K_m$ for some m and the projection maps from $\mathscr{D}_{K_m}$ into $\mathscr{D}(\Omega)$ are just the inclusion maps.

Let $\{\phi_m\}$ be a sequence in $\mathscr{D}(\Omega)$ converging to zero. Then every open set containing zero must also contain all but a finite number of the ϕ_m. We now show that we can find a single compact set K which contains the supports of all the ϕ_m. If not, there exists a sequence of points $\{x_m\}$ in Ω which does not converge in Ω and a sequence $\{\phi_m\}$ in $\mathscr{D}(\Omega)$ such that $\phi_m(x_m) \neq 0$. Let

$$W = \{\phi \in \mathscr{D}(\Omega) \mid |\phi(x_m)| < m^{-1}|\phi_m(x_m)|, \quad \text{for all } m\}.$$

Then as each K_m contains only a finite number of the x_l, $W \cap \mathscr{D}_{K_m}$ is open in $\mathscr{D}_{K_m}$. Hence W is open in $\mathscr{D}(\Omega)$ and must contain all but a finite number of the ϕ_m. But $\phi_m \notin W$ for any m, and we have a contradiction. Hence the existence of a compact set K as claimed is established. In the space $\mathscr{D}_K$ if W is any neighbourhood of zero, then W will contain all but a finite number of the ϕ_m. Hence $\phi_m \to 0$ in $\mathscr{D}_K$. Now using the norms $\|\cdot\|_N$ which generate the topology on $\mathscr{D}_K$ it follows that ϕ_m and all its derivatives converge uniformly to zero in K. Thus we have reconciled the definition of the topology on $\mathscr{D}(\Omega)$ given above with that described in Chapter 1 (cf. Definition 1.2.1).

Now let $T \in \mathscr{D}'(\Omega)$. We saw that (cf. Theorem 1.2.2) T restricted to $\mathscr{D}_K$ is also continuous for any compact set $K \subset \Omega$. Now if $a > 0$ is any number then $T^{-1}\{|x| < a\}$ is an open set in $\mathscr{D}_K$ containing the origin and hence a member of the local base. Thus there exists an N such that if $\phi \in \mathscr{D}_K$ and if $\|\phi\|_N < 1/N$, then $|T(\phi)| < a$. Hence for any $\phi \in \mathscr{D}_K$,

$$|T(\phi)| \leqslant Na\|\phi\|_N$$

which proves relation (1.2.15).

For more details on the topology of $\mathscr{D}(\Omega)$, see Rudin [2].

APPENDIX 3

Calculus in Banach Spaces

We will now describe the concept of integration and differentiation of functions which take values in a Banach space.

Let V be a Banach space and $\phi : [0, 1] \to V$ a continuous function. We would like to give a meaning to the expression

$$\int_0^1 \phi(t)\, dt.$$

If $V = \mathbb{R}$, then this is well known and can be interpreted as a limit of Riemann sums. A finite sum commutes with linear operations and so does the limit if the operation is continuous. It is this criterion which we will expect of integrals of vector-valued functions. More precisely, we define the integral as follows.

Definition A3.1 Let $\phi : [0, 1] \to V$ be a continuous function. The integral

$$\int_0^1 \phi(t)\, dt,$$

if it exists, is the unique vector $y \in V$ such that

$$f(y) = \int_0^1 f(\phi(t))\, dt, \tag{A3.1}$$

for every $f \in V^*$. ∎

Remark A3.1 Note that in (A3.1) the integrand is a real-valued function and so the integral can be defined in the usual way. Also as expressed in the above definition, we have no problem of uniqueness. If y_1 and y_2 satisfied (A3.1) for all $f \in V^*$, then $f(y_1) = f(y_2)$ for all such f and by the Hahn-Banach Theorem, it follows that $y_1 = y_2$. ∎

We now show that the integral exists for a continuous function ϕ. Let $n \geqslant 1$ be an integer and set

$$t_i^n = i2^{-n}, 0 \leqslant i \leqslant 2^n. \tag{A3.2}$$

Define the Riemann sum

$$y_n = 2^{-n} \sum_{i=0}^{2^n-1} \phi(t_i^n). \tag{A3.3}$$

We first show that $\{y_n\}$ is a Cauchy sequence in V.

Let $\epsilon > 0$. Since ϕ is uniformly continuous, there exists $\delta > 0$ such that $|t - s| < \delta$ implies that $\|\phi(t) - \phi(s)\| < \epsilon$. Now let n_0 be such that $2^{-n} < \delta$ for all $n \geqslant n_0$. Then for any $m \geqslant 0$,

$$y_n - y_{n+m} = 2^{-n} \sum_{i=0}^{2^n-1} \phi(t_i^n) - 2^{n+m} \sum_{i=0}^{2^{n+m}-1} \phi(t_i^{n+m})$$

But notice that $t_i^n = i2^{-n} = 2^m i 2^{-(n+m)} = t_{i2^m}^{n+m}$. Thus we may write

$$y_n - y_{n+m} = 2^{-n} \sum_{i=0}^{2^n-1} (\phi(t_i^n) - 2^{-m} \sum_{j=0}^{2^m-1} \phi(t_{i2^m+j}^{n+m}))$$

$$= 2^{-(n+m)} \sum_{i=0}^{2^n-1} \sum_{j=0}^{2^m-1} (\phi(t_i^n) - \phi(t_{i2^m+j}^{n+m}))$$

and since $|t_i^n - t_{i2^m+j}^{n+m}| < \delta$, we deduce that

$$\|y_{n+m} - y_n\| < \epsilon, \; n \geqslant n_0, \; m \geqslant 0.$$

Thus $\{y_n\}$ is a Cauchy sequence and has a limit y. Now if $f \in V^*$,

$$f(y) = \lim_{n\to\infty} f(y_n)$$

$$= \lim_{n\to\infty} 2^{-n} \sum_{i=0}^{2^n-1} f(\phi(t_i^n))$$

$$= \int_0^1 f(\phi(t))\, dt$$

by the usual definition of the Riemann integral of the continuous real-valued function $f\cdot\phi$. Thus y satisfies (A3.1) and is the integral of ϕ over $[0, 1]$.

The integral can be shown to exist for more general classes of functions. See Rudin [2] and the references therein.

Theorem A3.1 Let $\phi : [0, 1] \to V$ be continuous. Then

$$\left\| \int_0^1 \phi(t)\, dt \right\| \leqslant \int_0^1 \|\phi(t)\|\, dt. \tag{A3.4}$$

Proof: By the Hahn-Banach Theorem, there exists $f \in V^*$ such that $\|f\| \leqslant 1$ and $f(y) = \|y\|$ where

$$y = \int_0^1 \phi(t)\, dt.$$

Thus

$$\left\| \int_0^1 \phi(t)\, dt \right\| = f(y) = \int_0^1 f(\phi(t))\, dt$$

$$\leqslant \int_0^1 |f(\phi(t))|\, dt \leqslant \int_0^1 \|\phi(t)\|\, dt. \quad \blacksquare$$

We can generalize the relation (A3.1) as follows.

Theorem A3.2 Let $\phi : [0, 1] \to V$ be continuous. Let W be another Banach

space and let $A : V \to W$ be a bounded linear operator. Then

$$A\left(\int_0^1 \phi(t)\, dt\right) = \int_0^1 A(\phi(t))\, dt. \tag{A3.5}$$

Proof Let $f \in W^*$. Then $f \cdot A \in V^*$. Thus by (A3.1)

$$(f \cdot A)\left(\int_0^1 \phi(t)\, dt\right) = \int_0^1 (f \cdot A)(\phi(t))\, dt.$$

Thus

$$f\left(A \int_0^1 \phi(t)\, dt\right) = \int_0^1 f(A(\phi(t)))\, dt$$

and (A3.5) follows since $f \in W^*$ was arbitrary. ∎

We can also define the integral

$$\int_a^b \phi(t)\, dt$$

for a continuous function $\phi : [a, b] \to V$, $a < b$, $a, b \in \mathbb{R}$, in the obvious way. If either a or b (or both) were infinite, the integral is defined as the limit of definite integrals, if the limit exists. Thus for instance, if $\phi : [0, \infty] \to V$, we define

$$\int_0^\infty \phi(t)\, dt = \lim_{R\to\infty} \int_0^R \phi(t)\, dt, \tag{A3.6}$$

if the limit exists.

If $a < b < c$, $a, b, c \in \mathbb{R} \cup \{\pm\infty\}$ and if the relevant integrals exist, it is easy to see that

$$\int_a^c \phi(t)\, dt = \int_a^b \phi(t)\, dt + \int_b^c \phi(t)\, dt. \tag{A3.7}$$

We now briefly describe differentiation of functions which are defined on a Banach space and take values in another Banach space. Let V and W be Banach spaces and let $f : V \to W$ be a function.

Definition A3.2 We say that f is differentiable at a point $u \in V$ if there exists a bounded linear operator $A : V \to W$ such that

$$\lim_{\|h\|_V \to 0} \frac{\|f(u+h) - f(u) - Ah\|_W}{\|h\|_V} = 0. \tag{A3.8}$$

The operator A is called the derivative or Fréchet derivative of f at u and is denoted by $f'(u)$. If f is differentiable at all points in an open set $\mathcal{O} \subset V$, then we say that f is differentiable on $\mathcal{O}$. If $\mathcal{O} = V$ and the map $u \to f'(u)$ is a continuous map of V into $\mathcal{L}(V, W)$, the space of bounded linear maps of V into W, we say that f is C^1. ∎

Remark A3.2 If f is differentiable at u, it follows that f is continuous at u. Also we can talk of *the* derivative at u. If we had two operators A_1 and

A_2 satisfying (A3.8) then $A_1 = A_2$. For if

$$f(u+h) - f(u) - A_1 h = w_1(h)$$

and

$$f(u+h) - f(u) - A_2 h = w_2(h),$$

then $\|w_1(h)\|/\|h\|$ and $\|w_2(h)\|/\|h\| \to 0$ as $\|h\| \to 0$. Hence

$$A_1 h - A_2 h = w_2(h) - w_1(h)$$

and so

$$\sup_{\|h\|=\epsilon} \|(A_1 - A_2)h\|$$

can be made arbitrarily small as $\epsilon \to 0$, i.e. $\|A_1 - A_2\| = 0$ or $A_1 = A_2$. ∎

If $W = \mathbb{R}$ then the Fréchet derivative at a point, $f'(u)$, belongs to $\mathcal{L}(V, \mathbb{R}) = V^*$. If V is a Hilbert space, then we may assume that $f'(u) \in V$ itself. We now relate the concept of the Fréchet derivative to usual notions of the derivative on $\mathbb{R}$ or $\mathbb{R}^n$.

Example A3.1 Let $f: \mathbb{R} \to \mathbb{R}$. Then $f'(t) \in \mathcal{L}(\mathbb{R}, \mathbb{R}) \cong \mathbb{R}$. Indeed we have that

$$f'(t)(s) = \frac{df}{dt}(t) \cdot s, \tag{A3.9}$$

where $\dfrac{df}{dt}(t)$ is the usual differential coefficient of f at t, which is a real number. ∎

Example A3.2 Let $f: \mathbb{R}^n \to \mathbb{R}$. Then $f'(x) \in \mathcal{L}(\mathbb{R}^n, \mathbb{R}) \cong \mathbb{R}^n$. In fact,

$$(f'(x), y) = \sum_{i=1}^{n} \frac{\partial f}{\partial x_i}(x) y_i \tag{A3.10}$$

and we can thus write

$$f'(x) = \left(\frac{\partial f}{\partial x_1}(x), \ldots, \frac{\partial f}{\partial x_n}(x) \right) = \nabla f(x). \quad ∎$$

Example A3.3 If $V = \mathbb{R}^n$, $W = \mathbb{R}^m$, then $f'(x)$ is a matrix of order $m \times n$. If $f = (f_1, \ldots, f_m)$, then

$$f'(x) = \begin{bmatrix} \dfrac{\partial f_1}{\partial x_1}(x) \ldots \dfrac{\partial f_1}{\partial x_n}(x) \\ \vdots \qquad\qquad \vdots \\ \dfrac{\partial f_m}{\partial x_1}(x) \ldots \dfrac{\partial f_m}{\partial x_n}(x) \end{bmatrix}. \quad ∎ \tag{A3.11}$$

We now compute the Fréchet derivatives of some functionals in infinite dimensional spaces.

Example A3.4 Let V be a Hilbert space and $a : V\times V \to \mathbb{R}$ a symmetric bilinear form which is continuous (cf. Section 3.1). Define

$$J(v) = \tfrac{1}{2}a(v, v) - (f, v), \quad f \in V. \tag{A3.12}$$

Then

$$J(v + h) - J(v) = a(v, h) - f(h) + a(h, h)/2$$

and so if we set

$$(J'(v), h) = a(v, h) - (f, h) \tag{A3.13}$$

we get that

$$\|J(v + h) - J(v) - (J'(v), h)\| \leqslant M\|h\|^2, \quad h \in V,$$

and so (A3.13) defines the derivative of J. This was used in Chapter 3 where we showed that the solution of

$$a(u, v) = (f, v), \quad v \in V$$

minimized J if $a(\ldots)$ were elliptic. Indeed then u is a critical point of J. ∎

We have the usual rules of calculus valid for the Fréchet derivative. For instance if $f : V \to W$ and $g : W \to Y$ and if $f'(u)$ and $g'(f(u))$ exist, then $(g\cdot f)'(u)$ exists and

$$(g\cdot f)'(u) = g'(f(u))\cdot f'(u) \tag{A3.14}$$

We can also define higher order derivatives. If $f : V \to W$ then $f' : V \to \mathcal{L}(V, W)$. We can now discuss the differentiability of f'. If the derivative exists then

$$f'' : V \to \mathcal{L}(V, \mathcal{L}(V, W))$$

and the latter space can be identified with the space of symmetric bilinear maps on $V\times V$ with values in W.

Example A3.4 If V is a Hilbert space and J as in (A3.12) then it follows from (A3.13) that

$$(J'(v + h), w) - (J'(v), w) = a(h, w)$$

from which we see that

$$J''(v)(h, w) = a(h, w).$$

The second derivative in this case is independent of $v \in V$. ∎

We can continue in this fashion. The nth order derivative at a point will be a symmetric n-linear map on V. In the above example, $J''' = 0$ and in general $J^{(n)}(v) = 0$ for all $v \in V$ when $n \geqslant 3$.

An important result in the theory of differentiable functions of a real variable is the Mean Value Theorem or its generalization known as Taylor's theorem. If f is a C^1 function on $(a, b) \subset \mathbb{R}$ and continuous on $[a, b]$ we may write

$$f(b) - f(a) = f'(\xi)(b - a) \tag{A3.15}$$

where $\xi \in (a, b)$. However this is not possible in general when the range space is not $\mathbb{R}$. For instance if $f: [0, 2\pi] \to \mathbb{R}^2$ given by

$$f(t) = (\cos t, \sin t),$$

then $f'(t) = (-\sin t, \cos t)$. However $f(0) = f(2\pi)$ but we can never have $f'(t) = 0$ for $t \in (0, 2\pi)$. Thus (A3.15) will not be possible. However (A3.15) leads to the estimate

$$|f(b) - f(a)| \leqslant (\sup_{\xi \in (a,b)} |f'(\xi)|)(b - a)$$

and such an estimate is possible. In fact we have the following result.

Theorem A3.3 (Mean Value Theorem) Let $f: V \to W$ be continuously differentiable in an open set $\mathcal{O} \subset V$. Let $[u, u + h] \subset \mathcal{O}$ be a segment of the form

$$\{v \in \mathcal{O} \mid v = u + th, 0 \leqslant t \leqslant 1\}.$$

Then

$$\|f(u + h) - f(u)\|_W \leqslant \sup_{t \in (0,1)} \|f'(u + th)\|_{\mathcal{L}(V,W)} \|h\|_V. \quad ■ \qquad \text{(A3.16)}$$

More generally we can have Taylor's formula for functions which are differentiable in one of the following forms.

Theorem A3.4 Let $f: V \to W$ be $(k + 1)$ times continuously differentiable in an open set $\mathcal{O} \subset V$. Then if $[u, u + h] \subset \mathcal{O}$, we have

$$f(u + h) = f(u) + f'(u)h + \tfrac{1}{2} f''(u)(h, h) + \ldots + \|h\|^k \epsilon(h) \qquad \text{(A3.17)}$$

where $\lim_{\|h\| \to 0} \epsilon(h) = 0$. Also

$$f(u + h) = f(u) + f'(u)h + \tfrac{1}{2} f''(u)(h, h) + \ldots$$
$$+ \int_0^1 \frac{(1 - t)^k f^{(k+1)}}{k!} (u + th)(h)^{(k+1)}\, dt, \qquad \text{(A3.18)}$$

where $(h)^k = (h, \ldots h)$, the argument being repeated k times. ■

Remark A3.3 The formula (A3.18) is known as Taylor's formula with integral form of the remainder. ■

For details on the Fréchet derivative, see Cartan [1].

APPENDIX 4

Stampacchia's Theorem

Let $\Omega \subset \mathbb{R}^n$ be a bounded open set. Two of the important properties of functions in $W^{1,p}(\Omega)$, $1 < p < \infty$, were proved by Stampacchia [1] and can be stated as follows.

Theorem A4.1 Let $u \in W^{1,p}(\Omega)$. Then for any $t \in \mathbb{R}$, and for every index i such that $1 \leqslant i \leqslant n$, $\dfrac{\partial u}{\partial x_i} = 0$ a.e. on the set $\{x \mid u(x) = t\}$. ▮

Theorem A4.2 Let $f : \mathbb{R} \to \mathbb{R}$ be a Lipschitz continuous function. If $u \in W^{1,p}(\Omega)$, then

(i) $f(u) \in W^{1,p}(\Omega)$, where $f(u)(x) = f(u(x))$, $x \in \Omega$, and

(ii) if f is differentiable except atmost at a finite number of points, say, $\{t_1, \ldots, t_k\}$, then

$$\frac{\partial}{\partial x_i}(f(u))(x) = \begin{cases} f'(u)\dfrac{\partial u}{\partial x_i}(x), & \text{if} \quad u(x) \notin \{t_1, \ldots, t_k\} \\ 0, & \text{otherwise,} \end{cases} \tag{A4.1}$$

the derivatives being understood in the sense of distributions. ▮

The first part of Theorem A4.2 can be proved fairly easily in a variety of ways (cf. Theorem 2.2.5). The proof of the second part, as we shall indicate later for completeness, involves the result of Theorem A4.1. The proof of Theorem A4.1 usually given uses the notion of linear density and relies on some sophisticated measure-theoretic arguments (cf. Stampacchia [1]). It also uses the fact that functions in Sobolev spaces are differentiable a.e. in a more conventional sense than only in the sense of distributions.

We give below an elementary and short proof of Theorem A4.1. We do this by proving Theorem A4.2 for the case $f(t) = |t|$. Our starting point is the well-known (and easy to prove) fact that if $f \in C^1(\mathbb{R})$, then $f(u) \in W^{1,p}(\Omega)$ and that

$$\frac{\partial}{\partial x_i}(f(u)) = f'(u)\frac{\partial u}{\partial x_i}, \quad 1 \leqslant i \leqslant n. \tag{A4.2}$$

(cf. Theorem 2.2.3).

Lemma A4.1 Let $u \in H^1(\Omega)$. Then $|u| \in H^1(\Omega)$ and

$$\frac{\partial(|u|)}{\partial x_i} = \text{sgn}\,(u)\,\frac{\partial u}{\partial x_i}, \quad 1 \leqslant i \leqslant n. \tag{A4.3}$$

where

$$\text{sgn}\,(u)(x) = \begin{cases} +1 & \text{if} \quad u(x) > 0 \\ -1 & \text{if} \quad u(x) < 0 \\ 0 & \text{if} \quad u(x) = 0. \end{cases} \tag{A4.4}$$

Proof Let $\epsilon > 0$. Define $f_\epsilon(t) = \sqrt{t^2 + \epsilon}$. Clearly $f_\epsilon \in C^1(\mathbf{R})$. Thus if $u \in H^1(\Omega)$, $f_\epsilon(u) \in H^1(\Omega)$ and

$$\frac{\partial(f_\epsilon(u))}{\partial x_i} = \frac{u}{\sqrt{u^2 + \epsilon}}\,\frac{\partial u}{\partial x_i}, \quad 1 \leqslant i \leqslant n. \tag{A4.5}$$

Since Ω is bounded,

$$\int_\Omega |f_\epsilon(u)|^2 = \epsilon|\Omega| + \int_\Omega u^2 \leqslant C,$$

and

$$\int_\Omega \left|\frac{\partial}{\partial x_i}\,(f_\epsilon(u))\right|^2 = \int_\Omega \frac{u^2}{u^2 + \epsilon}\left|\frac{\partial u}{\partial x_i}\right|^2 \leqslant \int_\Omega \left|\frac{\partial u}{\partial x_i}\right|^2 \leqslant C,$$

C independent of ϵ. Thus $\{f_\epsilon(u)\}$ is bounded in $H^1(\Omega)$ and so there exists a subsequence $\epsilon_m \to 0$ (as $m \to \infty$) and a function $v \in H^1(\Omega)$ such that $f_{\epsilon_m}(u) \to v$ weakly in $H^1(\Omega)$ and, by Rellich's Theorem, strongly in $L^2(\Omega)$ and pointwise a.e. (by going to a further subsequence if necessary). But $f_\epsilon(t) \to |t|$ and so $v = |u|$. Thus $|u| \in H^1(\Omega)$.

Let $\phi \in \mathcal{D}(\Omega)$. Since $\dfrac{\partial(f_{\epsilon_m}(u))}{\partial x_i} \to \dfrac{\partial(|u|)}{\partial x_i}$ weakly in $L^2(\Omega)$, we have, for $1 \leqslant i \leqslant n$,

$$\int_\Omega \frac{\partial(|u|)}{\partial x_i}\,\phi = \lim_{m\to\infty} \int_\Omega \frac{u}{\sqrt{u^2 + \epsilon_m}}\,\frac{\partial u}{\partial x_i}\,\phi.$$

But the function $\dfrac{u}{\sqrt{u^2 + \epsilon_m}}\,\dfrac{\partial u}{\partial x_i}\,\phi \to \text{sgn}\,(u)\,\dfrac{\partial u}{\partial x_i}\,\phi$ pointwise and is dominated by $\left|\dfrac{\partial u}{\partial x_i}\right|\,|\phi|$ which is integrable. Thus by the Dominated Convergence Theorem, it follows that

$$\int_\Omega \frac{\partial(|u|)}{\partial x_i}\,\phi = \int_\Omega (\text{sgn}\,u)\,\frac{\partial u}{\partial x_i}\,\phi$$

which completes the proof. ∎

Proof of Theorem A4.1 If $u \geqslant 0$, then $|u| = u$. If $1 \leqslant i \leqslant n$, $\dfrac{\partial(|u|)}{\partial x_i}$ can be represented by two $L^2(\Omega)$-functions, viz., $\dfrac{\partial u}{\partial x}$ itself and sgn $(u)\,\dfrac{\partial u}{\partial x_i}$, by

Lemma A4.1. Thus, as these two must be equal a.e., we get that

$$\frac{\partial u}{\partial x_i}(x) = 0 \text{ a.e. on the set } \{x \mid u(x) = 0\}.$$

Now if $u \in H^1(\Omega)$ is an arbitrary function, we write it as

$$u = u^+ - u^-$$

where $u^+ = \frac{1}{2}(|u| + u)$ and $u^- = \frac{1}{2}(|u| - u)$. Then

$$\{x \mid u(x) = 0\} = \{x \mid u^+(x) = 0\} \cap \{x \mid u^-(x) = 0\}$$

and $\dfrac{\partial u^+}{\partial x_i} = 0$ a.e. on $\{x \mid u^+(x) = 0\}$ while $\dfrac{\partial u^-}{\partial x_i} = 0$ a.e. on $\{x \mid u^-(x) = 0\}$.

Thus as $\dfrac{\partial u}{\partial x_i} = \dfrac{\partial u^+}{\partial x_i} - \dfrac{\partial u^-}{\partial x_i}$, it follows that $\dfrac{\partial u}{\partial x_i} = 0$ a.e. on the set $\{x \mid u(x) = 0\}$.

For arbitrary $t \in \mathbb{R}$, replace u by $u - t$ which is in $H^1(\Omega)$, since Ω is bounded. This completes the proof. ∎

Let f be a Lipschitz continuous function with Lipschitz constant M. We will outline, for the sake of completeness, the proof of relation (A4.1). If $\rho \in \mathscr{D}(\mathbb{R})$ such that $\operatorname{supp}(\rho) \subset [-1, 1]$ and $\int_{\mathbb{R}} \rho = 1$, we set $\rho_\epsilon(x) = \dfrac{1}{\epsilon}\rho\left(\dfrac{x}{\epsilon}\right)$. These are the standard mollifiers. If we set $f_\epsilon = \rho_\epsilon * f$, then $f_\epsilon \in C^\infty(\mathbb{R})$ and

$$\left.\begin{array}{ll} f_\epsilon(t) \to f(t) & \text{for all } \quad t \in \mathbb{R}, \\ f_\epsilon'(t) \to f'(t) & \text{for all } \quad t \notin \{t_1, \ldots, t_k\} \\ |f_\epsilon'(t)| \leqslant M & \end{array}\right\} \tag{A4.6}$$

Thus $f_\epsilon(u) \in H^1(\Omega)$ for $u \in H^1(\Omega)$ and $f_\epsilon(u) \to f(u)$ weakly in $H^1(\Omega)$ exactly as in the proof of Lemma A4.1. If $\phi \in \mathscr{D}(\Omega)$,

$$\int_\Omega \frac{\partial}{\partial x_i}(f(u))\phi = \lim_{\epsilon \to 0} \int_\Omega \frac{\partial}{\partial x_i}(f_\epsilon(u))\phi = \lim_{\epsilon \to 0} \int_\Omega f_\epsilon'(u)\frac{\partial u}{\partial x_i}\phi. \tag{A4.7}$$

Now if $E_i = \{x \mid u(x) = t_i\}$ and $E = \bigcup_{i=1}^{k} E_i$, then $f_\epsilon'(u)\dfrac{\partial u}{\partial x_i} \to f'(u)\dfrac{\partial u}{\partial x_i}$ on $\Omega \setminus E$. On E, we have, by Theorem A4.1, $\dfrac{\partial u}{\partial x_i} = 0$ a.e. and so $f_\epsilon'(u)\dfrac{\partial u}{\partial x_i} = 0$ a.e. on E. Also by (A4.6)

$$\left| f_\epsilon(u)\frac{\partial u}{\partial x_i}\phi \right| \leqslant M\left|\frac{\partial u}{\partial x_i}\right| |\phi|$$

and we can deduce (A4.1) from (A4.7) by the Dominated Convergence Theorem.

We conclude by proving the results when $p \neq 2$. If $1 < p < \infty$, we split these into two cases:

(i) $p > 2$. Then as Ω is bounded, $u \in W^{1,p}(\Omega)$ implies that $u \in W^{1,2}(\Omega) = H^1(\Omega)$ and so $\frac{\partial u}{\partial x_i} = 0$ a.e. on the set $\{x \mid u(x) = t\}$, for any $t \in \mathbb{R}$.

(ii) $p < 2$. In this case $f_\epsilon(t) = \sqrt{t^2 + \epsilon}$. Since $|f_\epsilon'(t)| \leqslant 1$, we still have that $\left\{\frac{\partial}{\partial x_i}(f_\epsilon(u))\right\}$ is bounded in $L^p(\Omega)$. Also

$$\int_\Omega (u^2 + \epsilon)^{p/2} \leqslant \int_\Omega |u|^p + \epsilon^{p/2}|\Omega| \leqslant C$$

using the inequality $(a + b)^q \leqslant a^q + b^q$ for $0 < q < 1$ with $q = p/2$. Thus once again the proofs go through. Thus Theorem A4.1 is true for $u \in W^{1,p}(\Omega)$, $1 < p < \infty$.

We can also relax the condition that Ω be bounded. If $\Omega \subset \mathbb{R}^n$ is any open set we write

$$\Omega = \bigcup_{m=1}^{\infty} \Omega_m, \ \Omega_m = \Omega \cap B(0; m),$$

$B(0; m)$ being the open ball centred at the origin and of radius m. If $u \in W^{1,p}(\Omega)$, then $u \in W^{1,p}(\Omega_m)$ and so $\frac{\partial u}{\partial x_i} = 0$ a.e. on the set $\Omega_m \cap \{x \mid u(x) = t\}$ and so $\frac{\partial u}{\partial x_i} = 0$ a.e. on the set $\{x \mid u(x) = t\}$. Again Theorem A4.2 follows as usual. Only in order that $f(u) \in L^p(\Omega)$ we need that $f(0) = 0$. Thus for arbitrary domains Ω, Theorem A4.2 is true with the restriction that $f(0) = 0$, for any $u \in W^{1,p}(\Omega)$, $1 < p < \infty$.

References

Adams, R. A., Sobolev Spaces, Academic Press, 1975.

Agmon, S., Lectures on Elliptic Boundary Value Problems, Van Nostrand, 1965.

Agmon, S., Douglis, A. and Nirenberg, L., Estimates near the boundary for solutions of elliptic partial differential equations satisfying general boundary value conditions, I, *Comm. Pure Appl. Math.*, **12**, pp. 623–727, 1959.

Amann, H., Fixed point equations and nonlinear eigenvalue problems in ordered Banach spaces, *SIAM Review*, **18**, No. 4, pp. 620–709, 1976.

Ambrosetti, A. and Rabinowitz, P. H., Dual variational methods in critical point theory and applications, *J. Funct. Anal.*, **14**, pp. 349–381, 1973.

Aubin, T., Problèmes isopérimétriques et espaces de Sobolev, *J. Diff. Geom.*, **11**, pp. 573–598, 1976.

Bahri, A. and Berestycki, H., A perturbation method in critical point theory and applications, *Trans. A.M.S.*, **207**, No. 1, pp. 1–32, 1981.

Baouendi, M. S., Sur une classe d'opérateurs elliptiques dégénérés, *Bull. Soc. Math. de France*, **95**, pp. 45–87, 1967.

Baouendi, M. S. and Goulaouic, C., Sur l'étude de la régularité d'opérateurs elliptiques dégénérés, *C.R.A.S.*, Paris, Série A, t.266, pp. 336–338, 1968.

Berger, M. S., Nonlinearity and Functional Analysis, Academic Press, 1978.

Brezis, H., Analyse Fonctionnelle, Théorie et Applications, Masson, Paris, 1983.

———, Some variational problems with lack of compactness, Proceedings of the Berkeley Symposium on Nonlinear Functional Analysis, 1983 (to appear).

Brezis, H. and Nirenberg, L., Positive solutions of nonlinear elliptic equations involving critical Sobolev exponents, *Comm. Pure Appl. Math.*, **36**, pp. 437–477, 1983.

Cartan, H., Differential Calculus, Hermann, 1971.

Cerami, G., Fortunato, D. and Struwe, M., Bifurcation and multiplicity results for nonlinear elliptic problems involving critical Sobolev exponents, *Ann. Inst. H. Poincaré, Analyse Nonlinéaire*, **1**, pp. 341–350, 1984.

Choquet, G. and Deny, J., Sur quelques propriétés de moyenne caractéristiques des fonctions harmoniques et polyharmoniques, *Bull. Soc. Math. de France*, **72**, pp. 118–140, 1944.

Ciarlet, P. G., Lectures on the Finite Element Method, T.I.F.R. Lecture Notes Series, Bombay, 1975.

———, The Finite Element Method for Elliptic Problems, North-Holland, 1978.

Ciarlet, P. G. and Raviart, P. A., A mixed finite element method for the biharmonic equation, pp. 125–145 in Mathematical Aspects of Finite Elements in Partial Differential Equations, C. de Boor (ed.), Academic Press, 1974.

Clark, D. C., A variant of the Lusternik-Schnirelman theory, *Indiana Univ. Math. J.*, **22**, pp. 65–74, 1972.

Courant, R. and Hilbert, D., Methods of Mathematical Physics (Vols. I and II), Interscience, 1962.

Deimling, K., Nonlinear Functional Analysis, Springer-Verlag, 1985.

Deny, J. and Lions, J. L., Les espaces du type Beppo-Levi, *Ann. Inst. Fourier* (*Grenoble*), **5**, pp. 305–370, 1955.

Dieudonne, J., Foundations of Modern Analysis, Academic Press, 1969.

————, Treatise on Analysis, Vol. II, Academic Press, 1976.

Dirac, P. A. M., The physical interpretation of the quantum dynamics, Proc. of the Royal Soc., London, Sec. A, **113**, pp. 621–641, 1926–27.

Dugundji, J., Topology, Prentice-Hall, Indian Edition, 1975.

Dunford, N. and Schwartz, J. T., Linear Operators, Part I, General Theory, Interscience, 1958.

Duvaut, G. and Lions, J. L., Les Inéquations en Mécanique et en Physique, Dunod, 1972.

Folland, G. B., Lectures on Partial Differential Equations, T.I.F.R. Lecture Notes Series, Bombay, 1983.

Friedman, A., Partial Differential Equations, Holt, Reinhart and Winston, 1969.

Geymonat, G. and Grisvard, P., Problemi ai limiti ellitici negli spazi di Sobolev con peso, *Le Matematiche*, **XXII**, fasc. 2, 1967.

Gilbarg, D. and Trudinger, N., Elliptic Partial Differential Equations of Second Order, Springer-Verlag, 1977.

Girault, V. and Raviart, P.-A., Finite Element Approximation of the Navier-Stokes Equations, Springer Lecture Notes in Math., No. 749, 1979.

Giusti, E., Minimal Surfaces and Functions of Bounded Variation, Birkhauser, 1984.

Grisvard, P., Behaviour of the solution of an elliptic boundary value problem in a polygonal or polyhedral domain, pp. 207–274 in Numerical Solutions of Partial Differential Equations, III, SYNSPADE, Bert Hubbard Ed., 1975.

Heaviside, On operators in mathematical physics, Proc. of the Royal Soc., London, **52**, pp. 504–529, 1893.

————, On operators in mathematical physics, Proc. of the Royal Soc., London, **54**, pp. 105–143, 1894.

Hille, E. and Phillips, R. S., Functional Analysis and Semigroups, A.M.S., Colloquium Publ., Vol. 31, Providence, R.I., 1957.

Hormander, L., Linear Partial Differential Operators, Springer-Verlag, 3rd Revised Printing, 1969.

Horvath, J., An introduction to distributions, *Amer. Math. Monthly*, **77**, pp. 227–240, 1970.

John, F., Partial Differential Equations, Springer International Student Edition, 3rd Edition, 1978.

Joshi, M. C. and Bose, R. K., Some Topics in Nonlinear Functional Analysis, Wiley Eastern Ltd., 1985.

Kac, M., Can one hear the shape of a drum?, *Amer. Math. Monthly*, **73**, pp. 1–23, 1966.

Kannai, Y., An elementary proof of the no-retraction theorem, *Amer. Math. Monthly*, **88**, pp, 264–268, 1981.

Keller, H. B., T.I.F.R. Lecture Notes Series, Bombay (to appear).

Kellogg, R. B., Li, T. Y. and Yorke, J., A constructive proof of the Brouwer fixed point theorem and computational results, *SIAM J. Num. Anal.*, **4**, pp. 473–483, 1976.

Kesavan, S., La méthode de Kikuchi appliquée aux équations de Von Karman, *Numer. Math.*, **32**, pp. 209–232, 1979.

———, Existence of solutions by the Galerkin method for a class of nonlinear problems, *Applicable Anal.*, **16**, pp. 279–290, 1983.

Kinderlehrer, D. and Stampacchia, G., An Introduction to Variational Inequalities and Their Applications, Academic Press, 1980.

Kondrat'ev, V. A., Boundary value problems for elliptic equations in domains with conical or angular points, *Trudy Markov Mat. Obsc.*, **16**, pp. 209–292, 1967.

Lieb, E., Sharp constants in the Hardy-Littlewood-Sobolev and related inequalities, *Ann. Math.*, **118**, pp. 349–374, 1983.

Lions, J. L., Quelques Méthodes de Résolution des Problèmes aux Limites Nonlinéaires, Dunod Gauthier-Villars, 1969.

Lions, J. L. and Magenes, E., Non-Homogeneous Boundary Value Problems and Applications, Vol. I, Springer-Verlag, 1972.

Lions, P. L., On the existence of positive solutions of semilinear elliptic equations, *SIAM Review*, **24**, No. 4, pp. 441–467, 1982.

Lumer, G. and Phillips, R. S., Dissipative operators in a Banach space, *Pacific J. Math.*, **11**, pp. 679–698, 1961.

Maz'ja, V. G., Sobolev Spaces, Springer-Verlag, 1985.

Milnor, J., Morse Theory, Annals of Mathematics Studies, No. 51, Princeton, 1969.

Morrey, C. B., Functions of several variables and absolute continuity, II, *Duke J. Math.*, **6**, pp. 187–215, 1940.

Nirenberg, L., On elliptic partial differential equations, *Ann. Sc. Norm. Sup., Pisa*, **13**, pp. 116–162, 1959.

———, Topics in Nonlinear Functional Analysis, Lecture Notes, Courant Institute, New York, 1974.

———, Variational and topological methods in nonlinear problems, *Bull. (New Series) of the A.M.S.*, **4**, No. 3, pp. 267–302, 1981.

———, On some variational methods, pp. 169–176 in Bifurcation Theory, Mechanics and Physics, C. P. Bruter, A. Aragnol and A. Lichnerowicz (eds.), D. Reidel Publ. Co., 1983.

Pazy, A., Semigroups of Linear Operators and Applications to Partial Differential Equations, Springer-Verlag, *Applied Math. Sciences*, **44**, 1983.

Phillips, R. S., Dissipative operators and hyperbolic systems of partial differential equations, *Trans. Amer. Math. Soc.*, **90**, pp. 193–254, 1959.

Protter, M. and Weinberger, H., Maximum Principles in Differential Equations, Prentice-Hall, 1967.

Pucci, P. and Serrin, J., A mountain pass theorem, *J. Diff. Eqns.* **60**, No. 1, pp. 142–149, 1985.

Rabier, P., Personal communication.

Rabinowitz, P. H., Théorie du Degré Topologique et Applications à des Problèmes aux Limites Nonlinéaires, Notes by H. Berestycki, Univ. Paris, VI, 1975.

———, Applications of Bifurcation Theory, Academic Press, 1977.

Rudin, W., Real and Complex Analysis, 2nd Edition, Tata McGraw-Hill, 1974.

———, Functional Analysis, Tata McGraw-Hill, 1974.

Scarf, H., The approximation of fixed points of a continuous mapping, *SIAM J. Appl. Math.*, **15**, pp. 1328–1343, 1967.

Schaeffer, D. G., Topics in Bifurcation Theory, pp. 219–262 in Systems of Nonlinear Partial Differential Equations, J. M. Ball (ed.), D. Reidel Publ. Co., 1983.

Schwartz, J. T., Nonlinear Functional Analysis, Gordon and Breach, 1969.

Schwartz, L., Théorie des Distributions, Hermann, 1966.

Simmons G. F., Introduction to Topology and Modern Analysis, McGraw-Hill, Kogakusha, 1963.

———, Differential Equations, Tata McGraw-Hill, 1974.

Sobolev, S. L., On a theorem of functional analysis, *Mat. Sb.*, **46**, pp. 471–496, 1938.

Stampacchia, G., Equations Elliptiques du Second Ordre à Coefficients Discontinus, Les Presses de l'Université de Montréal, 1966.

Strang, G. and Fix, G. J., An Analysis of the Finite Element Method, Prentice-Hall, 1973.

Talenti, G., Best constants in Sobolev inequality, *Ann. Mat. Pura Appl.*, **110**, pp. 353–372, 1976.

Temam, R. and Strang, G., Functions of bounded deformation, *Arch. Rat. Mech. Anal.*, **75**, pp. 7–21, 1980.

Treves, J. F., Linear Partial Differential Equations with Constant Coefficients, Gordon and Breach, 1966.

———, Topological Vector Spaces, Distributions and Kernels, Academic Press, 1967.

———, Applications of distributions to PDE theory, *Amer. Math. Monthly*, **71**, pp. 241–248, 1970.

———, Basic Linear Partial Differential Equations, Academic Press, 1975.

Index

D1130504